Lecture Notes in Computer Science   9496

Commenced Publication in 1973
Founding and Former Series Editors:
Gerhard Goos, Juris Hartmanis, and Jan van Leeuwen

Editorial Board

More information about this series at http://www.springer.com/series/7410

Jens Groth (Ed.)

Cryptography and Coding

15th IMA International Conference, IMACC 2015
Oxford, UK, December 15–17, 2015
Proceedings

 Springer

Editor
Jens Groth
University College London
Cambridge
UK

ISSN 0302-9743   ISSN 1611-3349   (electronic)
Lecture Notes in Computer Science
ISBN 978-3-319-27238-2   ISBN 978-3-319-27239-9   (eBook)
DOI 10.1007/978-3-319-27239-9

Library of Congress Control Number: 2015957651

LNCS Sublibrary: SL4 – Security and Cryptology

Printed on acid-free paper

This Springer imprint is published by SpringerNature
The registered company is Springer International Publishing AG Switzerland

Preface

The International Conference of Cryptography and Coding is the biennial conference of the Institute of Mathematics and its Applications (IMA) on cryptography and coding theory. The conference series has been running for more than three decades and the 15th edition was held December 15–17, 2015, in inspirational and historical surroundings at St. Catherine's College at the University of Oxford.

We received 36 submissions from authors all over the world on a diverse set of topics both in cryptography and coding theory. The Program Committee selected 18 of the submissions for presentation at the conference. The review process was double-blind and rigorous: Each submission was reviewed independently by at least three reviewers in an individual review phase, and subsequently considered by the Program Committee in a discussion phase. Feedback from the reviews and discussions was given to the authors and their revised submissions are included in the proceedings.

The Program Committee selected one distinguished article for the best paper award. Congratulations to Kenneth G. Paterson, Jacob C.N. Schuldt, Dale L. Sibborn, and Hoeteck Wee for winning the award this year with their paper "Security Against Related Randomness Attacks via Reconstructive Extractors."

In addition to the presentations of accepted papers, the conference also featured four keynote talks by internationally leading scientists on their research in the interface of cryptography and coding theory. I am grateful to Sihem Mesnager, Allison Bishop, Alexander May, and Daniel Wichs for accepting our invitation and sharing the insights gathered from their exciting research. Sihem Mesnager kindly offered a companion paper to her talk, "On Existence (Based on an Arithmetical Problem) and Constructions of Bent Functions," co-authored by Gérard Cohen and David Madore.

Running a conference like IMACC requires the effort of many people and many thanks are due. I would like to thank the Steering Committee for their trust and support. I thank the authors for their submissions, and the Program Committee and the external reviewers for their effort in selecting the scientific program. Thanks also goes to the IACR and Shai Halevi for their cooperation and for letting us use the WebSubRev software to manage the submission and review process. I appreciate the assistance by Alfred Hofmann and Anna Kramer from Springer in the production of the proceedings. Finally, I am incredibly thankful to conference officer (general chair) Lizzi Lake and her colleagues at the Institute of Mathematics and its Applications for handling all the practical matters of the conference.

December 2015   Jens Groth

IMACC 2015

15th IMA International Conference on Cryptography and Coding

St. Catherine's College, University of Oxford, Oxford, UK
December 15–17, 2015

Sponsored by the Institute of Mathematics and its Applications (IMA)
In cooperation with the International Association for Cryptologic Research (IACR)

General Chair

Lizzi Lake   Institute of Mathematics and its Applications, UK

Program Chair

Jens Groth   University College London, UK

Steering Committee

Liqun Chen	HP Labs, UK
Bahram Honary	University of Lancaster, UK
Christopher Mitchell	Royal Holloway, University of London, UK
Matthew G. Parker	University of Bergen, Norway
Kenneth G. Paterson	Royal Holloway, University of London, UK
Fred Piper	Royal Holloway, University of London, UK
Martijn Stam	University of Bristol, UK

Program Committee

Martin Albrecht	Royal Holloway, University of London, UK
Marco Baldi	Università Politecnica delle Marche, Italy
Colin Boyd	NTNU, Norway
Claude Carlet	University of Paris 8, LAGA, France
Pascale Charpin	Inria Rocquencourt, France
Liqun Chen	HP Labs, UK
Nicolas Courtois	University College London, UK
Marten van Dijk	University of Connecticut, USA
Pooya Farshim	Queen's University Belfast, UK
Sebastian Faust	Ruhr University Bochum, Germany
Matthieu Finiasz	CryptoExperts, France

Contents

Invited Paper

On Existence (Based on an Arithmetical Problem) and Constructions of Bent Functions

Sihem Mesnager[1]([✉]), Gérard Cohen[2], and David Madore[2]

[1] Department of Mathematics, University of Paris VIII, University of Paris XIII,
LAGA, UMR 7539, CNRS, and Télécom ParisTech, Paris, France
smesnager@univ-paris8.fr
[2] Télécom ParisTech, UMR 5141, CNRS, Paris, France

Abstract. Bent functions are maximally nonlinear Boolean functions. They are wonderful creatures introduced by O. Rothaus in the 1960's and studied firstly by J. Dillon since 1974. Using some involutions over finite fields, we present new constructions of bent functions in the line of recent Mesnager's works. One of the constructions is based on an arithmetical problem. We discuss existence of such bent functions using Fermat hypersurface and Lang-Weil estimations.

Keywords: Boolean functions · Bent functions · Finite fields · Arithmetic and geometric tools

1 Introduction

Bent functions are maximally nonlinear Boolean functions with an even number of variables. They were introduced by Rothaus [36] in 1976 but already studied by Dillon [14] since 1974. For their own sake as interesting combinatorial objects, but also for their relations to coding theory (e.g. Reed-Muller codes, Kerdock codes), combinatorics (e.g. difference sets), design theory (any difference set can be used to construct a symmetric design), sequence theory, and applications in cryptography (design of stream ciphers and of S-boxes for block ciphers), they have attracted a lot of research for four decades. Yet, their classification is still elusive, therefore, not only their characterization, but also their generation are challenging problems. A non-exhaustive list of references dealing with constructions of binary bent Boolean functions is [1–4, 6, 8, 10, 11, 14–17, 22–25, 27–32, 35, 37]. Some open problems can be found in [7]. For a recent survey, see [9]. A book devoted especially to bent functions and containing a complete survey (including variations, generalizations and applications) is [34].

The paper was presented as a part of an invited talk entitled "Bent functions and their connections to coding theory and cryptography" at the fifteenth International Conference on Cryptography and Coding, Oxford, United Kingdom (IMACC 2015) given by S. Mesnager.

J. Groth (Ed.): IMACC 2015, LNCS 9496, pp. 3–19, 2015.
DOI: 10.1007/978-3-319-27239-9_1

Bent functions occur in pairs. In fact, given a bent function one can define its dual which is again bent. Computing the dual of a given bent function is not an easy task in general. Recently, the first author has derived in [31] several new infinite classes of bent functions defined over the finite field $\mathbb{F}_{2^n}$ with their duals. All these families are obtained by selecting three pairwise distinct bent functions from general classes and satisfying some conditions. In [32], the first author extends the results of [31] and exhibits several new infinite families of bent functions, together with their duals. Some of them are obtained via new infinite families of permutations that the author provides with their compositional inverses. In [32], secondary-like constructions of permutations leading to several families of bent functions have also been introduced. The paper is in the line of [31,32]. Our objective is to provide more primary constructions of bent functions defined over the finite field $\mathbb{F}_{2^{2m}} \simeq \mathbb{F}_{2^m} \times \mathbb{F}_{2^m}$ in bivariate representation in terms of the sum of the products of trace functions.

This paper is organized as follows. Formal definitions and necessary preliminaries are introduced in Sect. 2. In Sect. 3, we present an overview of the previous constructions of binary bent functions related to our work. Next, in the line of [31,32] based on special permutations, we investigate bent functions from involutions. We focus on monomial involutions and show how one can derive bent functions. A main result is given by Theorem 2. Finally, in Sect. 5, we study the existence of functions derived from Theorem 2. The problem of designing new primary bent functions turns out to be an arithmetical problem that we study by giving solutions using arithmetic and geometric tools.

2   Notation and Preliminaries

A Boolean function on the finite field $\mathbb{F}_{2^n}$ of order 2^n is a mapping from $\mathbb{F}_{2^n}$ to the prime field $\mathbb{F}_2$. It can be represented as a polynomial in one variable $x \in \mathbb{F}_{2^n}$ of the form $f(x) = \sum_{j=0}^{2^n-1} a_j x^j$ where the a_j's are elements of the field. Such a function f is Boolean if and only if a_0 and a_{2^n-1} belong to $\mathbb{F}_2$ and $a_{2j} = a_j^2$ for every $j \notin \{0, 2^n - 1\}$ (where $2j$ is taken modulo $2^n - 1$). This leads to a unique representation which we call the *polynomial form* (for more details, see e.g. [6]). First, recall that for any positive integers k, and r dividing k, the trace function from $\mathbb{F}_{2^k}$ to $\mathbb{F}_{2^r}$, denoted by Tr_r^k, is the mapping defined for every $x \in \mathbb{F}_{2^k}$ as:

$$Tr_r^k(x) := \sum_{i=0}^{\frac{k}{r}-1} x^{2^{ir}} = x + x^{2^r} + x^{2^{2r}} + \cdots + x^{2^{k-r}}.$$

In particular, we denote the *absolute trace* over $\mathbb{F}_2$ of an element $x \in \mathbb{F}_{2^n}$ by $Tr_1^n(x) = \sum_{i=0}^{n-1} x^{2^i}$. We make use of some known properties of the trace function such as $Tr_1^n(x) = Tr_1^n(x^2)$ and for every integer r dividing k, the mapping $x \mapsto Tr_r^k(x)$ is $\mathbb{F}_{2^k}$-linear.

The *bivariate representation* of Boolean functions makes sense only when n is an even integer. It plays an important role for defining bent functions and is defined as follows: we identify $\mathbb{F}_{2^n}$ (where $n = 2m$) with $\mathbb{F}_{2^m} \times \mathbb{F}_{2^m}$ and consider then the input to f as an ordered pair (x, y) of elements of $\mathbb{F}_{2^m}$. There exists a unique bivariate polynomial

$$\sum_{0 \leq i, j \leq 2^m - 1} a_{i,j} x^i y^j$$

over $\mathbb{F}_{2^m}$ such that f is the bivariate polynomial function over $\mathbb{F}_{2^m}$ associated to it. Then the algebraic degree of f equals $\max_{(i,j) \mid a_{i,j} \neq 0}(w_2(i) + w_2(j))$. The function f being Boolean, its bivariate representation can be written in the (non unique) form $f(x, y) = Tr_1^m(P(x, y))$ where $P(x, y)$ is some polynomial in two variables over $\mathbb{F}_{2^m}$. There exist other representations of Boolean functions not used in this paper (see e.g. [6]) in which we shall only consider functions in their bivariate representation.

If f is a Boolean function defined on $\mathbb{F}_{2^n}$, then the Walsh Hadamard transform of f is the discrete Fourier transform of the sign function $\chi_f := (-1)^f$ of f, whose value at $\omega \in \mathbb{F}_{2^n}$ is defined as follows:

$$\forall \omega \in \mathbb{F}_{2^n}, \quad \widehat{\chi_f}(\omega) = \sum_{x \in \mathbb{F}_{2^n}} (-1)^{f(x) + Tr_1^n(\omega x)}.$$

Bent functions can be defined in terms of the Walsh transform as follows.

Definition 1. *Let n be an even integer. A Boolean function f on $\mathbb{F}_{2^n}$ is said to be bent if its Walsh transform satisfies $\widehat{\chi_f}(a) = \pm 2^{\frac{n}{2}}$ for all $a \in \mathbb{F}_{2^n}$.*

The automorphism group of the set of bent functions (i.e., the group of permutations π on $\mathbb{F}_{2^n}$ such that $f \circ \pi$ is bent for every bent function f) is the general affine group, that is, the group of linear automorphisms composed by translations. The corresponding notion of equivalence between functions is called *affine equivalence*. Also, if f is bent and ℓ is affine, then $f + \ell$ is bent. A class of bent functions is called a *complete class* if it is globally invariant under the action of the general affine group and under the addition of affine functions. The corresponding notion of equivalence is called *extended affine equivalence*, in brief, *EA-equivalence*.

Bent functions occur in pair. In fact, given a bent function f over $\mathbb{F}_{2^n}$, we define its *dual function*, denoted by $\tilde{f}$, when considering the signs of the values of the Walsh transform $\widehat{\chi_f}(x)$ $(x \in \mathbb{F}_{2^n})$ of f. More precisely, $\tilde{f}$ is defined by the equation:

$$(-1)^{\tilde{f}(x)} 2^{\frac{n}{2}} = \widehat{\chi_f}(x). \tag{2.1}$$

Due to the involution law the Fourier transform is self-inverse. Thus the dual of a bent function is again bent, and $\tilde{\tilde{f}} = f$. A bent function is said to be self-dual if $\tilde{f} = f$.

Let us recall a fundamental class of Boolean bent functions. Bent functions from the Maiorana-McFarland construction are defined over $\mathbb{F}_{2^m} \times \mathbb{F}_{2^m}$ by (2.2):

$$f(x, y) = Tr_1^m(\phi(y)x) + g(y), \quad (x, y) \in \mathbb{F}_{2^m} \times \mathbb{F}_{2^m} \tag{2.2}$$

where m is some positive integer, ϕ is a function from $\mathbb{F}_{2^m}$ to itself and g stands for a Boolean function over $\mathbb{F}_{2^m}$. We have the following well-known result (e.g. see [6,34]).

Proposition 1. *Let m be a positive integer. Let g be a Boolean function defined over $\mathbb{F}_{2^m}$. Define f over $\mathbb{F}_{2^m} \times \mathbb{F}_{2^m}$ by (2.2). Then f is bent if and only if ϕ is a permutation of $\mathbb{F}_{2^m}$. Furthermore, its dual function $\tilde{f}$ is*

$$\tilde{f}(x,y) = Tr_1^m(y\phi^{-1}(x)) + g(\phi^{-1}(x)) \tag{2.3}$$

where ϕ^{-1} denotes the inverse mapping of the permutation ϕ.

The class of bent functions given by (2.2) is the so-called Maiorana-McFarland class. It has been widely studied because its Walsh transform can be easily computed and its elements are completely characterized (e.g. see [6]).

3   Related Previous Constructions of Bent Functions

In [5] a secondary construction of bent functions is provided (building new bent functions from already defined ones). It is proved there that if f_1, f_2 and f_3 are bent, then if $\psi := f_1 + f_2 + f_3$ is bent and if $\tilde{\psi} = \tilde{f}_1 + \tilde{f}_2 + \tilde{f}_3$, then $g(x) = f_1(x)f_2(x) + f_1(x)f_3(x) + f_2(x)f_3(x)$ is bent, and $\tilde{g} = \tilde{f}_1\tilde{f}_2 + \tilde{f}_1\tilde{f}_3 + \tilde{f}_2\tilde{f}_3$. Next, the first author has completed this result by proving in [31] that the converse is also true. The combined result is stated in the following theorem.

Theorem 1 *([31]). Let n be an even integer. Let f_1, f_2 and f_3 be three pairwise distinct bent functions over $\mathbb{F}_{2^n}$ such that $\psi = f_1 + f_2 + f_3$ is bent. Let g be a Boolean function defined by*

$$g(x) = f_1(x)f_2(x) + f_1(x)f_3(x) + f_2(x)f_3(x). \tag{3.1}$$

Then g is bent if and only if $\tilde{\psi} + \tilde{f}_1 + \tilde{f}_2 + \tilde{f}_3 = 0$. Furthermore, if g is bent, then its dual function $\tilde{g}$ is given by

$$\tilde{g}(x) = \tilde{f}_1(x)\tilde{f}_2(x) + \tilde{f}_1(x)\tilde{f}_3(x) + \tilde{f}_2(x)\tilde{f}_3(x), \ \forall x \in \mathbb{F}_{2^n}.$$

In [31,32], the first author has studied functions g of the shape (3.1) and derived several new primary constructions of bent functions.

To apply Theorem 1 to a 3-tuple of functions of the form (2.2) with $g = 0$, one has to choose appropriately the maps ϕ involved in their expressions. The following result is proven in [31].

Corollary 1. *Let m be a positive integer. Let ϕ_1, ϕ_2 and ϕ_3 be three permutations of $\mathbb{F}_{2^m}$. Then,*

$$g(x,y) = Tr_1^m(x\phi_1(y))Tr_1^m(x\phi_2(y)) + Tr_1^m(x\phi_1(y))Tr_1^m(x\phi_3(y))$$
$$+ Tr_1^m(x\phi_2(y))Tr_1^m(x\phi_3(y))$$

is bent if and only if

1. $\psi = \phi_1 + \phi_2 + \phi_3$ is a permutation,
2. $\psi^{-1} = \phi_1^{-1} + \phi_2^{-1} + \phi_3^{-1}$.

Furthermore, its dual function $\tilde{g}$ is given by

$$\tilde{g}(x,y) = Tr_1^m(\phi_1^{-1}(x)y)Tr_1^m(\phi_2^{-1}(x)y) + Tr_1^m(\phi_1^{-1}(x)y)Tr_1^m(\phi_3^{-1}(x)y)$$
$$+ Tr_1^m(\phi_2^{-1}(x)y)Tr_1^m(\phi_3^{-1}(x)y).$$

Permutations satisfying $(\mathcal{A}_m)$ were introduced by the first author in [32].

Definition 2. *Let m be a positive integer. Three permutations ϕ_1, ϕ_2 and ϕ_3 of $\mathbb{F}_{2^m}$ are said to satisfy $(\mathcal{A}_m)$ if the following two conditions hold.*

1. *Their sum $\psi = \phi_1 + \phi_2 + \phi_3$ is a permutation of $\mathbb{F}_{2^m}$.*
2. *$\psi^{-1} = \phi_1^{-1} + \phi_2^{-1} + \phi_3^{-1}$.*

Several new bent functions have been exhibited from monomial permutations (see [31]) and from more families of new permutations of $\mathbb{F}_{2^m}$ (see [32]). Firstly, we list below the constructions obtained by the first author in [31].

1. Bent functions obtained by selecting Niho bent functions:
   - $f(x) = Tr_1^m(\lambda x^{2^m+1}) + Tr_1^n(ax)Tr_1^n(bx)$; $x \in \mathbb{F}_{2^n}$, $n = 2m$, $\lambda \in \mathbb{F}_{2^m}^\star$ and $(a,b) \in \mathbb{F}_{2^n}^\star \times \mathbb{F}_{2^n}^\star$ such that $a \neq b$ and $Tr_1^n(\lambda^{-1}b^{2^m}a) = 0$.

   $$\tilde{f}(x) = Tr_1^m(\lambda^{-1}x^{2^m+1}) + \left(Tr_1^m(\lambda^{-1}a^{2^m+1}) + Tr_1^n(\lambda^{-1}a^{2^m}x)\right) \times$$
   $$\left(Tr_1^m(\lambda^{-1}b^{2^m+1}) + Tr_1^n(\lambda^{-1}b^{2^m}x)\right) + 1.$$

   - $g(x) = Tr_1^m(x^{2^m+1}) + Tr_1^n\left(\sum_{i=1}^{2^{r-1}-1} x^{(2^m-1)\frac{i}{2^r}+1}\right) + Tr_1^n(\lambda x)Tr_1^n(\mu x)$; $x \in \mathbb{F}_{2^n}$, $n = 2m$, $(\lambda, \mu) \in \mathbb{F}_{2^m}^\star \times \mathbb{F}_{2^m}^\star$ ($\lambda \neq \mu$).

   $$\tilde{g}(x) = Tr_1^m\left((u(1+x+x^{2^m})+u^{2^{n-r}}+x^{2^m})(1+x+x^{2^m})^{\frac{1}{2^r-1}}\right) \times Tr_1^m\left((\lambda + \mu)(1+x+x^{2^m})^{\frac{1}{2^r-1}}\right) + Tr_1^m\left((u(1+x+x^{2^m})+u^{2^{n-r}}+x^{2^m}+\lambda)(1+x+x^{2^m})^{\frac{1}{2^r-1}}\right) \times Tr_1^m\left((u(1+x+x^{2^m})+u^{2^{n-r}}+x^{2^m}+\mu)(1+x+x^{2^m})^{\frac{1}{2^r-1}}\right);$$
   where $u \in \mathbb{F}_{2^n}$ satisfying $u + u^{2^m} = 1$.

2. Bent functions obtained by selecting bent Boolean functions of Maiorana-McFarland's class:
   - $f(x,y) = Tr_1^m(a_1y^dx)Tr_1^m(a_2y^dx) + Tr_1^m(a_1y^dx)Tr_1^m(a_3y^dx) + Tr_1^m(a_2y^dx)Tr_1^m(a_3y^dx)$; where $(x,y) \in \mathbb{F}_{2^m} \times \mathbb{F}_{2^m}$, d is a positive integer which is not a power of 2 and $\gcd(d, 2^m-1) = 1$, a_i's are pairwise distinct such that $b := a_1+a_2+a_3 \neq 0$ and $a_1^{-e}+a_2^{-e}+a_3^{-e} = b^{-e}$ where $e = d^{-1}$ (mod 2^m-1).

   $$\tilde{f}(x,y) = Tr_1^m(a_1^{-e}x^ey)Tr_1^m(a_2^{-e}x^ey) + Tr_1^m(a_1^{-e}x^ey)Tr_1^m(a_3^{-e}x^ey)$$
   $$+ Tr_1^m(a_2^{-e}x^ey)Tr_1^m(a_3^{-e}x^ey).$$

- $g(x, y) = Tr_1^m(a^{-11}x^{11}y)Tr_1^m(a^{-11}c^{-11}x^{11}y)Tr_1^m(a^{-11}x^{11}y)Tr_1^m(c^{11}a^{-11}$
  $x^{11}y) + Tr_1^m(a^{-11}c^{-11}x^{11}y)Tr_1^m(c^{11}a^{-11}x^{11}y)$; where $(x, y) \in \mathbb{F}_{2^m} \times \mathbb{F}_{2^m}$,
  $a \in \mathbb{F}_{2^n}^{\star}$ with $n = 2m$ is a multiple of 4 but not of 10, $c \in \mathbb{F}_{2^m}$ is such that
  $c^4 + c + 1 = 0$.
  $\tilde{g}(x, y) = Tr_1^m(ay^dx)Tr_1^m(acy^dx) + Tr_1^m(ay^dx)Tr_1^m(ac^{-1}y^dx) + Tr_1^m(acy^dx)$
  $Tr_1^m(ac^{-1}y^dx)$; with $d = 11^{-1}$ (mod $2^n - 1$).
- $h(x, y) = (Tr_1^m(a_1y^dx) + g_1(y))(Tr_1^m(a_2y^dx) + g_2(y)) + (Tr_1^m(a_1y^dx) + g_1(y))(Tr_1^m(a_3y^dx) + g_3(y)) + (Tr_1^m(a_2y^dx) + g_2(y))(Tr_1^m(a_3y^dx) + g_3(y))$;
  where $m = 2r$, $\gcd(d, 2^m - 1) = 1$, a_1, a_2 and a_3 are three pairwise distinct
  elements of $\mathbb{F}_{2^m}$ such that $b := a_1 + a_2 + a_3 \neq 0$ and $a_1^{-e} + a_2^{-e} + a_3^{-e} = b^{-e}$
  and for $i \in \{1, 2, 3\}$, $g_i \in \mathcal{D}_m := \{g : \mathbb{F}_{2^m} \to \mathbb{F}_2 \mid g(ax) = g(x), \forall(a, x) \in \mathbb{F}_{2^r} \times \mathbb{F}_{2^m}\}$.
  $\tilde{h}(x, y) = (Tr_1^m(a_1^{-e}x^ey) + g_1(x^e))(Tr_1^m(a_2^{-e}x^ey) + g_2(x^e)) + (Tr_1^m(a_1^{-e}x^ey) + g_1(x^e))(Tr_1^m(a_3^{-e}x^ey) + g_3(x^e)) + (Tr_1^m(a_2^{-e}x^ey) + g_2(x^e))(Tr_1^m(a_3^{-e}x^ey) + g_3(x^e))$ where $e = d^{-1}$ (mod $2^m - 1$).

3. Self-dual bent functions obtained by selecting functions from Maiorana-McFarland completed class[1]:

   - $g(x) = Tr_1^{4k}(a_1x^{2^k+1})Tr_1^{4k}(a_2x^{2^k+1}) + Tr_1^{4k}(a_1x^{2^k+1})Tr_1^{4k}(a_3x^{2^k+1}) + Tr_1^{4k}(a_2x^{2^k+1})Tr_1^{4k}(a_3x^{2^k+1})$; where $x \in \mathbb{F}_{2^{4k}}$, $k \geq 2$, a_1, a_2, a_3 be three pairwise
     distinct nonzero solutions in $\mathbb{F}_{2^{4k}}$ of the equation $\lambda^{2^{3k}} + \lambda = 1$ such that
     $a_1 + a_2 + a_3 \neq 0$.

4. Bent functions obtained by selecting functions from PS_{ap}:

   - $f(x, y) = Tr_1^m(a_1y^{2^m-2}x)Tr_1^m(a_2y^{2^m-2}x) + Tr_1^m(a_1y^{2^m-2}x)Tr_1^m(a_3y^{2^m-2}x) + Tr_1^m(a_2y^{2^m-2}x)Tr_1^m(a_3y^{2^m-2}x)$; where $(x, y) \in \mathbb{F}_{2^m} \times \mathbb{F}_{2^m}$,
     the a_i's are pairwise distinct in $\mathbb{F}_{2^m}$ such that $a_1 + a_2 + a_3 \neq 0$.

   $$\tilde{f}(x, y) = f(y, x).$$

5. Bent functions obtained by combining Niho bent functions and self-dual bent functions:

   - $f(x) = Tr_1^{2k}(x^{2^{2k}+1}) + Tr_1^{4k}(ax)Tr_1^{2k}(x^{2^{2k}+1}) + Tr_1^{4k}(ax)Tr_1^{4k}(\lambda_2(x + \beta)^{2^k+1}) + Tr_1^{4k}(ax)$; where $x \in \mathbb{F}_{2^{4k}}$ $(k \geq 2)$, $\lambda_2 \in \mathbb{F}_{2^{4k}}$ such that $\lambda_2 + \lambda_2^{2^{3k}} = 1$, $a \in \mathbb{F}_{2^{4k}}^{\star}$ is a solution of $a^{2^{2k}} + \lambda_2^{2^{-k}}a^{2^{-k}} + \lambda_2 a^{2^k} = 0$ and $\beta \in \mathbb{F}_{2^{4k}}$
     such that $Tr_1^{4k}(\beta a) = Tr_1^{2k}(a^{2^{2k}+1}) + Tr_1^{4k}(\lambda_2 a^{2^k+1})$.

   $$\tilde{f}(x) = Tr_1^{2k}(x^{2^{2k}+1}) + \left(Tr_1^{2k}(x^{2^{2k}+1}) + Tr_1^{4k}(\lambda_2 x^{2^k+1}) + Tr_1^{4k}(\beta x)\right) \times \left(Tr_1^{4k}(a^{2^k}x) + Tr_1^{2k}(a^{2^{2k}+1})\right).$$

Secondly, we list below the infinite families of bent functions from new permutations and their duals provided by the first author in [32].

---

[1] The Maiorana-McFarland completed class is the smallest class containing the class of Maiorana-McFarland which is globally invariant under the action of the general affine group and under the addition of affine functions.

1. Let m be a positive integer. Let L be a linear permutation on $\mathbb{F}_{2^m}$. Let f be a Boolean function over $\mathbb{F}_{2^m}$ such that $\mathcal{L}_f^0 := \{\alpha \in \mathbb{F}_{2^m} \mid D_\alpha f = 0\}$ is of dimension at least two over $\mathbb{F}_2$. Let $(\alpha_1, \alpha_2, \alpha_3)$ be any 3-tuple of pairwise distinct elements of $\mathcal{L}_f^0$ such that $\alpha_1 + \alpha_2 + \alpha_3 \neq 0$. Then the Boolean function g defined in bivariate representation on $\mathbb{F}_{2^m} \times \mathbb{F}_{2^m}$ by $g(x, y) = Tr_1^m(xL(y)) + f(y)\Big(Tr_1^m(L(\alpha_1)x)Tr_1^m(L(\alpha_2)x) + Tr_1^m(L(\alpha_1)x)Tr_1^m(L(\alpha_3)x) + Tr_1^m(L(\alpha_2)x)$ $Tr_1^m(L(\alpha_3)x)\Big)$ is bent and its dual function $\tilde{g}$ is given by $\tilde{g}(x, y) = Tr_1^m(L^{-1}$ $(x)y) + f(L^{-1}(x))\Big(Tr_1^m(\alpha_1 y)Tr_1^m(\alpha_2 y) + Tr_1^m(\alpha_1 y)Tr_1^m(\alpha_3 y) + Tr_1^m(\alpha_2 y)Tr_1^m$ $(\alpha_3 y)\Big)$.

2. Let $m = 2k$. Let $a \in \mathbb{F}_{2^k}$ and $b \in \mathbb{F}_{2^m}$ such that $b^{2^k+1} \neq a^2$. Set $\alpha = b^{2^k+1} + a^2$ and $\rho = a + b^{2^k}$. Let g_1, g_2 and g_3 be three Boolean functions over $\mathbb{F}_{2^k}$. Then the Boolean function h defined in bivariate representation on $\mathbb{F}_{2^m} \times \mathbb{F}_{2^m}$ by

$$h(x, y) = Tr_1^m(axy + bxy^{2^k}) + Tr_1^m(xg_1(Tr_k^m(\rho y)))Tr_1^m(xg_2(Tr_k^m(\rho y)))$$
$$+ Tr_1^m(xg_1(Tr_k^m(\rho y)))Tr_1^m(xg_3(Tr_k^m(\rho y)))$$
$$+ Tr_1^m(xg_2(Tr_k^m(\rho y)))Tr_1^m(xg_3(Tr_k^m(\rho y)))$$

is bent and its dual function $\tilde{h}$ is given by

$$\tilde{h}(x, y) = Tr_1^m\Big(\alpha^{-1}(axy + bx^{2^k}y)\Big)$$
$$+ Tr_1^m\Big(\alpha^{-1}(a + b)yg_1(Tr_k^m(x))\Big)Tr_1^m\Big(\alpha^{-1}(a + b)yg_2(Tr_k^m(x))\Big)$$
$$+ Tr_1^m\Big(\alpha^{-1}(a + b)yg_1(Tr_k^m(x))\Big)Tr_1^m\Big(\alpha^{-1}(a + b)yg_3(Tr_k^m(x))\Big)$$
$$+ Tr_1^m\Big(\alpha^{-1}(a + b)yg_2(Tr_k^m(x))\Big)Tr_1^m\Big(\alpha^{-1}(a + b)yg_3(Tr_k^m(x))\Big).$$

3. Let n be a multiple of m where m is a positive integer and $n \neq m$. Let ϕ_1, ϕ_2 and ϕ_3 be three permutations over $\mathbb{F}_{2^m}$ satisfying $(\mathcal{A}_m)$. Let (a_1, a_2, a_3) be a 3-tuple of $\mathbb{F}_{2^m}^{\star}$ such that $a_1 + a_2 + a_3 \neq 0$. Set

$$g(x, y) = Tr_1^n(x\phi_1(y))Tr_1^n(x\phi_2(y)) + Tr_1^n(x\phi_1(y))Tr_1^n(x\phi_3(y))$$
$$+ Tr_1^n(x\phi_2(y))Tr_1^n(x\phi_3(y))$$

if $(x, y) \in \mathbb{F}_{2^n} \times \mathbb{F}_{2^m}$ and

$$g(x, y) = Tr_1^n(a_1 xy^{2^n-2})Tr_1^n(a_2 xy^{2^n-2}) + Tr_1^n(a_1 xy^{2^n-2})Tr_1^n(a_3 xy^{2^n-2})$$
$$+ Tr_1^n(a_2 xy^{2^n-2})Tr_1^n(a_3 xy^{2^n-2})$$

if $(x, y) \in \mathbb{F}_{2^n} \times \mathbb{F}_{2^m} \setminus \mathbb{F}_{2^m}$. Then g is bent and its dual function $\tilde{g}$ is defined by

$$\tilde{g}(x, y) = Tr_1^n(\phi_1^{-1}(x)y)Tr_1^n(\phi_2^{-1}(x)y) + Tr_1^n(\phi_1^{-1}(x)y)Tr_1^n(\phi_3^{-1}(x)y)$$
$$+ Tr_1^n(\phi_2^{-1}(x)y)Tr_1^n(\phi_3^{-1}(x)y)$$

if $(x, y) \in \mathbb{F}_{2^m} \times \mathbb{F}_{2^n}$ and

$$\tilde{g}(x, y) = Tr_1^n(a_1 x^{2^n - 2} y) Tr_1^n(a_2 x^{2^n - 2} y) + Tr_1^n(a_1 x^{2^n - 2} y) Tr_1^n(a_3 x^{2^n - 2} y)$$
$$+ Tr_1^n(a_2 x^{2^n - 2} y) Tr_1^n(a_3 x^{2^n - 2} y)$$

if $(x, y) \in \mathbb{F}_{2^n} \setminus \mathbb{F}_{2^m} \times \mathbb{F}_{2^n}$.

4. Let n be a multiple of m where m is a positive integer and $n \neq m$. Let ϕ_1, ϕ_2 and ϕ_3 be three permutations over $\mathbb{F}_{2^m}$ satisfying $(\mathcal{A}_m)$. Let $a \in \mathbb{F}_{2^m}^\star$ and $c \in \mathbb{F}_{2^n}$ such that $c^4 + c + 1 = 0$. Let d be the inverse of 11 modulo $2^n - 1$. Set

$$g(x, y) = Tr_1^n(x\phi_1(y)) Tr_1^n(x\phi_2(y)) + Tr_1^n(x\phi_1(y)) Tr_1^n(x\phi_3(y))$$
$$+ Tr_1^n(x\phi_2(y)) Tr_1^n(x\phi_3(y))$$

if $(x, y) \in \mathbb{F}_{2^n} \times \mathbb{F}_{2^m}$ and

$$g(x, y) = Tr_1^n(axy^d) Tr_1^n(acxy^d) + Tr_1^n(axy^d) Tr_1^n(ac^{-1}xy^d)$$
$$+ Tr_1^n(acxy^d) Tr_1^n(ac^{-1}xy^d)$$

if $(x, y) \in \mathbb{F}_{2^n} \times \mathbb{F}_{2^n} \setminus \mathbb{F}_{2^m}$. Then g is bent and its dual function $\tilde{g}$ is defined by

$$\tilde{g}(x, y) = Tr_1^n(\phi_1^{-1}(x)y) Tr_1^n(\phi_2^{-1}(x)y) + Tr_1^n(\phi_1^{-1}(x)y) Tr_1^n(\phi_3^{-1}(x)y)$$
$$+ Tr_1^n(\phi_2^{-1}(x)y) Tr_1^n(\phi_3^{-1}(x)y)$$

if $(x, y) \in \mathbb{F}_{2^m} \times \mathbb{F}_{2^n}$ and

$$\tilde{g}(x, y) = Tr_1^n(a^{-11} x^{11} y) Tr_1^n(a^{-11} c^{-11} x^{11} y) + Tr_1^n(a^{-11} x^{11} y) Tr_1^n(a^{-11} c^{11} x^{11} y)$$
$$+ Tr_1^n(a^{-11} c^{-11} x^{11} y) Tr_1^n(a^{-11} c^{11} x^{11} y)$$

if $(x, y) \in \mathbb{F}_{2^n} \setminus \mathbb{F}_{2^m} \times \mathbb{F}_{2^n}$.

5. Let n be a multiple of m where m is a positive integer and $n \neq m$. Let ϕ_1, ϕ_2 and ϕ_3 be three permutations over $\mathbb{F}_{2^m}$ satisfying $(\mathcal{A}_m)$. Let $\alpha \in \mathbb{F}_{2^m}^\star$. Let d be a positive integer such that d and $2^n - 1$ are coprime. Denote by e the inverse of d modulo $2^n - 1$. Set

$$g(x, y) = Tr_1^n(x\phi_1(y)) Tr_1^n(x\phi_2(y)) + Tr_1^n(x\phi_1(y)) Tr_1^n(x\phi_3(y))$$
$$+ Tr_1^n(x\phi_2(y)) Tr_1^n(x\phi_3(y))$$

if $(x, y) \in \mathbb{F}_{2^n} \times \mathbb{F}_{2^m}$ and $g(x, y) = Tr_1^n(\alpha xy^d)$ if $(x, y) \in \mathbb{F}_{2^n} \times \mathbb{F}_{2^n} \setminus \mathbb{F}_{2^m}$. Then g is bent and its dual function $\tilde{g}$ is defined by

$$\tilde{g}(x, y) = Tr_1^n(\phi_1^{-1}(x)y) Tr_1^n(\phi_2^{-1}(x)y) + Tr_1^n(\phi_1^{-1}(x)y) Tr_1^n(\phi_3^{-1}(x)y)$$
$$+ Tr_1^n(\phi_2^{-1}(x)y) Tr_1^n(\phi_3^{-1}(x)y)$$

if $(x, y) \in \mathbb{F}_{2^m} \times \mathbb{F}_{2^n}$ and $\tilde{g}(x, y) = Tr_1^n(\alpha^{-e} x^e y)$ if $(x, y) \in \mathbb{F}_{2^n} \setminus \mathbb{F}_{2^m} \times \mathbb{F}_{2^n}$.

6. Let $n = 2m$ where m is a positive integer. Let ϕ_1, ϕ_2 and ϕ_3 be three permutations over $\mathbb{F}_{2^m}$ satisfying $(\mathcal{A}_m)$. Let d be a positive integer such that $d+1$ and $2^n - 1$ are coprime. Let $\lambda \in \mathbb{F}_{2^m}^\star$. Set

$$g(x,y) = Tr_1^n(x\phi_1(y))Tr_1^n(x\phi_2(y)) + Tr_1^n(x\phi_1(y))Tr_1^n(x\phi_3(y))$$
$$+ Tr_1^n(x\phi_2(y))Tr_1^n(x\phi_3(y))$$

if $(x,y) \in \mathbb{F}_{2^n} \times \mathbb{F}_{2^m}$ and $g(x,y) = Tr_1^n\left(\lambda xy\left(Tr_m^n(y)\right)^d\right)$ if $(x,y) \in \mathbb{F}_{2^n} \times \mathbb{F}_{2^n} \setminus \mathbb{F}_{2^m}$. Then g is bent and its dual function $\tilde{g}$ is defined by

$$\tilde{g}(x,y) = Tr_1^n(\phi_1^{-1}(x)y)Tr_1^n(\phi_2^{-1}(x)y) + Tr_1^n(\phi_1^{-1}(x)y)Tr_1^n(\phi_3^{-1}(x)y)$$
$$+ Tr_1^n(\phi_2^{-1}(x)y)Tr_1^n(\phi_3^{-1}(x)y)$$

if $(x,y) \in \mathbb{F}_{2^m} \times \mathbb{F}_{2^n}$ and $\tilde{g}(x,y) = Tr_1^n\left(\lambda^{-\frac{1}{d+1}}x\left(Tr_m^n(x)\right)^{-\frac{d}{d+1}}y\right)$ if $(x,y) \in \mathbb{F}_{2^n} \setminus \mathbb{F}_{2^m} \times \mathbb{F}_{2^n}$.

4 More Constructions of Bent Functions

In this section, we provide from classes of involutions more primary constructions of bent functions in the line of [31, 32].

An *involution* is a special permutation, but the involution property includes the bijectivity as it appears in the classical definition.

Definition 3. *Let F be any function over $\mathbb{F}_{2^n}$. We say that F is an involution if $F \circ F(x) = x$, for all $x \in \mathbb{F}_{2^n}$.*

In a recent work, Charpin, Mesnager and Sarkar [12] have provided a mathematical study of these involutions. The authors have considered several classes of polynomials and characterized when they are involutions (especially monomials as well as linear involutions) and presented several constructions. New involutions from known ones have also been derived. The following result is an easy consequence of Theorem 1 showing that one can derive bent functions from involutions.

Corollary 2. *Let m be a positive integer. Let ϕ_1, ϕ_2 and ϕ_3 be three involutions of $\mathbb{F}_{2^m}$. Then,*

$$g(x,y) = Tr_1^m(x\phi_1(y))Tr_1^m(x\phi_2(y)) + Tr_1^m(x\phi_1(y))Tr_1^m(x\phi_3(y))$$
$$+ Tr_1^m(x\phi_2(y))Tr_1^m(x\phi_3(y))$$

is bent if and only if $\psi = \phi_1 + \phi_2 + \phi_3$ is an involution. Furthermore, its dual function $\tilde{g}$ is given by $\tilde{g}(x,y) = g(y,x)$.

Remark 1. Notice that this gives a very handy way to compute the dual (namely, transpose the two arguments), in stark contrast with the univariate case.

Using a monomial involution (see [12]), a first construction of a new family of bent functions is given by the following statement.

Theorem 2. *Let n be an integer. Let d be a positive integer such that $d^2 \equiv 1$ (mod $2^n - 1$). Let Φ_1, Φ_2 and Φ_3 be three mappings from $\mathbb{F}_{2^n}$ to $\mathbb{F}_{2^n}$ defined by $\Phi_i(x) = \lambda_i x^d$ for all $i \in \{1,2,3\}$, where the $\lambda_i \in \mathbb{F}_{2^n}^\star$ are pairwise distinct such that $\lambda_i^{d+1} = 1$ and $\lambda_0^{d+1} = 1$, where $\lambda_0 := \lambda_1 + \lambda_2 + \lambda_3$. Let g be the Boolean function defined over $\mathbb{F}_{2^n} \times \mathbb{F}_{2^n}$ by*

$$g(x,y) = Tr_1^n(\Phi_1(y)x)Tr_1^n(\Phi_2(y)x) + Tr_1^n(\Phi_2(y)x)Tr_1^n(\Phi_3(y)x)$$
$$+ Tr_1^n(\Phi_1(y)x)Tr_1^n(\Phi_3(y)x). \tag{4.1}$$

Then the Boolean function g defined over $\mathbb{F}_{2^n} \times \mathbb{F}_{2^n}$ by (4.1) is bent and its dual is given by $\tilde{g}(x,y) = g(y,x)$.

Proof. Set $f_i(x,y) := Tr_1^n(\Phi_i(y)x)$ for all $i \in \{1,2,3\}$. The function f_i belongs to Maiorana-McFarland's class. Moreover, $\Phi_i(y) = \lambda_i y^d$ is a polynomial over $\mathbb{F}_{2^m}$ which is an involution if and only if $\lambda_i^{d+1} = 1$ and $d^2 \equiv 1$ (mod $2^n - 1$). Indeed, we have $\Phi_i(\Phi_i(y)) = \lambda_i^{d+1} y^{d^2}$, hence $\lambda_i^{d+1} y^{d^2} = y$ if and only if $\lambda_i^{d+1} \equiv 1$ and $y^{d^2} \equiv y$ (mod $y^{2^n} + y$), that is, $d^2 \equiv 1$ (mod $2^n - 1$). Using the same arguments, $\sum_{i=1}^3 \Phi_i$ is an involution since we have $(\lambda_1 + \lambda_2 + \lambda_3)^{d+1} = 1$ by hypothesis. Now, since Φ_i (resp. $\sum_{i=1}^3 \Phi_i$) is in particular a permutation over $\mathbb{F}_{2^n}$, for every $i \in \{1,2,3\}$ the Boolean function f_i (resp. $\psi := \sum_{i=1}^3 f_i$) is bent whose dual function equals $\tilde{f}_i$ (resp. $\tilde{\psi}$) defined by $\tilde{f}_i(x,y) = Tr_1^n(y\Phi_i^{-1}(x)) = Tr_1^n(y\Phi_i(x))$, $\forall (x,y) \in \mathbb{F}_{2^n} \times \mathbb{F}_{2^n}$ (resp. $\tilde{\psi}(x,y) = Tr_1^n(y(\Phi_1 + \Phi_2 + \Phi_3)^{-1}(x)) = Tr_1^n(y(\Phi_1 + \Phi_2 + \Phi_3)(x)))$. Therefore, the condition of bentness given in Theorem 1 holds, which completes the proof.

Remark 2. Note that if we multiply λ_1, λ_2, λ_3 by a same non-zero constant a say, $\lambda_i = \frac{1}{a}\mu_i$ for all $i \in \{1,2,3\}$, then the functions g constructed via the λ_i and those h constructed via the μ_i are linked by the relation $h(x,y) = g(ax,y)$. Therefore the functions g and h are affinely equivalent.

The existence of bent functions given in Theorem 2 is a non-trivial arithmetical problem and is discussed in the next session.

Using similar arguments as previously, we derive in Propositions 2 and 3 more constructions of bent functions based on some involutions of $\mathbb{F}_{2^n}$ (see [12]) as application of Corollary 2.

Proposition 2. *Let $n = rk$ be an integer with $k > 1$ and $r > 1$. For $i \in \{1,2,3\}$, let γ_i be an element of $\mathbb{F}_{2^n}^\star$ such that $Tr_k^n(\gamma_i) = 0$ and Φ_i be a mapping defined over $\mathbb{F}_{2^n}$ by*

$$\Phi_i(x) = x + \gamma_i Tr_k^n(x).$$

Then the Boolean function g defined over $\mathbb{F}_{2^n} \times \mathbb{F}_{2^n}$ by (4.1) is bent and its dual function is given by $\tilde{g}(x,y) = g(y,x)$.

Proposition 3. *Let $n = 2m$ be an even integer. Let h_1, h_2, h_3 be three linear mappings from $\mathbb{F}_{2^m}$ to itself. For $i \in \{1, 2, 3\}$, let Φ_i be a mapping from $\mathbb{F}_{2^n}$ to itself defined by*

$$\Phi_i(x) = h_i(Tr_m^n(x)) + x.$$

Then the Boolean function g defined over $\mathbb{F}_{2^n} \times \mathbb{F}_{2^n}$ by (4.1) is bent and its dual function is given by $\tilde{g}(x, y) = g(y, x)$.

Remark 3. Set $\Phi_i'(x) = h_i(Tr_m^n(x)) + x^{2^m}$. Let g' be the Boolean function derived from (4.1) using the Φ_i''s. Then g' is bent and its dual is given by $\tilde{g}'(x, y) = g'(y, x)$. Clearly the functions g (given by the previous theorem) and g' are affinely equivalent. We point out that very recently, Mesnager has exhibited in [33] several new constructions of bent functions employing involutions.

5   Finding Primary Bent Functions from Theorem 2

5.1   Discussion

We now turn to the question of finding values n, d and λ_i which can be used in Theorem 2 and further satisfying certain "non-obviousness" conditions to be laid out below. In other words, we are looking for n, d such that $d^2 \equiv 1 \pmod{2^n - 1}$ and $\lambda_i \in \mathbb{F}_{2^n}^\star$ such that $\lambda_i^{d+1} = 1$ with $\lambda_0 + \lambda_1 + \lambda_2 + \lambda_3 = 0$ and perhaps some additional constraints such as $\lambda_i \neq \lambda_j$ for $i \neq j$. We further refine the problem by introducing the quantity $e := \operatorname{lcm}(d+1, N)/(d+1) = N/\gcd(d+1, N)$ where $N := 2^n - 1$; the significance of this quantity is that for λ_i to be a $(d+1)$-st root of unity in $\mathbb{F}_{2^n}$, a necessary and sufficient condition is that λ_i be a nonzero e-th power, say $\lambda_i = Z_i^e$ (because there are $\gcd(r, N)$ solutions to $rx = 0$ in $\mathbb{Z}/N\mathbb{Z}$, namely the multiples of $N/\gcd(r, N)$).

So, discussing on the value of e, we now have two problems: the *arithmetical problem*, namely, finding for which values of n, d we have $d^2 \equiv 1 \pmod{2^n - 1}$ with $N/\gcd(d+1, N) = e$; and the *algebraic problem*, namely, finding $Z_0, \ldots, Z_3$ nonzero such that $Z_0^e + Z_1^e + Z_2^e + Z_3^e = 0$ (and perhaps some additional constraints for non-obviousness).

In the sequel, we shall denote by $G(e) \leq \mathbb{F}_{2^n}^\star$ the cyclic group of e-th powers.

5.2   The Arithmetical Problem

Given an odd positive integer e, we ask upon what conditions we can find n, d such that $d^2 \equiv 1 \pmod{2^n - 1}$ with $N/\gcd(d+1, N) = e$ for $N := 2^n - 1$.

Let us temporarily forget about N being $2^n - 1$ (except that it is odd). Now if $N = p_1^{v_1} \cdots p_s^{v_s}$ where the p_i are distinct odd primes, finding d such that $d^2 \equiv 1 \pmod{N}$ amounts, by the Chinese remainder theorem, to choosing $\varepsilon_i \in \{\pm 1\}$, and taking $d \equiv \varepsilon_i \pmod{p_i^{v_i}}$ (thus, there are 2^s possible values of d with $d^2 \equiv 1 \pmod{N}$). Then clearly $N/\gcd(d+1, N)$ is the product of the $p_i^{v_i}$ where i ranges over those indices such that $\varepsilon_i = +1$. So if we fix e (a positive odd integer) and look for appropriate values of N, we find that there exists a d (necessarily unique) such that $d^2 \equiv 1 \pmod{N}$ and $N/\gcd(d+1, N) = e$ iff N is the product of e by a positive odd integer prime to it, in other words, N odd and $N \equiv te \pmod{e^2}$ where t is prime to e (and defined modulo e).

Now if we fix an odd positive integer e, and if we choose for t one of the $\varphi(e)$ invertible classes mod e (where φ is Euler's totient function), we are interested in those n such that $2^n \equiv 1 + te \pmod{e^2}$. Not much more can be said about this in general unless we know something about the multiplicative order of 2 mod e^2, but at least we can discuss the small values of e:

Proposition 4. – *For $e = 3$: there exists d such that $d^2 \equiv 1 \pmod{2^n - 1}$ with $N/\gcd(d + 1, N) = e$ (again with $N := 2^n - 1$) iff $n \equiv 2$ or $n \equiv 4$ (mod 6).*
– *For $e = 5$: there exists d such that $d^2 \equiv 1 \pmod{2^n - 1}$ with $N/\gcd(d+1, N) = e$ (again with $N := 2^n - 1$) iff n is congruent mod 20 to one of the following values: $4, 8, 12, 16$.*
– *For $e = 7$: there exists d such that $d^2 \equiv 1 \pmod{2^n - 1}$ with $N/\gcd(d+1, N) = e$ (again with $N := 2^n - 1$) iff n is congruent mod 21 to one of the following values: $3, 6, 9, 12, 15, 18$.*

Proof. In each case, we compute the order of 2 mod e^2, namely 6 for $e = 3$, resp. 20 for $e = 5$, and 21 for $e = 7$, and we then simply compute 2^n mod e^2 for each value of n modulo this order, keeping those which are congruent to $1 + te$ for t prime to e.

5.3   The Algebraic Problem: Generalities

We now turn to the "algebraic problem": given e a positive odd integer and n such that e divides $N := 2^n - 1$, we wish to find $Z_0, \ldots, Z_3$ nonzero such that $Z_0^e + Z_1^e + Z_2^e + Z_3^e = 0$.

The latter equation defines (in 3-dimensional projective space $\mathbb{P}^3_{\mathbb{F}_{2^n}}$) a smooth algebraic surface of a class known as *Fermat hypersurfaces*, which have been studied from the arithmetic and geometric points of view (see, e.g., [13, Sect. 2.14]). The equation has obvious solutions: if $\{i_0, i_1, i_2, i_3\} = \{0, 1, 2, 3\}$ is a labeling of the indices and ω, ω' two e-th roots of unity, then any solution to $\omega Z_{i_0} + Z_{i_1} = 0$ and $\omega' Z_{i_2} + Z_{i_3} = 0$ satisfies $Z_0^e + Z_1^e + Z_2^e + Z_3^e = 0$: these are known as the *standard lines* on the Fermat surface, corresponding to cases where two of the λ_i are equal. Solutions which do not lie on one of the lines are known as nonobvious solutions. We now comment on their existence and explicitly construct some.

5.4   Using the Lang-Weil Estimates

Assume $e \geq 3$ (some odd integer) is arbitrary but fixed. We show that nonobvious solutions exist for n large enough, albeit in a nonconstructive way.

The polynomial $Z_0^e + Z_1^e + Z_2^e + Z_3^e$ is irreducible over the algebraic closure of $\mathbb{F}_2$. (Indeed, if it could be written as PQ with P, Q nonconstant, then all its partial derivatives would vanish where $P = Q = 0$, and nontrivial such points would exist because elementary dimension theory, e.g. [19, Theorem I.7.2], guarantees that over an algebraically closed field, r homogeneous polynomials in $> r$ variables always have a nontrivial common zero. But on the other hand it is clear that the partial derivatives of $Z_0^e + Z_1^e + Z_2^e + Z_3^e$ never all vanish unless all the Z_i vanish. In geometric terms, what we are saying is that a smooth projective hypersurface is geometrically irreducible.)

Because of this, we can apply the Lang-Weil estimates [20, Theorem 1], and conclude that the number of solutions to $Z_0^e + Z_1^e + Z_2^e + Z_3^e = 0$ (in projective 3-space, i.e., up to multiplication by a common constant) over $\mathbb{F}_{2^n}$ is $q^2 + O(q^{3/2})$ where $q := 2^n$ and the constant implied by $O(q^{3/2})$ is absolute. Even if we deduct the at most $O(q)$ points located on each of the curves $Z_i = 0$ and standard lines, we are still left with the same estimate for the number of solutions. This proves:

Proposition 5. *For any odd $e \geq 3$, there exists n_0 such that if $n \geq n_0$, there exist $Z_0, \ldots, Z_3 \in \mathbb{F}_{2^n}$ all nonzero and not located on the standard lines $(\omega Z_{i_0} + Z_{i_1} = 0) \wedge (\omega' Z_{i_2} + Z_{i_3} = 0)$, such that $Z_0^e + Z_1^e + Z_2^e + Z_3^e = 0$.*

In particular, if d is such that $d^2 \equiv 1 \pmod{2^n - 1}$ and $(2^n - 1)/\gcd(d + 1, 2^n - 1) = e$, and if we let $\lambda_i = Z_i^e$, Theorem 2 applies, and no two of the λ_i are equal.

5.5   A Lower Bound on the Number of Solutions

Denote by $N(s, e, g)$ the number of solutions of

$$x_1^e + \cdots x_s^e = g, x_i \in \mathbb{F}_{2^n}, g \in \mathbb{F}_{2^n}^\star.$$

By Theorem 5.22 in [21] (see also [38]), we have:

$$N(s, e, g) \geq 2^{n(s-1)} - (e - 1)^s 2^{n(s-1)/2}.$$

In particular, in the cases of interest to us, namely $s = 2, 3, g \in G(e)$:

– $N(2, e, g) \geq 2^n - (e - 1)^2 2^{n/2} > 0$, for $2^n > (e - 1)^4$.
– $N(3, e, g) \geq 2^{2n} - (e - 1)^3 2^n > 0$, for $2^n > (e - 1)^3$.

Since we are interested only in nontrivial solutions, we should substract at most $2e$ from $N(2, e, g)$ and $3e^2 2^n$ from $N(3, e, g)$ respectively. Once we know there are solutions, there exist deterministic algorithms for finding them, running in polynomial time in terms of e and n (see Theorem A3 in [39]).

5.6   A Semi-explicit Construction

Proposition 6. *If $N > e(2e + 1)$, there exist non trivial zero sums of 4 terms in $G(e)$.*

Proof. Consider all the $M := \binom{|G(e)|}{2}$ pairs $\{a, b\}$ of elements in $G(e)$. If $M > N$, two different pairs must have the same sum, providing a non-trivial 4-term 0-sum of elements of $G(e)$. This occurs as soon as $N > e(2e + 1)$.

Remark 4. Let $c^i + a = c^j + b$ be such a sum; upon normalization, we get: $c + c^{j-i+1} + a' + b' = 0$. That is, we can fix freely one element (c) in the sum.

5.7   From Three to Four e-powers

Let $a + b + c = 0$ be a non-trivial zero sum of 3 elements of $G(e)$ (e-th powers). By cubing this equation, we get: $c^3 = a^3 + b^3 + ab(a + b) = a' + b' + abc$, i.e., a non-trivial zero sum of 4 elements in $G(e)$!

Remark 5. This generalizes to any characteristic $p \neq 3$, but since now we have: $-c^3 = a^3 + b^3 - 3abc$, we need -1 and 3 to be e-th powers (a sufficient condition being that e does not divide $(p - 1)$, in which case all elements of F_p are e-th powers).

6   The Case $e = 3$

We now specialise to the case $e = 3$ and delve further into the study of explicit solutions.

6.1   Explicit parametrization in the case $e = 3$

if $e = 3$, the equation $Z_0^3 + Z_1^3 + Z_2^3 + Z_3^3 = 0$ defines a smooth *cubic surface* (here, a diagonal one), and the 27 sets of simultaneous equations $\ell_{\omega, i_0, i_1 | \omega', i_2, i_3} :=$ $(\omega Z_{i_0} + Z_{i_1} = 0) \wedge (\omega' Z_{i_2} + Z_{i_3} = 0)$ (with $\{i_0, i_1, i_2, i_3\} = \{0, 1, 2, 3\}$ and ω, ω' any two cube roots of unity) define the 27 lines on that cubic surface. We refer to [19, V. Sect. 4] as well as [26, Chap. IV] and the references therein for general background on cubic surfaces and their configuration of 27 lines.

Geometrically (i.e., over an algebraically closed field), a smooth cubic surface is isomorphic to the *blowup* of the projective plane in six points in general position: see [19, *loc. cit.*] or [18, p. 480 & 545]: in practice, this means that the points on the cubic surface correspond to points on the projective plane, except for the six exceptional points which must be replaced by their set of tangent directions (and correspond to six pairwise skew lines on the cubic surface); in particular, the cubic surface is *rational*, meaning that its points can be (almost bijectively) parametrized by rational functions. The same analysis can be performed for a cubic surface over an arbitrary field provided we can find six pairwise skew lines which are (collectively) defined over the base field. This is the case for $Z_0^3 + Z_1^3 + Z_2^3 + Z_3^3 = 0$ over any field, as we can simultaneously "blow down" the two lines $\ell_{\omega_0, 0, 1 | \omega_0, 2, 3}$, for ω_0 ranging over the two primitive cube roots of unity, and their image under cyclic permutations of (Z_1, Z_2, Z_3), all six of which are pairwise skew. Explicitly, in characteristic two, if we blow them down to the points $(1 : \omega_0 : 1)$ and corresponding cyclic permutation of the coordinates $(U : V : W)$, we get the parametrization:

$$Z_0 = UV^2 + VW^2 + WU^2$$
$$Z_1 = U^2 V + V^2 W + W^2 U + V^3 + W^3$$
$$Z_2 = U^2 V + V^2 W + W^2 U + U^3 + W^3$$
$$Z_3 = U^2 V + V^2 W + W^2 U + U^3 + V^3$$

satisfying $Z_0^3 + Z_1^3 + Z_2^3 + Z_3^3 = 0$, whose inverse is given (projectively, i.e., up to constants) by

$$U = Z_0^2 + Z_1^2 + Z_1 Z_2 + Z_2 Z_3 + Z_3^2$$
$$V = Z_1^2 + Z_0 Z_2 + Z_2^2 + Z_0 Z_3 + Z_3^2$$
$$W = Z_0 Z_1 + Z_0 Z_2 + Z_1 Z_2 + Z_0 Z_3 + Z_1 Z_3 + Z_2 Z_3 + Z_3^2$$

(or any one obtained by cyclically permuting both Z_1, Z_2, Z_3 and U, V, W).

The gist of the above explanations is that, if over any field of characteristic two, we substitute *any* values U, V, W other than the six exceptional points $(1 : \omega_0 : 1)$, $(1 : 1 : \omega_0)$, $(\omega_0 : 1 : 1)$ in the first set of equations above, we obtain a solution to $Z_0^3 + Z_1^3 + Z_2^3 + Z_3^3 = 0$; if furthermore the point $(U : V : W)$ is not located on one of the fifteen plane lines through two of the exceptional points (e.g., $U = V$, $V = W$, $U = W$, etc.) or one of the six conics through five of them, the resulting (Z_0, Z_1, Z_2, Z_3) will not be on one of the lines of the cubic surface (i.e., it will be *nonobvious* in the terminology used above), and if $(U : V : W)$ is furthermore chosen outside of the plane cubics $UV^2 + VW^2 + WU^2 = 0$ etc. (given by the equations for the Z_i themselves), the point will have nonzero coordinates so we can use it in construction given in Theorem 2.

(The equations themselves can be checked without any appeal to the machinery of algebraic geometry: for example, using symetries, it is straightforward that, in characteristic two, $(UV^2 + VW^2 + WU^2)^3 + (U^2 V + V^2 W + W^2 U + V^3 + W^3)^3 + (U^2 V + V^2 W + W^2 U + U^3 + W^3)^3 + (U^2 V + V^2 W + W^2 U + U^3 + V^3)^3 = 0$; and one can similarly check that substituting the first set of equations in the second recovers U, V, W up to a common factor, namely $U^4 V + UV^4 + U^2 V^2 W + UVW^3 + W^5$).

6.2   An explicit example

We present an explicit example with $n = 10$ and $d = 340$. To this end, we represent $\mathbb{F}_{2^{10}}$ modulo the minimal polynomial $m(x) := x^{10} + x^6 + x^5 + x^3 + x^2 + x + 1$. Let $\xi \in \mathbb{F}_{2^{10}}$ be the class of x mod $m(x)$. Then for example taking $U = 1$, $V = \xi$, $W = 1 + \xi$ in the equations above gives $Z_0 = \xi^3 + \xi^2 + 1$, $Z_1 = \xi^3 + \xi^2$, $Z_2 = \xi^2 + 1$ and $Z_3 = \xi$, whose cubes, viz., $\lambda_0 = \xi^9 + \xi^8 + \xi^7 + \xi^4 + \xi^3 + \xi^2 + 1$, $\lambda_1 = \xi^9 + \xi^8 + \xi^7 + \xi^6$, $\lambda_2 = \xi^6 + \xi^4 + \xi^2 + 1$ and $\lambda_3 = \xi^3$ all satisfy $\lambda_i^{341} = 1$ (and sum up to 0).

Acknowledgments. The first author thanks Jens Groth (Program Chair of the international conference IMACC 2015) for his nice invitation.

References

1. Budaghyan, L., Carlet, C., Helleseth, T., Kholosha, A., Mesnager, S.: Further results on Niho bent functions. IEEE Trans. Inf. Theor. **58**(11), 6979–6985 (2012)
2. Canteaut, A., Charpin, P., Kyureghyan, G.: A new class of monomial bent functions. Finite Fields Appl. **14**(1), 221–241 (2008)

3. Carlet, C.: Two new classes of bent functions. In: Helleseth, T. (ed.) EUROCRYPT 1993. LNCS, vol. 765, pp. 77–101. Springer, Heidelberg (1994)
4. Carlet, C.: A construction of bent function. In: Proceedings of the Third International Conference on Finite Fields and Applications, pp. 47–58. Cambridge University Press (1996)
5. Carlet, C.: On bent and highly nonlinear balanced/resilient functions and their algebraic immunities. In: Fossorier, M.P.C., Imai, H., Lin, S., Poli, A. (eds.) AAECC 2006. LNCS, vol. 3857, pp. 1–28. Springer, Heidelberg (2006)
6. Carlet, C.: Boolean functions for cryptography and error correcting codes. In: Crama, Y., Hammer, P.L. (eds.) Boolean Models and Methods in Mathematics, Computer Science, and Engineering, pp. 257–397. Cambridge University Press, Cambridge (2010)
7. Carlet, C.: Open problems on binary bent functions. In: Proceeding of the conference Open problems in mathematical and computational sciences, Sept. 18-20, 2013, in Istanbul, Turkey, pp. 203–241. Springer (2014)
8. Carlet, C., Mesnager, S.: On Dillon's class H of bent functions, Niho bent functions and o-polynomials. J. Comb. Theor. Ser. A **118**(8), 2392–2410 (2011)
9. Carlet, C., and Mesnager, S.: Four decades of research on bent functions. Journal Designs, Codes and Cryptography (to appear)
10. Charpin, P., Gong, G.: Hyperbent functions, Kloosterman sums and Dickson polynomials. In: ISIT 2008, pp. 1758–1762 (2008)
11. Charpin, P., Kyureghyan, G.: Cubic monomial bent functions: a subclass of $\mathcal{M}$. SIAM J. Discrete Math. **22**(2), 650–665 (2008)
12. Charpin, P., Mesnager, S., Sarkar, S.: On involutions of finite fields. In: Proceedings of 2015 IEEE International Symposium on Information Theory, ISIT (2015)
13. Debarre, O.: Higher-Dimensional Algebraic Geometry. Universitext. Springer, New York (2001)
14. Dillon, J.: Elementary Hadamard difference sets. Ph.D. thesis, University of Maryland (1974)
15. Dillon, J., Dobbertin, H.: New cyclic difference sets with Singer parameters. Finite Fields Appl. **10**(3), 342–389 (2004)
16. Dobbertin, H., Leander, G., Canteaut, A., Carlet, C., Felke, P., Gaborit, P.: Construction of bent functions via Niho power functions. J. Comb. Theor. Ser. A **113**, 779–798 (2006)
17. Gold, R.: Maximal recursive sequences with 3-valued recursive crosscorrelation functions. IEEE Trans. Inf. Theor. **14**(1), 154–156 (1968)
18. Griffiths, P., Harris, J.: Principles of Algebraic Geometry. Wiley, New York (1978)
19. Hartshorne, R.: Algebraic geometry. In: Hartshorne, R. (ed.) GTM, vol. 52. Springer, New York (1977)
20. Lang, S., Weil, A.: Number of points on varieties in finite fields. Amer. J. Math. **76**, 819–827 (1954)
21. Lidl, R., Niederreiter, H.: Finite Fields, Encyclopedia Mathematics Applications, vol. 20. Addison-Wesley, Reading (1983)
22. Leander, G.: Monomial bent functions. IEEE Trans. Inf. Theor. **52**(2), 738–743 (2006)
23. Leander, G., Kholosha, A.: Bent functions with 2^r Niho exponents. IEEE Trans. Inf. Theor. **52**(12), 5529–5532 (2006)
24. Li, N., Helleseth, T., Tang, X., Kholosha, A.: Several new classes of bent functions from Dillon exponents. IEEE Trans. Inf. Theor. **59**(3), 1818–1831 (2013)
25. McFarland, R.L.: A family of noncyclic difference sets. J. Comb. Theor. Ser. A **15**, 1–10 (1973)

26. Manin, Y.L.: Cubic Forms: Algebra, Geometry, Arithmetic. North-Holland, Amsterdam (1974)
27. Mesnager, S.: A new family of hyper-bent boolean functions in polynomial form. In: Parker, M.G. (ed.) Cryptography and Coding 2009. LNCS, vol. 5921, pp. 402–417. Springer, Heidelberg (2009)
28. Mesnager, S.: Hyper-bent boolean functions with multiple trace terms. In: Hasan, M.A., Helleseth, T. (eds.) WAIFI 2010. LNCS, vol. 6087, pp. 97–113. Springer, Heidelberg (2010)
29. Mesnager, S.: Bent and hyper-bent functions in polynomial form and their link with some exponential sums and Dickson polynomials. IEEE Trans. Inf. Theor. **57**(9), 5996–6009 (2011)
30. Mesnager, S.: A new class of bent and hyper-bent boolean functions in polynomial forms. Des. Codes Crypt. **59**(1–3), 265–279 (2011)
31. Mesnager, S.: Several new infinite families of bent functions and their duals. IEEE Trans. Inf. Theor. **60**(7), 4397–4407 (2014)
32. Mesnager, S.: Further constructions of infinite families of bent functions from new permutations and their duals. Journal of Cryptography and Communications (CCDS). Springer (to appear)
33. Mesnager, S.: A note on constructions of bent functions from involutions. Cryptology ePrint Archive: Report 2015/982 (2015)
34. Mesnager, S.: Bent functions: fundamentals and results. Springer, New York (2016, to appear)
35. Mesnager, S., Flori, J.P.: Hyper-bent functions via Dillon-like exponents. IEEE Trans. Inf. Theor. **59**(5), 3215–3232 (2013)
36. Rothaus, O.S.: On "bent" functions. J. Comb. Theor. Ser. A **20**, 300–305 (1976)
37. Yu, N.Y., Gong, G.: Construction of quadratic bent functions in polynomial forms. IEEE Trans. Inf. Theor. **52**(7), 3291–3299 (2006)
38. Winterhof, A.: On Waring's problem in finite fields. Acta Arithmetica LXXXVII.2 **87**, 171–177 (1998)
39. van de Woestijne, C.E.: Deterministic equation solving over finite fields. Ph.D. Thesis, Math. Inst. Univ. Leiden (2006)

Best Paper Award

Security Against Related Randomness Attacks via Reconstructive Extractors

Kenneth G. Paterson[1], Jacob C.N. Schuldt[2]([☒]), Dale L. Sibborn[1], and Hoeteck Wee[3]

[1] Royal Holloway, University of London, Surrey, UK
[2] National Institute of Advanced Industrial Science and Technology, Tokyo, Japan
jacob.schuldt@aist.go.jp
[3] École Normale Supérieure, Paris, France

Abstract. This paper revisits related randomness attacks against public key encryption schemes as introduced by Paterson, Schuldt and Sibborn (PKC 2014). We present a general transform achieving security for public key encryption in the related randomness setting using as input any secure public key encryption scheme in combination with an auxiliary-input reconstructive extractor. Specifically, we achieve security in the function-vector model introduced by Paterson *et al.*, obtaining the first constructions providing CCA security in this setting. We consider instantiations of our transform using the Goldreich-Levin extractor; these outperform the previous constructions in terms of public-key size and reduction tightness, as well as enjoying CCA security. Finally, we also point out that our approach leads to an elegant construction for Correlation Input Secure hash functions, which have proven to be a versatile tool in diverse areas of cryptography.

Keywords: Public-key encryption · Related randomness attacks · Auxiliary-inputs · Reconstructive extractors · CIS hash functions

1 Introduction

In recent work, and motivated by numerous practical attacks involving diverse kinds of randomness failure, Paterson, Schuldt and Sibborn [22] introduced *related randomness attacks* against public key encryption schemes. In such an attack, the adversary is able to control the randomness and public keys used during encryption; the security target is that messages encrypted under an honestly generated public key should still remain hidden from the adversary to the maximum extent that this is possible. In the model of Paterson *et al.* [22], the adversary is able to force the encryption scheme to use random values that are

---

The first and third authors were supported by EPSRC grant EP/L018543/1, the second author is supported by JSPS KAKENHI Grant Number 15K16006, and the last author is partially supported by ANR Project EnBiD (ANR-14-CE28-0003).

© Springer International Publishing Switzerland 2015
J. Groth (Ed.): IMACC 2015, LNCS 9496, pp. 23–40, 2015.
DOI: 10.1007/978-3-319-27239-9_2

related to one another in ways that are specified by functions acting on the randomness space of the scheme. This modelling is inspired by practical attacks like those by Ristenpart and Yilek in [25], which exploit randomness generation in virtual machines, and extends the *Reset Attack (RA)* setting considered by Yilek in [28]. As demonstrated in [22], it is also connected to other research topics such as security against related key (RKA) attacks and leakage resilience[1].

1.1 The RRA Setting

In the Related Randomness Attack (RRA) setting, the adversary can not only force the reuse of existing random values as in the RA setting, but can also force the use of *functions of* those random values. The extra adversarial power in the RRA setting allows the modelling of reset attacks in which the adversary does not have an exact reset capability, but where the randomness used after a reset is in some way related to that used on previous resets. Such behaviours were observed in the experimental work by Ristenpart *et al.* [25], for example. Via access to an **Enc** oracle, the RRA adversary is able to get arbitrary messages encrypted under arbitrary public keys, using functions ϕ of an initial set of well-distributed but unknown random values. The public keys can even be maliciously generated, and hence, the adversary might know the corresponding private keys. The adversary is tasked with winning an indistinguishability-style game defined via a left-or-right oracle, **LR**, which consistently returns the encryption of either the first or second message of message pairs submitted to the oracle. The encryptions are with respect to an honestly generated target public key pk^*, but again where the adversary can force the use of functions ϕ of the initial random values. When the functions ϕ are limited to coming from some set Φ, we speak of a Φ-restricted adversary.

Because the adversary may know all but one of the private keys, it can check that its challenger is behaving correctly with respect to its encryption queries. Moreover, these queries concern public keys that are outside the control of the challenger. This makes achieving security in the RRA setting technically quite challenging, while practically relevant.

1.2 Previous Results

Paterson *et al.* [22] gave a variety of security models and constructions for PKE secure under related randomness attacks (RRA) in the CPA and CCA settings. As a first contribution, they explored the use of the Random Oracle Model, obtaining necessary and sufficient conditions on the function set Φ that are required to obtain RRA security (these being collision-resistance and output-unpredictability of Φ). They also showed how to transform any PKE scheme PKE

---

[1] See also [22] for an extended discussion of the practical motivation for studying related randomness attacks based on the attack literature as represented by [2,6,7, 9,11–13,15–17,19,21,25].

into a new PKE scheme Hash-PKE that is RRA-secure for Φ-restricted adversaries, simply by hashing the random input together with the public key and message during encryption. This construction is closely related to approaches in [3, 25].

In the standard model, Paterson *et al.* were able to show that any Φ-restricted related key attack-secure PRF (RKA-PRF) can be used to build a RRA-secure PKE scheme for Φ-restricted adversaries, thus transferring security from the RKA setting for PRFs to the RRA setting for PKE. Using the RKA-PRFs currently available in the literature [1, 4, 20] to instantiate this construction, schemes secure for function families Φ consisting of polynomials of bounded degree, can be achieved. However, providing security for function families Φ not enjoying such a convenient algebraic structure would be much more relevant to practical related randomness attacks. But this is a challenging task: in fact, the results by Wichs [27] imply that, for a large class of encryption schemes[2], security for arbitrary function families Φ cannot be shown via a black-box reduction, based on any cryptographic game involving a single-stage adversary (e.g. computational assumptions like DDH, IND-CCA security of a public key encryption scheme, etc.). To obtain further constructions, Paterson *et al.* considered weakened security models; the weakening taking place along two independent dimensions: the degree of control that the adversary enjoys over the public keys under which it can force encryptions for related random values, and the degree of adaptivity it has in the selection of functions $\phi \in \Phi$. More specifically, they considered the situations where:

- The public keys are all honestly generated at the start of the security game, the public keys and all but one of the private keys are then given to the adversary, and the adversary can adaptively specify the functions $\phi \in \Phi$ involved in its queries. This is called the honest-key, related randomness attack (HK-RRA) setting in [22].
- There is no restriction on public keys, but instead of letting the adversary adaptively choose the functions $\phi \in \Phi$, the security game itself is parametrised by a vector of functions $\boldsymbol{\phi} = (\phi_1, \ldots, \phi_q)$ that will be used in the attack, and security is required to hold for all choices of $\boldsymbol{\phi}$ from some set $\boldsymbol{\Phi}$. This is called the function-vector, related randomness attack (FV-RRA) setting in [22]. The difference between this setting and the (adaptive) HK-RRA setting is subtle, but note that in the FV-RRA setting, the adversary's choice of ϕ_i cannot depend on the oracles' outputs for the previously used functions $\phi_1, \ldots, \phi_{i-1}$, whereas in the HK-RRA setting, it may.

In the first of these two settings, Paterson *et al.* obtained a generic construction for a scheme achieving HK-RRA security based on combining any PKE scheme with a Correlated-Input Secure (CIS) hash function [14]. However, the then-known instantiations of CIS hash functions only enabled them to obtain selective, HK-RRA security for Φ-restricted adversaries where Φ is a large class

---

[2] Specifically, Wichs' results apply to encryption schemes which are injective with respect to the used randomness.

of polynomial functions. In view of recent results on RKA-PRFs [1], this construction now appears to be superseded by their earlier generic construction using Φ-restricted RKA-PRFs.

In the second of the two settings, they gave a direct construction for a PKE scheme that is FV-RRA-CPA secure solely under the DDH assumption, assuming the component functions ϕ_i of ϕ are simultaneously hard to invert on a random input. The scheme is based on a specific PKE scheme of Boneh *et al.* [8] that is secure in the so-called *auxiliary input setting*, wherein the adversary is given a hard-to-invert function of the secret key as part of its input. However, in this setting, only a CPA-secure scheme was given in [22].

1.3   Our Contributions

In this paper, we give a new, general transform for achieving FV-RRA-ATK security for hard-to-invert function families from standard IND-ATK security, where ATK $\in \{\text{CPA}, \text{CCA}\}$. In fact, the transform works for a stronger notion of FV-RRA-ATK security than was originally introduced in [22]: we will allow an adversary to also manipulate the randomness used for the **LR** queries, instead of being restricted to using only the identity function in such queries. Furthermore, besides yielding schemes secure in the CCA setting, which was left as an open problem in [22], we show that this transform allows us to construct encryption schemes that have tighter security reductions (and are more efficient) than the single FV-RRA-CPA secure scheme that was presented in [22]. As motivation for considering the class of hard-to-invert functions, note that achieving FV-RRA-ATK security for this class would be relevant in modelling the one-way state evolution of a PRNG which has exhausted its entropy pool but which doesn't receive new entropy.

Auxiliary-Input Reconstructive Extractors: Our transform makes use of a technical tool called an *auxiliary-input reconstructive extractor*. Classically, an *extractor* is a function Ext, which, given an input and a seed, produces an output that is statistically indistinguishable from elements chosen uniformly at random from some set Σ, provided the input is chosen from a distribution with sufficient min-entropy and the seed is chosen uniformly at random. A *reconstructive extractor* is an extractor with the additional property that, roughly speaking, allows the efficient reconstruction of the input x from any distinguisher $\mathcal{D}$ that successfully distinguishes the output of the extractor from random. This is formalised in terms of the existence of an oracle machine Rec outputting x. Then an *auxiliary-input* reconstructive extractor is a reconstructive extractor in which the output still remains indistinguishable when the distinguisher $\mathcal{D}$ is also given access to the output of a leakage function $h(\cdot)$ on input x. Our actual definition (Definition 6) extends this idea further still: the distinguisher $\mathcal{D}$ is given either a set of uniformly random values or the set of outputs of the extractor when evaluated on $\phi(x)$ for all $\phi \in \boldsymbol{\phi}$, where $\boldsymbol{\phi}$ is a vector of functions defined by the game.

Our Transform (Intuition): Equipped with an auxiliary-input reconstructive extractor, our transform to achieve FV-RRA-ATK security is conceptually simple:

- We append a uniformly random extractor seed s to each public key, resulting in a new public key denoted $\hat{pk}$.
- The encryption algorithm consumes a random value r from some set of bit strings; this is fed into the extractor to create a value $K \leftarrow \text{Ext}(r, s)$. This value K used as a key for a Pseudorandom Function (PRF) F to compute $r' \leftarrow F_K(\hat{pk}\|m)$ where m is the message to be encrypted. Finally, r' is used as the actual randomness for encryption, and we simply encrypt with the original encryption algorithm.
- Decryption works exactly as in the original decryption algorithm.

Details of the construction are given in Fig. 4 in Sect. 4.

Intuitively, a challenge encryption constructed using randomness value $\phi(r)$ remains secure, since the extractor guarantees an output indistinguishable from random, even when the adversary gains access to encryptions under the related randomness values $\phi'(r)$. Hence, the PRF, which uses the extractor output as a key, will guarantee that independent randomness values are used for different public key and message pairs. In turn, this implies that the adversary is forced to break the security of the underlying PKE scheme to learn anything about the encrypted challenge messages. That this approach attains FV-RRA-ATK security is formally proven in Theorem 1.

Instantiations. In Sect. 5, we consider the instantiation of our transform using the Goldreich-Levin extractor. This provides a particularly neat construction of FV-RRA-ATK-secure PKE in which we start with an IND-ATK-secure scheme and augment it with a simple inner-product computation to prepare the key for the PRF. However, we stress that, given the limited strength of known results for the security of the Goldreich-Levin extractor in the auxiliary input setting [10], our results using this extractor are in turn limited to the original FV-RRA-ATK security model of [22] (i.e. in which the adversary is restricted to using the identity function in its **LR** queries). Still, the schemes obtained from using our transform with this extractor have significant benefits compared to the single concrete FV-RRA-CPA-secure scheme from [22]. For example, we obtain shorter public keys and a tighter security reduction compared to the scheme from [22]. Most importantly, we obtain FV-RRA-CCA security in a completely generic way.

Connection to CIS Hash Functions: As a final contribution, in Sect. 6, we explore the connections between auxiliary-input reconstructive extractors and Correlated-Input Secure (CIS) hash functions. The latter were introduced by Goyal *et al.* in [14] and have proven useful in a variety of cryptographic constructions including password-based login, efficient searches on encrypted data and RKA-PRFs. We will show that any reconstructive extractor can be used to construct a secure CIS hash function of a certain type. Specifically, our security definition for CIS hash functions involves functions that are selected from pre-specified sets, as opposed to being adaptively selected as in the strongest (but mostly unachieved) definitions in [14]. Using the Goldreich-Levin extractor once more provides a construction for CIS hash functions that is exquisitely simple: given key $c \in \mathbb{Z}_p^n$ and input $r \in H^n$ (where H is an arbitrary subset of $\mathbb{Z}_p$), the CIS hash function output is simply:

$$h_c(r) := \langle r, c \rangle$$

where the inner product is evaluated over $\mathbb{Z}_p$.

2 Preliminaries

Notation. Throughout the paper we will use $\lambda \in \mathbb{N}$ to denote the security parameter, which will sometimes be written in its unary representation, 1^λ. We denote by $y \leftarrow x$ the assignment of y to x, and by $s \leftarrow_\$ S$ we denote the selection of an element s uniformly at random from the set S. The notation $[n]$ represents the set $\{1, 2, \ldots, n\}$. For an algorithm A, we denote by $y \leftarrow A(x; r)$ that A is run with input x and random coins r, and that the output is assigned to y.

All our security games and proofs will utilise code-based games and the associated language. Here we briefly recall the basic definitions from Bellare *et al.* in [5]. A game consists of at least two procedures. We begin with **Initialise**, which assigns starting values to all variables and then gives outputs, if there are any, to the adversary. The adversary $\mathcal{A}$ may then submit queries to the oracle procedures, and when $\mathcal{A}$ halts (and possibly outputs a value) the **Finalise** procedure begins. **Finalise** will take the output from $\mathcal{A}$ (if there is one) as its input and will output its own value. The value output by **Finalise** is defined to be the output of the game. We write $\mathbb{P}[G^{\mathcal{A}} \Rightarrow b]$ to denote the probability that game G outputs bit b when run with $\mathcal{A}$. For brevity, in what follows ATK will denote either CPA or CCA, where theorems or statements apply to both games. Any proofs or figures will refer to the CCA setting, but may be easily modified to the CPA case.

Public Key Encryption. We denote a specific PKE scheme by PKE $=$ (PKE.K, PKE.E, PKE.D). All three algorithms are polynomial-time. The randomised key generation algorithm PKE.K takes the security parameter as its input and outputs a key pair (pk, sk). The encryption algorithm, on input a message $m \in \mathcal{M}$ and a public key pk chooses random coins from Rnd and uses these coins to output a ciphertext c. The decryption algorithm is deterministic. Its inputs are a private key sk and a ciphertext c. The algorithm either outputs a message m or an error symbol $\perp$. We require the scheme PKE to satisfy the correctness property. That is, for all $\lambda \in \mathbb{N}$, all pairs (pk, sk) output by the key generation algorithm, and all messages $m \in \mathcal{M}$, we require that PKE.D$(sk, $PKE.E$(pk, m)) = m$.

Definition 1. *The advantage of an* IND-ATK *adversary* $\mathcal{A}$ *against a scheme* PKE *is*

$$\mathbf{Adv}_{\mathrm{PKE}, \mathcal{A}}^{\mathrm{ind\text{-}atk}}(\lambda) := 2 \cdot \mathbb{P}[\text{IND-ATK}_{\mathrm{PKE}}^{\mathcal{A}}(\lambda) \Rightarrow 1] - 1$$

where game IND-ATK *is shown in Fig. 1. A scheme* PKE *is* IND-ATK *secure if the advantage of any polynomial-time adversary is negligible in the security parameter* λ.

proc. Initialise(λ):	proc. LR(m_0, m_1):	proc. Dec(c):
$b \leftarrow_\$ \{0,1\}$;	$c \leftarrow_\$ \text{PKE.E}(pk, m_b)$	if $c \in \mathcal{S}$, return $\perp$
$(pk, sk) \leftarrow_\$ \text{PKE.K}(1^\lambda)$;	$\mathcal{S} \leftarrow \mathcal{S} \cup \{c\}$	else return PKE.D(sk, c)
$\mathcal{S} \leftarrow \emptyset$;	return c	
return pk		proc. Finalise(b'):
		If $b = b'$, return 1

Fig. 1. Game IND-ATK for PKE. (If ATK = CPA, the adversary's access to **proc. Dec** is removed.)

proc. Initialise(λ):	proc. Initialise(λ):
$K \leftarrow_\$ \text{Keys}_\lambda$	FunTab $\leftarrow \emptyset$
proc. Function(x):	proc. Function(x):
return $F(K, x)$	if FunTab[x] $= \perp$,
	$\quad$ FunTab[x] $\leftarrow_\$ \text{Rng}_\lambda$
proc. Finalise(b):	return FunTab[x]
return b	
	proc. Finalise(b):
	return b

Fig. 2. Games for PRF security. Game PRFReal is on the left, PRFRand on the right.

Pseudorandom Functions. We recall the standard definition of pseudorandom functions:

Definition 2. *Let* F : $\text{Keys}_\lambda \times \text{Dom}_\lambda \rightarrow \text{Rng}_\lambda$ *be a family of functions. The advantage of a* PRF *adversary* $\mathcal{A}$ *against* F *is*

$$\mathbf{Adv}^{\text{prf}}_{F,\mathcal{A}}(\lambda) := \mathbb{P}[\text{PRFReal}^{\mathcal{A}}_F(\lambda) \Rightarrow 1] - \mathbb{P}[\text{PRFRand}^{\mathcal{A}}_\$(\lambda) \Rightarrow 1]$$

where the games PRFReal *and* PRFRand *are defined in Fig. 2. We say* F *is a secure* PRF *family if the advantage of any polynomial-time adversary is negligible in the security parameter* λ.

3   Function Vector Related Randomness Security

In this section we recall the FV-RRA-ATK notion of security from [22], and then slightly strengthen this definition to encompass a more general attack.

The FV-RRA-ATK game is designed to capture related randomness attacks, in which the adversary is allowed to obtain challenge encryptions, as well as encryptions for maliciously chosen keys, using related randomness values. This is achieved by giving the adversary access to an encryption oracle **Enc** which enables the adversary to manipulate the random values used for the encryption. More specifically, the standard FV-RRA-ATK security game is parametrised

proc. Initialise(λ):	**proc. LR**(m_0, m_1, i, j):	**proc. Enc**(pk, m, i, j):
$b \leftarrow_\$ \{0, 1\}$;	If CoinTab$[i] = \perp$,	if CoinTab$[i] = \perp$,
$(pk^*, sk^*) \leftarrow_\$$ PKE.K(1^λ);	$\quad$ CoinTab$[i] \leftarrow_\$$ Rnd	$\quad$ CoinTab$[i] \leftarrow_\$$ Rnd
CoinTab $\leftarrow \emptyset$; $\mathcal{S} \leftarrow \emptyset$;	$r_i \leftarrow$ CoinTab$[i]$	$r_i \leftarrow$ CoinTab$[i]$
return pk^*	$c \leftarrow$ PKE.E$(pk^*, m_b; \phi_j(r_i))$	$c \leftarrow$ PKE.E$(pk, m; \phi'_j(r_i))$
	$\mathcal{S} \leftarrow \mathcal{S} \cup \{c\}$	return c
proc. Dec(c):	return c	
if $c \in \mathcal{S}$, then return $\perp$		**proc. Finalise**(b'):
else return PKE.D(sk^*, c)		if $b = b'$, return 1

Fig. 3. Game (ϕ, ϕ')-FV-RRA-ATK, where $\phi = (\phi_1, \dots, \phi_q)$ and $\phi' = (\phi'_1, \dots, \phi'_{q'})$. (If ATK = CPA, then the adversary's access to **proc. Dec** is removed.)

by a vector of functions $\phi = (\phi_1, \dots, \phi_q)$, where $q := q(\lambda)$ is polynomial in the security parameter λ, and the adversary may request encryption queries by submitting a tuple of the form (pk, m, i, j) to its **Enc** oracle. This tuple consists of a public key pk, a message m, an index i selecting the random value r_i with which to encrypt, and an index j that selects the function ϕ_j that modifies the randomness r_i before encryption. Hence, the adversary will receive the response PKE.E$(pk, m; \phi_j(r_i))$, where the values r_i are uniform and independent. The adversary may furthermore query a Left or Right (**LR**) oracle with a tuple (m_0, m_1, i). The response of this oracle will be PKE.E$(pk^*, m_b; r_i)$, where pk^* is the target public key and b is a bit, both of which are chosen uniformly and independently during the initialisation stage of the security game. Note that the randomness values r_i used to respond to **LR** queries are uniformly chosen random values. In the CCA version of the game, an adversary can additionally submit ciphertexts c to a decryption oracle **Dec**. The decryption oracle will return PKE.D(sk^*, c) as long as the ciphertext c was not returned by the **LR** oracle. When the adversary has made all its (polynomially many) queries, it will submit a bit b' to a **Finalise** procedure, which represents the adversary's guess for the bit b. The **Finalise** procedure will output 1 (representing an adversarial win) if $b = b'$. The security game for this notion is given in Fig. 3.

We will now introduce some new definitions that slightly strengthen the FV-RRA-ATK notion from [22] outlined above. Our strengthening allows an adversary to manipulate the randomness used for the **LR** queries, instead of being restricted to using only the identity function. The security game for our new notion is given in Fig. 3. The major difference from the definition of [22] is that the game is parametrised by two sets of functions, ϕ and ϕ'. An adversary may only use functions from ϕ in its **LR** queries, and the functions in ϕ' may only be used for **Enc** queries. Notice that if $\phi = \{\text{id}\}$, then this definition recovers the corresponding FV-RRA-ATK security game and notion from [22]. While our generic transform is proven secure in the stronger model shown in Fig. 3, we stress that, because of the limitations of currently known reconstructive extractors, our concrete instantiation of the transform will be secure only in the weaker model of [22].

The following definition has been adapted from [22] for our purposes. The definition captures natural restrictions which must be placed on an adversary with the capability of controlling the randomness of the challenge encryptions in an IND-ATK style security game. This is reminiscent of the restrictions put in place in the security definition for deterministic encryption (e.g. see [24]).

Definition 3. *Let $\mathcal{A}$ be an adversary in Game (ϕ, ϕ')-FV-RRA-ATK that queries r different randomness indices to its **LR** and **Enc** oracles and makes $q_{i,\phi}$ queries to its **LR** oracle with index i and function $\phi \in \phi$. Let $(m_0^{i,\phi,1}, m_1^{i,\phi,1})$, $\ldots, (m_0^{i,\phi,q_{i,\phi}}, m_1^{i,\phi,q_{i,\phi}})$ be $\mathcal{A}$'s **LR** queries for index $i \in [r]$ and $\phi \in \phi$. Suppose that for all pairs $(i, \phi) \in [r] \times \phi$ and for all $j \neq k \in [q_{i,\phi}]$, we have:*

$$m_0^{i,\phi,j} = m_0^{i,\phi,k} \text{ iff } m_1^{i,\phi,j} = m_1^{i,\phi,k}.$$

Then we say that $\mathcal{A}$ is equality-pattern respecting.

Note that any adversary that is not equality-pattern respecting can trivially win the game in Fig. 3. More specifically, the adversary can simply query its **LR** oracle with the tuples (m_0, m_1, i, j) and (m_0, m_2, i, j), where m_0, m_1 and m_2 are all distinct. The values i and j can be an arbitrary values from the appropriate domain. If the bit b is equal to 0, the adversary will receive identical ciphertexts, whereas the ciphertexts will differ if b equals 1. This results in a trivial win for an adversary. In contrast, an equality-respecting adversary cannot exploit the available oracles in this particular way, and is forced to mount a non-trivial attack against the scheme to win the security game.

With the above definition in place, we can now formally define FV-RRA-ATK security.

Definition 4. *Let $\phi = (\phi_1, \ldots, \phi_q)$ and $\phi' = (\phi'_1, \ldots, \phi'_{q'})$ be vectors of $q := q(\lambda)$ and $q' := q'(\lambda)$ functions respectively. We define the advantage of an equality-pattern respecting, (ϕ, ϕ')-FV-RRA-ATK adversary $\mathcal{A}$ against a PKE scheme PKE to be:*

$$\mathbf{Adv}_{\mathsf{PKE},\mathcal{A}}^{(\phi,\phi')\text{-fv-rra-atk}}(\lambda) := 2 \cdot \Pr[(\phi, \phi')\text{-FV-RRA-ATK}_{\mathsf{PKE}}^{\mathcal{A}}(\lambda) \Rightarrow 1] - 1.$$

If Φ and Φ' are sets of vectors of functions, then a PKE scheme PKE is said to be (Φ, Φ')-FV-RRA-ATK secure if, for all $\phi \in \Phi$ and for all $\phi' \in \Phi'$, the advantage of any equality-pattern respecting, (ϕ, ϕ')-FV-RRA-ATK adversary against PKE that runs in polynomial time is negligible in the security parameter λ.

Similar to the notion defined in [22], it is possible to reduce the above defined FV-RRA-ATK security to a simpler notion in which the security game involves only a single uniformly chosen random value used in all oracle queries. The following lemma follows easily from Lemma 1 of [22] and is therefore presented without a proof.

Lemma 1. *Consider an equality-pattern respecting, (ϕ, ϕ')-FV-RRA-ATK adversary $\mathcal{A}$ that queries q_r distinct randomness indices and makes at most q_{LR}*

LR queries. Then there exists an equality-pattern respecting, (ϕ, ϕ')-FV-RRA-ATK adversary $\mathcal{B}$ that queries at most 1 randomness index and makes at most q_{LR} LR queries such that

$$\mathbf{Adv}_{\mathrm{PKE},\mathcal{A}}^{(\phi,\phi')\text{-fv-rra-atk}}(\lambda) \leq q_r \cdot \mathbf{Adv}_{\mathrm{PKE},\mathcal{B}}^{(\phi,\phi')\text{-fv-rra-atk}}(\lambda),$$

where $\mathcal{B}$ runs in approximately the same time as $\mathcal{A}$. In the CCA setting, $\mathcal{B}$ makes the same number of decryption queries as $\mathcal{A}$.

4 Obtaining FV-RRA Security from Auxiliary-Input Reconstructive Extractors

In this section we present the main result of the paper. Recall that this result improves upon the work of Paterson et al. [22] by proposing a transform that converts any IND-ATK scheme into an FV-RRA-ATK scheme via the use of an auxiliary-input reconstructive extractor. Recall also that the authors of [22] only provided a single concrete instantiation of a FV-RRA-CPA secure scheme. In the later sections we will provide instantiations of our transform that are not only able to meet the stronger FV-RRA-CCA notion, but also provide shorter public keys and a tighter security reduction compared to the scheme from [22].

Before introducing the extractors we utilise in our transform, we first need to define the notion of a vector of functions being δ-hard-to-compute with respect to another vector of functions.

Definition 5. *Let $\phi = (\phi_1, \ldots, \phi_q)$ and $\phi' = (\phi'_1, \ldots, \phi'_{q'})$ denote vectors of functions on a set $\mathtt{Rnd}_\lambda$, where $q := q(\lambda)$ and $q' := q'(\lambda)$ are polynomial in the security parameter λ. Let $\delta(\lambda)$ be a function. We say that ϕ is $\delta(\lambda)$-hard-to-compute with respect to ϕ' if, for all polynomial time algorithms $\mathcal{A}$ and all sufficiently large λ, we have:*

$$\Pr[\phi_i(r) \leftarrow \mathcal{A}(\phi'_1(r), \ldots, \phi'_q(r)) : r \leftarrow_\$ \mathtt{Rnd}_\lambda] \leq \delta(\lambda),$$

for all $i \in \{1, \ldots, q\}$. We say that a set of vectors of functions $\boldsymbol{\Phi}$ is δ-hard-to-compute with respect to $\boldsymbol{\Phi}'$ if each vector $\phi \in \boldsymbol{\Phi}$ is δ-hard-to-compute with respect to every vector in $\boldsymbol{\Phi}'$ (note that the vectors in such a set $\boldsymbol{\Phi}$ need not all be of the same dimension, but we assume they each have dimension that is polynomial in λ). If $\delta = \mathrm{negl}(\lambda)$, then we simply say that $\boldsymbol{\Phi}$ is hard-to-compute with respect to $\boldsymbol{\Phi}'$.

A natural question to ask is: what functions satisfy this notion of being δ-hard-to-compute? For simplicity, consider the scenario where $\boldsymbol{\Phi} = \{\mathrm{id}\}$ (in which case we simply say that $\boldsymbol{\Phi}'$ is δ-hard-to-invert, cf. Definition 14 of [22]), and assume that $\boldsymbol{\Phi}'$ consists of only one function, say ϕ'. In this scenario, an obvious example of a δ-hard-to-invert function is a function that fixes certain bits of the output e.g. a function ϕ' that takes a bit-string of length n as input, and returns a string consisting of k zero bits followed by the least significant $n - k$ bits of the input (for $0 \leq k \leq n$). No information is leaked about the first k bits of the input, and hence no algorithm can invert ϕ' with probability greater than 2^{-k} when the input string is uniformly random. Therefore, if $k \geq -\log_2 \delta$, the function ϕ' (and, consequently, $\boldsymbol{\Phi}'$) is δ-hard-to-invert. This example can naturally be extended to the case where $\boldsymbol{\Phi}'$ contains multiple vectors of functions and $\boldsymbol{\Phi} \neq \{\text{id}\}$.

We now introduce our generalised definition of an auxiliary-input reconstructive extractor.

Definition 6. *An $(\epsilon, \delta, \boldsymbol{\Phi}, \boldsymbol{\Phi}')$-auxiliary-input reconstructive extractor is a pair of functions* (Ext, Rec) *such that* Ext *is an extractor that maps from $\{0,1\}^n \times \{0,1\}^d$ to Σ, and* Rec *is an oracle machine that on input $(1^n, 1/\epsilon)$ runs in time* $\text{poly}(n, 1/\epsilon, \log(|\Sigma|))$. *Furthermore, for every $x \in \{0,1\}^n$, every $\phi = (\phi_1, \ldots, \phi_q) \in \boldsymbol{\Phi}$, every $\phi' \in \boldsymbol{\Phi}'$, and every function $\mathcal{D}$ such that*

$$\left| \Pr_{s \leftarrow_\$ \{0,1\}^d}[\mathcal{D}(s, \{\text{Ext}(\phi_i(x), s)\}_{i \in \{1,\ldots,q\}}, \phi'(x)) = 1] \right.$$

$$\left. - \Pr_{\substack{s \leftarrow_\$ \{0,1\}^d \\ \sigma_i \leftarrow_\$ \Sigma}}[\mathcal{D}(s, \{\sigma_i\}_{i \in \{1,\ldots,q\}}, \phi'(x)) = 1] \right| \geq \epsilon$$

we require that

$$\Pr[\text{Rec}^{\mathcal{D}}(1^n, 1/\epsilon, \phi'(x)) - \phi_i(x)] \geq \delta$$

for some $i \in \{1, \ldots, q\}$, where $\phi = (\phi_1, \ldots, \phi_q)$, $q := q(\lambda)$ is polynomial, and the probability is over the coin tosses of Rec. *If, for every $\mathcal{D}$ with non-negligible ϵ,* Rec *reconstructs $\phi_i(x)$ with non-negligible probability, we may simply say that* (Ext, Rec) *is a $(\boldsymbol{\Phi}, \boldsymbol{\Phi}')$-auxiliary-input reconstructive extractor.*

Armed with this new definition of an auxiliary-input reconstructive extractor, we are ready to state the main result of this paper. We show that any extractor satisfying Definition 6 can be used in conjunction with an IND-ATK secure PKE scheme and a PRF to meet the FV-RRA-ATK security notion in Fig. 3. The encryption scheme that achieves this result is in Fig. 4. The algorithm works by appending a uniformly random extractor seed to each public key, but leaving the private key unmodified. The encryption algorithm generates a uniformly random r, which is then fed into the extractor (using the seed from the public key). The output of the extractor is used as a key for a PRF, and the input to the PRF is the public key appended with the message. Finally, the output of the PRF is used as the new randomness for encryption, and then we simply encrypt with the standard encryption algorithm.

Theorem 1. *If $\boldsymbol{\Phi}$ is hard-to-compute with respect to $\boldsymbol{\Phi}'$ and* (Ext, Rec) *is an $(\boldsymbol{\Phi}, \boldsymbol{\Phi}')$-auxiliary-input reconstructive extractor, then the PKE scheme* EXT-PKE *in Fig. 4 is $(\boldsymbol{\Phi}, \boldsymbol{\Phi}')$-FV-RRA-ATK secure when instantiated with a secure PRF and an IND-ATK secure PKE scheme* PKE. *More precisely, consider any polynomial-size vectors of functions $\phi \in \boldsymbol{\Phi}$ and $\phi' \in \boldsymbol{\Phi}'$, any $(\epsilon, \delta, \boldsymbol{\Phi}, \boldsymbol{\Phi}')$-auxiliary-input reconstructive extractor* (Ext, Rec), *and any equality-pattern respecting, (ϕ, ϕ')-FV-RRA-ATK adversary $\mathcal{A}$ against* EXT-PKE. *Suppose $\mathcal{A}$ makes q_{LR} **LR** queries*

Alg. EXT-PKE.K(1^λ):	Alg. EXT-PKE.E($\hat{pk}, m$):	Alg. EXT-PKE.D($\hat{sk}, c$):
$(pk, sk) \leftarrow$ PKE.K(1^λ)	$r \leftarrow_\$ $ Rnd	$m \leftarrow$ PKE.D(sk, c)
$s \leftarrow$ seeds	$K \leftarrow$ Ext(r, s)	return m
$\hat{pk} \leftarrow (pk, s)$	$r' \leftarrow F_K(\hat{pk} \| m)$	
$\hat{sk} \leftarrow (sk)$	$c \leftarrow$ PKE.E($pk, m; r'$)	
return $\hat{pk}$	return c	

Fig. 4. Scheme EXT-PKE built from a reconstructive extractor, a PKE scheme PKE, and a PRF F.

and uses q_r randomness indices. Then, either Φ is not δ-hard-to-compute with respect to Φ', or there exists a PRF adversary $\mathcal{B}$, and an IND-ATK adversary $\mathcal{C}$, all running in polynomial time, such that:

$$\mathbf{Adv}_{\text{EXT-PKE},\mathcal{A}}^{(\phi,\phi')\text{-fv-rra-atk}}(\lambda) < 2q_r \cdot q \cdot \mathbf{Adv}_{F,\mathcal{B}}^{\text{prf}}(\lambda) + q_r \cdot q_{LR} \cdot \mathbf{Adv}_{\text{PKE},\mathcal{C}}^{\text{ind-atk}}(\lambda) + 2q_r\epsilon.$$

The proof of the above theorem can be found in the full version of the paper.

5   Instantiation of an Auxiliary-Input Reconstructive Extractor

Given Theorem 1, it now remains to see what extractors exist that satisfy Definition 6. The strongest extractor we are aware of is the Goldreich-Levin extractor, whose properties are analysed in [10, Theorem 1]. That theorem states the following (with the notation changed to remain consistent with ours):

Theorem 2. Let p be a prime, and let H be an arbitrary subset of $\mathbb{Z}_p$. Let $f : H^n \to \{0, 1\}^*$ be any (possibly randomised) function. If there is a distinguisher $\mathcal{D}$ that runs in time t such that

$$\left| \Pr[\mathbf{r} \leftarrow H^n, y \leftarrow f(\mathbf{r}), \mathbf{s} \leftarrow \mathbb{Z}_p^n : \mathcal{D}(y, \mathbf{s}, \langle \mathbf{r}, \mathbf{s} \rangle) = 1] \right.$$

$$\left. - \Pr[\mathbf{r} \leftarrow H^n, y \leftarrow f(\mathbf{r}), \mathbf{s} \leftarrow \mathbb{Z}_p^n, u \leftarrow \mathbb{Z}_p : \mathcal{D}(y, \mathbf{s}, u) = 1] \right| = \epsilon$$

then there is an inverter $\mathcal{A}$ that runs in time $t' = t \cdot \text{poly}(n, |H|, 1/\epsilon)$ such that[3]

$$\Pr[\mathbf{r} \leftarrow H^n, y \leftarrow f(\mathbf{r}) : \mathcal{A}(y) = \mathbf{r}] \geq \frac{\epsilon^3}{512 \cdot n \cdot p^3}. \tag{1}$$

This theorem can be used to obtain an auxiliary-input reconstructive extractor. Specifically, consider the extractor Ext that maps from $H^n \times \mathbb{Z}_p^n$ to $\mathbb{Z}_p$ (where H is a subset of $\mathbb{Z}_p$) defined as

$$\text{Ext}(r, s) = \langle r, s \rangle.$$

---

[3] The bound quoted in [10] had the denominator $512np^2$. However, we believe the bound has a slight error and should in fact be $512np^3$, as given here. The bound in Eq. (1) was also used by Paterson et al. in [23].

Alg. **EXT-PKE.K**(1^λ):	Alg. **EXT-PKE.E**$(\hat{pk}, m)$:	Alg. **EXT-PKE.D**$(\hat{sk}, c)$:
$(pk, sk) \leftarrow \text{PKE.K}(1^\lambda)$	$r \leftarrow_\$ H^\lambda$	$m \leftarrow \text{PKE.D}(sk, c)$
$s \leftarrow \mathbb{Z}_p^\lambda$	$K \leftarrow \langle r, s \rangle$	return m
$\hat{pk} \leftarrow (pk, s)$	$r' \leftarrow F_K(\hat{pk}\|m)$	
$\hat{sk} \leftarrow (sk)$	$c \leftarrow \text{PKE.E}(pk, m; r')$	
return $\hat{pk}$	return c	

Fig. 5. Scheme **EIP-PKE** (Euclidean Inner Product) built from a PKE scheme **PKE**, and a PRF F. Here, H denotes a subset of $\mathbb{Z}_q$.

Matching the notation of Theorem 2 with Definition 6, **Rec** is now $\mathcal{A}$, $\boldsymbol{\Phi} = \{\text{id}\}$, $\boldsymbol{\Phi}'$ is the set of δ-hard-to-invert vectors of functions, ϕ' is the function f, and the extractor **Ext** is easily seen to be an $(\epsilon, \delta, \text{id}, \boldsymbol{\Phi}')$-auxiliary-input reconstructive extractor, where

$$\epsilon = \sqrt[3]{512\delta\lambda p^3}.$$

Note that, in the proof of [10], the theorem is stated with one function f. However, we now use a vector of functions $(\phi'_1, \ldots, \phi'_q)$ in our proof. Fortunately this is not problematic, since we can simply interpret f (whose output is in $\{0, 1\}^*$) as a vector of functions. That is, we can set $f(r) = (\phi'_1(r), \ldots, \phi'_q(r))$.

By combining Theorem 2 with Theorem 1, we easily obtain the following theorem.

Theorem 3. *Let $\boldsymbol{\Phi}'$ be a set of hard-to-invert vectors of functions on $\{0, 1\}^\lambda$. Then PKE scheme* **EIP-PKE** *in Fig. 5 is* (id, $\boldsymbol{\Phi}'$)-*FV-RRA-ATK secure. More precisely, consider any polynomial-size vector of functions $\phi' \in \boldsymbol{\Phi}'$ which are δ-hard-to-invert, and any equality-pattern respecting,* (id, ϕ')-*FV-RRA-ATK adversary $\mathcal{A}$ against* **EIP-PKE**. *Suppose $\mathcal{A}$ makes q_{LR}* **LR** *queries and uses q_r randomness indices. Then there exists a PRF adversary $\mathcal{B}$ and an IND-ATK adversary $\mathcal{C}$, all running in polynomial time, such that:*

$$\mathbf{Adv}^{(\text{id},\phi')\text{-fv-rra-atk}}_{\text{EIP-PKE},\mathcal{A}}(\lambda) < 2q_r \cdot \mathbf{Adv}^{\text{prf}}_{F,\mathcal{B}}(\lambda) + q_r \cdot q_{LR} \cdot \mathbf{Adv}^{\text{ind-atk}}_{\text{PKE},\mathcal{C}}(\lambda) + 2q_r \sqrt[3]{512\delta\lambda p^3}.$$

While the above theorem limits the challenge functions modifying the input to the extractor to being the identity function, the schemes resulting from our transform using the above reconstructive extractor still enjoy several advantages over the single FV-RRA-CPA-secure scheme that was presented in [22]. Most notably, [22] only gave one concrete scheme, which is only secure in the CPA version of the FV-RRA-ATK game. Our theorem not only shows how to achieve CCA security (which was left as an open problem in [22]), but also shows how to convert *any* IND-CCA scheme into an FV-RRA-CCA secure scheme. Furthermore, the security bound of our theorem is tighter than that of [22], and our theorem facilitates the use of much smaller public keys. For comparison, when using our transform with the above Goldreich-Levin extractor, the public key of the underlying PKE scheme is modified to include λ additional components from $H \subset \mathbb{Z}_q$. Hence, transforming, for example, the PKE scheme by Kurosawa and

Desmedt [18], yields a scheme with public keys consisting of $\lambda+4$ group elements and a hash function key. In contrast, the modified BHHO scheme presented in [22] requires public keys consisting of $2 \cdot k(\lambda)$ group elements (where k is polynomial). Furthermore, the loss of security in the security reduction of the modified BHHO scheme includes the component $\sqrt[3]{512\delta kp^4}$, which originates from the reduction to the δ-hard-to-invert functions. In comparison, the corresponding loss of security obtained from applying our transform is $\sqrt[3]{512\delta\lambda p^3}$, which leads to a weaker requirement on the δ-hard-to-invert functions.

It remains an open question whether there exists extractors that will enable stronger notions of FV-RRA-ATK security to be shown for schemes like EXT-PKE (Fig. 4), or alternative extractors that have, for example, shorter seeds. However, this seems difficult at present. A standard technique to obtain an (ϵ, δ)-auxiliary-input reconstructive extractor is to use complexity-leveraging with a standard reconstructive extractor [26]. Unfortunately, this technique does not appear to work in the FV-RRA-ATK setting. More specifically, if we wish to use complexity-leveraging, we require the range of the auxiliary function to be smaller than the domain. However, for our FV-RRA-ATK game to make sense, we require that for each ϕ we have $\mathcal{D}(\phi) = \mathcal{R}(\phi) = \texttt{Rnd}$. Hence, complexity-leveraging seems to be incompatible with the FV-RRA-ATK model.

6   Connections with CIS Hash Functions

We will now briefly explore the connections between $(\epsilon, \delta, \boldsymbol{\Phi}, \boldsymbol{\Phi}')$-auxiliary-input reconstructive extractors and correlated-input secure (CIS) hash functions. In particular, we will show that any reconstructive extractor can be used to construct a secure CIS hash function. Correlated-input secure hash functions were first studied by Goyal et al. in [14]. They introduced several definitions of security, but the one we shall be concerned with is the pseudorandomness notion. Intuitively, a hash function is (pseudorandom) correlated-input secure if the challenge output of the hash function is indistinguishable from random even when an adversary is allowed to see outputs on correlated inputs. That is, an adversary can submit correlation functions ϕ to its oracle and will receive $h(\phi(r))$, where h is the (possibly keyed) hash function, and r is a uniformly random input chosen at the beginning of the security game. The adversary may submit multiple oracle queries, and finally forwards a challenge function ϕ^* to the oracle. The game will return either $h(\phi^*(r))$ or z, where z is chosen uniformly at random from the range of the hash function. The hash function is (adaptively) secure if the adversary has negligible advantage in distinguishing the outputs.

As noted in [14], CIS hash functions have applications to password-based login and efficient searches on encrypted data. Furthermore, they share interesting connections with Related-Key Attack secure primitives. However, the CIS hash function construction presented in [14] only achieves selective security for correlation functions ϕ corresponding to polynomials of bounded degree, which limits its usefulness in the above mentioned applications. Constructing adaptive CIS hash functions for a wide class of functions is a challenging task,

proc. Initialise(λ):	proc. Challenge(j):	proc. Query(i):
$b \leftarrow_\$ \{0,1\}$;	if $b = 0$,	return $h_c(\phi_i'(r))$
$h_c \leftarrow_\$ \mathcal{H}$	$z \leftarrow_\$ \mathcal{R}(h_c)$	
$r \leftarrow_\$ \mathcal{D}(h_c)$	return z	proc. Finalise(b'):
return h_c	else,	If $b = b'$, return 1
	return $h_c(\phi_j(r))$	

Fig. 6. The (ϕ, ϕ')-CIS hash game, where $\phi = (\phi_1, \ldots, \phi_q)$ and $\phi' = (\phi_1', \ldots, \phi_{q'}')$.

in particular for non-algebraic function classes. This is evidenced by the results of Wichs [27], which show that injective CIS hash functions cannot be proved secure for arbitrary correlation functions ϕ via a black-box reduction, based on any cryptographic game. However, here we show that auxiliary-input reconstructive extractors can be used to construct a specific kind of CIS hash functions. To explore this connection, we must consider a variant of the CIS hash security game that was presented in [14]. The security game is shown in Fig. 6, while our definition of security is given below.

Definition 7. *The advantage of an adversary $\mathcal{A}$ against a family of hash functions $\mathcal{H}$ in the (ϕ, ϕ')-CIS game (Fig. 6) is defined to be*

$$\mathbf{Adv}_{\mathcal{H},\mathcal{A}}^{(\phi,\phi')\text{-cis}}(\lambda) := 2 \cdot \Pr[(\phi, \phi')\text{-CIS}_{\mathcal{H}}^{\mathcal{A}}(\lambda) \Rightarrow 1] - 1.$$

Definition 8. *A family of hash functions $\mathcal{H}$ is said to be $(\boldsymbol{\Phi}, \boldsymbol{\Phi}')$-pseudorandom correlated-input secure if, for all $\phi \in \boldsymbol{\Phi}$, all $\phi' \in \boldsymbol{\Phi}'$, and all polynomial time adversaries $\mathcal{A}$, we have*

$$\mathbf{Adv}_{\mathcal{H},\mathcal{A}}^{(\phi,\phi')\text{-cis}}(\lambda) \leq \mathrm{negl}(\lambda).$$

Notice that in our new definition, instead of letting the adversary adaptively choose the functions as in [14], the security game itself is parametrised with function vectors ϕ and ϕ', and security is required to hold for all choices of $\phi \in \boldsymbol{\Phi}$ and $\phi' \in \boldsymbol{\Phi}'$. It is worth stressing that there is a subtle difference between the two approaches to defining security for CIS hash functions, and the definition used here implies that the function vectors ϕ and ϕ' will be independent of the chosen hash function (i.e. the hash function key c).

With these definitions and notions in place we can define our hash function family $\mathcal{H}$ from an extractor as follows:

$$h_c(r) := \mathtt{Ext}(r, c).$$

The following theorem establishes the security of the hash function, based on the security of the underlying auxiliary-input reconstructive extractor.

Theorem 4. *Let $\mathtt{Ext}$ be an $(\epsilon, \delta, \boldsymbol{\Phi}, \boldsymbol{\Phi}')$-auxiliary-input reconstructive extractor, and let $\boldsymbol{\Phi}$ be δ-hard-to-compute with respect to $\boldsymbol{\Phi}'$. Consider the hash function* *family* $\mathcal{H}$ *defined by the hash functions* $h_c(r) := \texttt{Ext}(c,r)$. *Then, for any* $\phi \in \Phi$, *any* $\phi' \in \Phi'$, *and all polynomial time adversaries* $\mathcal{A}$, *we have*

$$\mathbf{Adv}_{\mathcal{H},\mathcal{A}}^{(\phi,\phi')\text{-cis}}(\lambda) < \epsilon.$$

We will sketch the proof of the above theorem.

Proof (Sketch). If an adversary $\mathcal{A}$ has advantage greater than or equal to ϵ, we would be able to build an extractor adversary $\mathcal{D}$ that distinguishes the outputs of the extractor with probability ϵ. This in turn would allow us to build a function Rec that recovers r with probability greater than δ (cf. Definition 6), which is not possible by assumption. Hence, we have a contradiction, so the advantage of the adversary $\mathcal{A}$ must be less than ϵ.   □

A concrete instantiation of such a CIS hash is possible via Theorem 1 of [10]. If we define

$$h_c(r) := \langle r, c \rangle, \tag{2}$$

where $c \in \mathbb{Z}_p^\lambda$ and $r \in H^\lambda$ for $H \subset \mathbb{Z}_p$, then the following corollary is obvious.

Corollary 1. *Consider the hash function family* $\mathcal{H}$ *defined by Eq. 2, and let* Φ' *be a set of* δ*-hard-to-invert functions. Then, for all* $\phi' \in \Phi'$, *and all polynomial time adversaries* $\mathcal{A}$, *we have*

$$\mathbf{Adv}_{\mathcal{H},\mathcal{A}}^{(\text{id},\phi')\text{-cis}}(\lambda) < \sqrt[3]{512\delta\lambda p^3}.$$

As highlighted above, CIS hash functions share interesting connections with RKA-secure primitives. In fact, [14] proposed a general approach for obtaining RKA-security via a CIS hash function. For example, consider a standard signature scheme given by algorithms $\{\texttt{KeyGen}, \texttt{Sign}, \texttt{Verify}\}$, and a (id, Φ')-pseudorandom correlated-input secure hash function h for which we assume a key c is publicly available. To obtain a RKA-secure signature scheme for functions Φ', simply replace the random coins r used by $\texttt{KeyGen}$ with $h_c(r)$, and the signing key with r. Furthermore, since the signing key of the original scheme is no longer stored, the algorithm $\texttt{Sign}$ must regenerate this from r using h_c and $\texttt{KeyGen}$. As shown in [14], the resulting signature scheme will be RKA-secure for functions Φ'.

Note that, in this approach, only a (id, Φ')-pseudorandom correlated-input secure hash function is required. Hence, by using the CIS hash function from Corollary 1 in the above sketched transformation, we can obtain a RKA-secure signature scheme for hard-to-invert functions. As far as the authors are aware, this is the first construction of a RKA-secure signature scheme for this class of functions. Furthermore, a similar result can be obtained for any primitive for which the above transformation applies. However, note that due to the properties of the above described security model for CIS hash functions, which implies that the functions Φ' are independent of the hash function key, we only obtain selective RKA-security.

References

1. Abdalla, M., Benhamouda, F., Passelègue, A., Paterson, K.G.: Related-key security for pseudorandom functions beyond the linear barrier. In: Garay, J.A., Gennaro, R. (eds.) CRYPTO 2014, Part I. LNCS, vol. 8616, pp. 77–94. Springer, Heidelberg (2014)
2. Becherer, A., Stamos, A., Wilcox, N.: Cloud computing security: raining on the trendy new parade. In: BlackHat USA (2009)
3. Bellare, M., Brakerski, Z., Naor, M., Ristenpart, T., Segev, G., Shacham, H., Yilek, S.: Hedged public-key encryption: how to protect against bad randomness. In: Matsui, M. (ed.) ASIACRYPT 2009. LNCS, vol. 5912, pp. 232–249. Springer, Heidelberg (2009)
4. Bellare, M., Cash, D.: Pseudorandom functions and permutations provably secure against related-key attacks. In: Rabin, T. (ed.) CRYPTO 2010. LNCS, vol. 6223, pp. 666–684. Springer, Heidelberg (2010)
5. Bellare, M., Rogaway, P.: The security of triple encryption and a framework for code-based game-playing proofs. In: Vaudenay, S. (ed.) EUROCRYPT 2006. LNCS, vol. 4004, pp. 409–426. Springer, Heidelberg (2006)
6. Bendel, M.: Hackers describe PS3 security as epic fail, gain unrestricted access (2011). http://www.exophase.com/20540/hackers-describe-ps3-security-as-epic-fail-gain-unrestricted-access/
7. Bernstein, D.J., Chang, Y.-A., Cheng, C.-M., Chou, L.-P., Heninger, N., Lange, T., van Someren, N.: Factoring RSA keys from certified smart cards: coppersmith in the wild. Cryptology ePrint Archive, report 2013/599 (2013). http://eprint.iacr.org/
8. Boneh, D., Halevi, S., Hamburg, M., Ostrovsky, R.: Circular-secure encryption from decision diffie-hellman. In: Wagner, D. (ed.) CRYPTO 2008. LNCS, vol. 5157, pp. 108–125. Springer, Heidelberg (2008)
9. Debian Security Advisory DSA-1571-1: OpenSSL - predictable random number generator (2008). http://www.debian.org/security/2008/dsa-1571
10. Dodis, Y., Goldwasser, S., Tauman Kalai, Y., Peikert, C., Vaikuntanathan, V.: Public-key encryption schemes with auxiliary inputs. In: Micciancio, D. (ed.) TCC 2010. LNCS, vol. 5978, pp. 361–381. Springer, Heidelberg (2010)
11. Dodis, Y., Pointcheval, D., Ruhault, S., Vergnaud, D., Wichs, D.: Security analysis of pseudo-random number generators with input: /dev/random is not robust. IACR Cryptology ePrint Archive 2013:338 (2013)
12. Dorrendorf, L., Gutterman, Z., Pinkas, B.: Cryptanalysis of the random number generator of the Windows operating system. ACM Trans. Inf. Syst. Secur. **13**(1) (2009)
13. Goldberg, I., Wagner, D.: Randomness and the Netscape browser (1996). http://www.drdobbs.com/windows/184409807
14. Goyal, V., O'Neill, A., Rao, V.: Correlated-input secure hash functions. In: Ishai, Y. (ed.) TCC 2011. LNCS, vol. 6597, pp. 182–200. Springer, Heidelberg (2011)
15. Gutterman, Z., Malkhi, D.: Hold your sessions: an attack on java session-id generation. In: Menezes, A. (ed.) CT-RSA 2005. LNCS, vol. 3376, pp. 44–57. Springer, Heidelberg (2005)
16. Gutterman, Z., Pinkas, B., Reinman, T.: Analysis of the linux random number generator. In: IEEE Symposium on Security and Privacy, pp. 371–385. IEEE Computer Society (2006)

17. Heninger, N., Durumeric, Z., Wustrow, E., Alex Halderman, J.: Mining your Ps and Qs: detection of widespread weak keys in network devices. In: USENIX Security Symposium, August 2012
18. Kurosawa, K., Desmedt, Y.G.: A new paradigm of hybrid encryption scheme. In: Franklin, M. (ed.) CRYPTO 2004. LNCS, vol. 3152, pp. 426–442. Springer, Heidelberg (2004)
19. Lenstra, A.K., Hughes, J.P., Augier, M., Bos, J.W., Kleinjung, T., Wachter, C.: Public keys. In: Safavi-Naini, R., Canetti, R. (eds.) CRYPTO 2012. LNCS, vol. 7417, pp. 626–642. Springer, Heidelberg (2012)
20. Lucks, S.: Ciphers secure against related-key attacks. In: Roy, B., Meier, W. (eds.) FSE 2004. LNCS, vol. 3017, pp. 359–370. Springer, Heidelberg (2004)
21. Michaelis, K., Meyer, C., Schwenk, J.: Randomly failed! the state of randomness in current java implementations. In: Dawson, E. (ed.) CT-RSA 2013. LNCS, vol. 7779, pp. 129–144. Springer, Heidelberg (2013)
22. Paterson, K.G., Schuldt, J.C.N., Sibborn, D.L.: Related randomness attacks for public key encryption. In: Krawczyk, H. (ed.) PKC 2014. LNCS, vol. 8383, pp. 465–482. Springer, Heidelberg (2014)
23. Paterson, K.G., Schuldt, J.C.N., Sibborn, D.L.: Related randomnessattacks for public key encryption. IACR Cryptology ePrint Archive 2014:337 (2014)
24. Raghunathan, A., Segev, G., Vadhan, S.: Deterministic public-key encryption for adaptively chosen plaintext distributions. In: Johansson, T., Nguyen, P.Q. (eds.) EUROCRYPT 2013. LNCS, vol. 7881, pp. 93–110. Springer, Heidelberg (2013)
25. Ristenpart, T., Yilek, S.: When good randomness goes bad: virtual machine reset vulnerabilities and hedging deployed cryptography. In: NDSS. The Internet Society (2010)
26. Wee, H.: Public key encryption against related key attacks. In: Fischlin, M., Buchmann, J., Manulis, M. (eds.) PKC 2012. LNCS, vol. 7293, pp. 262–279. Springer, Heidelberg (2012)
27. Wichs, D.: Barriers in cryptography with weak, correlated and leaky sources. In: Kleinberg, R.D. (ed.) ITCS, pp. 111–126. ACM (2013)
28. Yilek, S.: Resettable public-key encryption: how to encrypt on a virtual machine. In: Pieprzyk, J. (ed.) CT-RSA 2010. LNCS, vol. 5985, pp. 41–56. Springer, Heidelberg (2010)

Authentication

MI-T-HFE, A New Multivariate Signature Scheme

Wenbin Zhang$^{(\boxtimes)}$ and Chik How Tan

Temasek Laboratories, National University of Singapore, Singapore, Singapore
{tslzw,tsltch}@nus.edu.sg

Abstract. In this paper, we propose a new multivariate signature scheme named MI-T-HFE as a competitor of QUARTZ. The core map of MI-T-HFE is of an HFEv type but more importantly has a specially designed trapdoor. This special trapdoor makes MI-T-HFE have several attractive advantages over QUARTZ. First of all, the core map and the public map of MI-T-HFE are both surjective. This surjectivity property is important for signature schemes because any message should always have valid signatures; otherwise it may be troublesome to exclude those messages without valid signatures. However this property is missing for a few major signature schemes, including QUARTZ. A practical parameter set is proposed for MI-T-HFE with the same length of message and same level of security as QUARTZ, but it has smaller public key size, and is more efficient than (the underlying HFEv- of) QUARTZ with the only cost that its signature length is twice that of QUARTZ.

Keywords: Post-quantum cryptography · Multivariate signature scheme · QUARTZ · HFEv

1   Introduction

Multivariate public key cryptosytems (MPKCs) are constructed using polynomials and their public keys are represented by a polynomial map $F = (f_1, \ldots, f_m)$: $\mathbb{F}_q^n \to \mathbb{F}_q^m$ where $\mathbb{F}_q$ is the field of q elements and each f_i is a polynomial. The security of MPKCs relies on the following MP problem:

MP Problem. Solve the system $f_1(\mathbf{x}) = 0$, $\ldots$, $f_m(\mathbf{x}) = 0$, where each f_i is a polynomial in $\mathbf{x} = (x_1, \ldots, x_m) \in \mathbb{F}_q^n$ and all coefficients are in $\mathbb{F}_q$.

This problem is usually called the MQ problem if the degree of the system is two; namely each f_i is a quadratic polynomial. The MP problem is NP-hard if the degree is at least two [GJ79]. Especially the MQ problem is also NP-hard in general. Based on this NP-hardness and along with its computational efficiency, MPKCs is considered as a potential candidate for post-quantum cryptography.

To use polynomial maps $F = (f_1, \ldots, f_m) : \mathbb{F}_q^n \to \mathbb{F}_q^m$ for public key cryptography, one needs to design trapdoors in the polynomial maps. Currently the most common construction of such a trapdoor is of the following bipolar form [DY09]:

© Springer International Publishing Switzerland 2015
J. Groth (Ed.): IMACC 2015, LNCS 9496, pp. 43–56, 2015.
DOI: 10.1007/978-3-319-27239-9_3

$$\bar{F} = L \circ F \circ R : \mathbb{F}_q^n \xrightarrow{R} \mathbb{F}_q^n \xrightarrow{F} \mathbb{F}_q^m \xrightarrow{L} \mathbb{F}_q^m$$

where L, R are invertible affine maps and $F = (f_1, \ldots, f_m)$ is a polynomial map. The public key is $\bar{F}$ while the secret key usually consists of L, R, F. It should be efficient to invert the central map F but infeasible to invert $\bar{F}$ unless one knows L, R, F.

In MPKCs multivariate polynomials can be used for both encryption schemes and signature schemes, and encryption schemes can often be converted to signature schemes, but here we shall focus on signature schemes only. The public key of a multivariate signature scheme is a specially designed polynomial map $F : \mathbb{F}_q^n \to \mathbb{F}_q^m$, a message is a vector $\mathbf{y} \in \mathbb{F}_q^m$ and a signature is a vector $\mathbf{x} \in \mathbb{F}_q^n$. Given any message $\mathbf{y}$, the signer need to solve the equation $F(\mathbf{x}) = \mathbf{y}$ using the trapdoor to find a solution as a signature $\mathbf{x}$. The verifier verifies if a signature $\mathbf{x}$ is valid by checking if it satisfies the equation $F(\mathbf{x}) = \mathbf{y}$. Notice that any message should have valid signatures in general. Hence F should be a surjective map, or otherwise there should be a good control on those invalid messages, i.e., those messages having no valid signatures. However having a good control on invalid messages may be troublesome, so it is preferred to have F being surjective.

Since the famous Matsumoto-Imai (MI) cryptosystem [MI88] was proposed in 1980's, various multivariate encryption and signature schemes have been constructed. The MI cryptosystem was broken by Patarin in 1995 [Pat95], but it has influenced many important variants. A few of them are to modify the MI cryptosystem by simple methods, such as FLASH for signature [PCG99] and Ding's internal perturbation of MI for encryption [Din04]. However all these simple modification of MI turned out to insecure. In 1996, Patarin [Pat96] proposed the famous Hidden Field Equation (HFE) encryption scheme which has been developed into a big family. Though the original HFE has been thoroughly broken [KS99, GJS06, BFP13], some of its variants still survive until now, such as HFEv for encryption and HFEv- for signature, especially QUARTZ as an instance of HFEv- [PCG01]. Inspired by the linearization attack to the MI cryptosystem, Patarin proposed the Oil-Vinegar (OV) signature scheme [Pat97]. OV was broken soon, but its variant Unbalanced Oil-Vinegar signature scheme [KPG99] and Rainbow [DS05b] survive until now. There were also many other schemes intended for signatures, but major signature schemes that remain secure are HFEv, HFEv-, QUARTZ, UOV, Rainbow, etc. However, the public map of HFEv, HFEv- generally cannot be surjective because their central polynomials are chosen randomly with restriction only on the degree. For UOV and Rainbow, it is not guaranteed that any message do have a valid signature though the failure probability is very small. So to implement these schemes in practice, one still has to handle those invalid messages.

In this paper, we propose a new multivariate signature scheme, named MI-T-HFE, to resolve the problem on surjectivity while maintaining efficiency and security. The core map of MI-T-HFE is a definitely surjective polynomial map, indeed an HFEv polynomial, and thus its public map is also surjective. The design of MI-T-HFE is motivated by the idea of [ZT14] where they propose a double perturbation of the MI cryptosystem by two perturbation methods, triangular perturbation and dual perturbation. Here we also modify the MI cryptosystem by two maps, an extended version of triangular maps and a special type of HFEv polynomials. The final map of this modification is an HFEv polynomial which has a large number of vinegar variables. This construction can also be viewed as an HFEv polynomial with a trapdoor embedded in its vinegar variables. In the name MI-T-HFE, MI, T and HFE stand for the MI cryptosystem, triangular perturbation and HFE polynomials respectively. Compared to QUARTZ, the signature generation of MI-T-HFE can be performed much faster, and MI-T-HFE can have smaller public key size. We examine the security of this construction against current main attacks in multivariate public key cryptography, and show that it can have the same level of security as QUARTZ.

This paper is organized as follows. Section 2 is a brief review of some previous results to be used in this paper. Our new signature scheme MI-T-HFE is then constructed in Sect. 3. Section 4 is devoted to the cryptanalysis of MI-T-HFE, then followed by a practical example given in Sect. 5. Finally Sect. 6 concludes this paper.

2 Preliminaries

In this section, we shall briefly review a few previous results which will be used in the rest of this paper.

2.1 The Matsumoto-Imai Cryptosystem

We first recall the Matsumoto-Imai (MI) cryptosystem [MI88] as follows. Let q be a power of 2, $\mathbb{K}$ a degree n extension of $\mathbb{F}_q$ and $\phi : \mathbb{K} \to \mathbb{F}_q^n$ the standard $\mathbb{F}_q$-linear map

$$\phi(a_0 + a_1 x + \cdots + a_{n-1} x^{n-1}) = (a_0, a_1, \ldots, a_{n-1}).$$

Let θ be an integer such that, $0 < \theta < n$ and $\gcd(q^\theta + 1, q^n - 1) = 1$. Define the following simple polynomial

$$\tilde{F} : \mathbb{K} \to \mathbb{K}, \quad \tilde{F}(X) = X^{1+q^\theta}.$$

This polynomial $\tilde{F}$ is invertible and its inverse is $\tilde{F}^{-1}(Y) = Y^\eta$ where $\eta(1 + q^\theta) \equiv 1 \bmod q^n - 1$.

The MI cryptosystem uses $F = \phi \circ \tilde{F} \circ \phi^{-1} : \mathbb{F}_q^n \to \mathbb{F}_q^n$ as the central map and its public map is constructed from F by composing two invertible affine transformation at the two ends $\bar{F} = L \circ F \circ R$. Since F is invertible, the MI cryptosystem is an encryption scheme. For convenience, we shall call such an F an MI map.

2.2 HFE

After breaking the MI cryptosystem [Pat95], Patarin then proposed Hidden Field Equations (HFE) for encryption in 1996 [Pat96] which significantly influences the development of multivariate public key cryptography.

Let q be a power of a prime (odd or even) and $\mathbb{K}$ a degree n extension of $\mathbb{F}_q$. HFE uses the following type of polynomials over $\mathbb{K}$ as the central map

$$H(X) = \sum a_{ij} X^{q^i + q^j} + \sum b_i X^{q^i} + c.$$

where the coefficients are randomly chosen in $\mathbb{K}$ and the degree of H is bounded by a relatively small number D. We shall call such an F an HFE map (polynomial).

The parameter D determines the efficiency and security level of HFE. $H(X) = Y$ can be solved by Berlekamp's algorithm and the complexity is known as

$$O(nD^2 \log_q D + D^3)$$

So it can be efficient if $\deg(H) \le D$ is small enough. However, it is first found that D cannot be too small otherwise it can be broken by attacks [KS99, Cou01, FJ03], and later on HFE was thoroughly broken by [GJS06, BFP13].

2.3 HFEv

Though HFE has been broken, some simple modification can make it secure against those attacks to HFE: HFEv which adds vinegar variables and HFEv- which deletes a few components from the public map.

HFEv uses the following type of polynomials as the central map

$$H(X, V) = \sum a_{ij} X^{q^i + q^j} + \sum b_{ij} X^{q^i} V^{q^j} + \sum c_{ij} V^{q^i + q^j} + \sum d_i X^{q^i} + \sum e_i V^{q^i} + f$$

where the degree of X is bounded by a relatively small parameter D but the degree of V can be arbitrary high. In addition, V varies only in a certain subspace of $\mathbb{K}$ of dimension v corresponding to the subspace $\mathbb{F}_q^v$ of $\mathbb{F}_q^n$. To invert H, one first assign a random value to V and then H is reduced to an HFE polynomial and thus can be solved by Berlekamp's algorithm. If HFEv is used for encryption, the parameter v should be small so that decryption won't be too slow.

HFEv- is HFEv with a few components deleted from the public map. It is intended for signature schemes. The most famous example of HFEv- is QUARTZ [PCG01] which has parameters $(q, D, n, v, r) = (2, 129, 103, 4, 3)$ where r is the number of components deleted.

The central polynomials of HFE, HFEv and HFEv- are randomly chosen with only one restriction on the degree, so the probability that are surjective is very small. Additional effort is then necessary to take care of those messages without valid signatures when using them for signature schemes. This could be quite troublesome, so a signature scheme with the public map being surjective is still preferred.

2.4 Triangular Maps and Perturbation

Triangular maps are of the following form

$$G(\mathbf{x}) = \begin{pmatrix} x_1 \\ x_2 + g_1(x_1) \\ \vdots \\ x_n + g_{n-1}(x_1, \ldots, x_{n-1}) \end{pmatrix}$$

where $g_1, \ldots, g_s$ are randomly chosen polynomials. The great advantage of this triangular structure is that G is bijective and it is very easy to solve $G(\mathbf{x}) = \mathbf{y}$ inductively.

In [ZT14], triangular maps are turned into a modification method, called triangular perturbation. Their method is to add to the central map the following triangular map

$$G(\mathbf{x}) = G(\mathbf{x}_1, \mathbf{x}_2) = \begin{pmatrix} x_{n+1} + g_1(\mathbf{x}_1) \\ x_{n+2} + g_2(\mathbf{x}_1, x_{n+1}) \\ \vdots \\ x_{n+s} + g_s(\mathbf{x}_1, x_{n+1}, \ldots, x_{n+s-1}) \end{pmatrix}$$

Namely, the modified central map is

$$F'(\mathbf{x}) = F(\mathbf{x}_1) + S \cdot G(\mathbf{x}_1, \mathbf{x}_2)$$

where S is a randomly chosen $m \times s$ matrix. Triangular perturbation can preserve the efficiency and surjectivity of the original scheme, because $G(\mathbf{x}_1, \mathbf{x}_2) = \mathbf{y}$ always has a solution $\mathbf{x}_2 = (x_{n+1}, \ldots, x_{n+s})$ for any $\mathbf{x}_1, \mathbf{y}$ and $x_{n+1}, \ldots, x_{n+s}$ can be computed straightforward by induction. However it cannot enhance the security if it is applied alone as its triangular structure is vulnerable to high rank attack.

In [ZT14], they also propose another modification method, called dual perturbation, and a new signature scheme by combining the two methods. They claim that the two methods can protect each other to resist current attacks. However we find that their scheme is indeed insecure. The reason is that their dual perturbation can be simplified as adding a random polynomial only on the second part of the variables after a linear transformation on the variables, and thus can be removed, contradicting their claim on the security.

3 The New Multivariate Signature Scheme MI-T-HFE

Though the construction of [ZT14] is insecure due to the failure of dual perturbation, we find that their idea of double perturbation, i.e., using two maps to protect each other remains interesting. In this section, we will apply their idea to embed a trapdoor into HFEv and thus construct a new signature scheme, named MI-T-HFE.

3.1   Preparation

Before giving the construction of MI-T-HFE, we shall first introduce two types of polynomial maps. The first type of polynomial map is an extended version of triangular maps,

$$
G(\mathbf{x}) = G(\mathbf{x}_1, \mathbf{x}_2) = \begin{pmatrix} \phi_1(x_{n+1}) + g_1(\mathbf{x}_1) \\ \phi_2(x_{n+2}) + g_2(\mathbf{x}_1, x_{n+1}) \\ \vdots \\ \phi_s(x_{n+s}) + g_s(\mathbf{x}_1, x_{n+1}, \ldots, x_{n+s-1}) \end{pmatrix}
$$

where $g_1, \ldots, g_s$ are randomly chosen polynomials and $\phi_i : \mathbb{F}_q \to \mathbb{F}_q$ are invertible polynomials, which can be easily inverted. If we want G to be quadratic, then choose g_i, ϕ_i to be quadratic. For example, if $k > 1$, $\mathbb{F}_{2^k} \to \mathbb{F}_{2^k}$, $x \mapsto x^2$ has an inverse $y \mapsto y^{2^{k-1}}$. Then each $\phi_i : \mathbb{F}_{2^k} \to \mathbb{F}_{2^k}$ can be chosen as $\phi_i(x) = a_i x^2$ where $a_i \in \mathbb{F}_{2^k}$ and $a_i \neq 0$. This type of maps with each $\phi_i(x) = x^2$ appears in [PG97]. We make the convention that if $q = 2$, we choose each $\phi_i(x) = x$ and if $q > 2$, we choose each $\phi_i(x) = a_i x^2$ for a constant $a_i \neq 0$.

Like the triangular perturbation [ZT14], extended triangular maps can also be used as a modification method, called extended triangular perturbation. It also preserves the efficiency and surjectivity of the original scheme, but is insecure against high rank attack. To protect (extended) triangular perturbation, the triangular structure should be hidden by adding a large amount of quadratic terms and cross terms of $\mathbf{x}_1, \mathbf{x}_2$.

Next we propose a special type of HFEv polynomials. Let $\mathbb{K} = \mathbb{F}_q[x]/(g(x))$ be a degree t extension of $\mathbb{F}_q$ where $g(x) \in \mathbb{F}_q[x]$ is a degree s irreducible polynomial. Let $\phi : \mathbb{K} \to \mathbb{F}_q^t$ be the standard $\mathbb{F}_q$-linear map

$$
\phi(a_0 + a_1 x + \cdots + a_{t-1} x^{t-1}) = (a_0, a_1, \ldots, a_{t-1}).
$$

Define the following type of polynomial over $\mathbb{K}$:

$$
H(X_1, X_2) = \sum_{0 \leq i < t} \sum_{1 \leq q^j \leq D} a_{ij} X_1^{q^i} X_2^{q^j} + \sum_{1 \leq q^i + q^j \leq D} b_{ij} X_2^{q^i + q^j} + \sum_{1 \leq q^j \leq D} c_j X_2^{q^j}.
$$

Here D is a relatively small number. Fixing a value of X_1, $H(X_1, X_2)$ is then an HFE polynomial of X_2, so X_2 can be solved efficiently from $H(X_1, X_2) = 0$ with a given X_1. Notice that this equation always has the zero solution $X_2 = 0$, but a nonzero solution is preferred. We can apply Berlekamp's algorithm to solve it and among those solutions, we pick a nonzero solution as X_2. We shall accept the zero solution $X_2 = 0$ if there is only the zero solution. It would be ideal that there is a nonzero solution for most values of X_1.

For $\mathbf{x}_1, \mathbf{x}_2 \in \mathbb{F}_q^t$, define the following map to be used next

$$
\bar{H} : \mathbb{F}_q^t \times \mathbb{F}_q^t \to \mathbb{F}_q^t, \quad \bar{H}(\mathbf{x}_1, \mathbf{x}_2) = \phi(H(\phi^{-1}(\mathbf{x}_1), \phi^{-1}(\mathbf{x}_2))).
$$

3.2   Construction of MI-T-HFE

Let q be a power of 2, $F : \mathbb{F}_q^n \to \mathbb{F}_q^n$ an MI map, $1 \leq s \leq n$ and $1 \leq t \leq n$. Combining the extended triangular map G and the HFEv map H defined above, we define the following trapdoor function for $\mathbf{x}_1 \in \mathbb{F}_q^n$, $\mathbf{x}_2 \in \mathbb{F}_q^s$, $\mathbf{x}_3 \in \mathbb{F}_q^t$,

$$F' : \mathbb{F}_q^{n+s+t} \to \mathbb{F}_q^n,$$

$$F'(\mathbf{x}_1, \mathbf{x}_2, \mathbf{x}_3) = F(\mathbf{x}_1) + S \cdot G(\mathbf{x}_1, \mathbf{x}_2) + T_2 \cdot \bar{H}(T_1 \cdot (\mathbf{x}_1, \mathbf{x}_2), \mathbf{x}_3) \qquad (3.1)$$

where S is an $n \times s$ matrix, T_1 an $t \times (n + s)$ matrix and T_2 an $n \times t$ matrix. This trapdoor function will serve as the central map of MI-T-HFE.

It should be noted that F' is indeed an HFEv map with $(\mathbf{x}_1, \mathbf{x}_2)$ as the $n + s$ vinegar variables. In addition, it is also a scheme obtained from the MI cryptosystem by perturbing it using an extended triangular map and an HFEv map just like the situation in [ZT14].

Randomly choose two invertible affine transformations $L_1 : \mathbb{F}_q^{n+s+t} \to \mathbb{F}_q^{n+s+t}$ and $L_2 : \mathbb{F}_q^n \to \mathbb{F}_q^n$. Then the public map of MI-T-HFE is

$$P(x_1, \ldots, x_{n+s+t}) = L_2 \circ F' \circ L_1 : \mathbb{F}_q^{n+s+t} \to \mathbb{F}_q^n.$$

The signature scheme MI-T-HFE is described as follows.

Public Key: The public key of MI-T-HFE consists of
1. The finite field $\mathbb{F}_q$.
2. The n polynomials in $P(x_1, \ldots, x_{n+s})$.

Private Key: The private key of MI-T-HFE consists of
1. The θ of the MI map F.
2. The extended triangular map G.
3. The matrix S.
4. The polynomial H.
5. The two matrices T_1, T_2.
6. The two invertible affine transformations L_1, L_2.

Signature Verification: For a given a message $\mathbf{y} \in \mathbb{F}_q^n$, a signature $\mathbf{x} \in \mathbb{F}_q^{n+s+t}$ will be accepted if it satisfies $\bar{F}'(\mathbf{x}) = \mathbf{y}$.

Signature Generation: For a given message $\mathbf{y} \in \mathbb{F}_q^n$, a valid signature is generated in the following procedure:
1. Compute $\mathbf{y}' = L_2^{-1}(\mathbf{y})$.
2. Randomly choose $\mathbf{u} = (u_1, \ldots, u_s) \in \mathbb{F}_q^s$, then solve $F(\mathbf{x}_1) = \mathbf{y}' - S \cdot \mathbf{u}$ to get a solution $\mathbf{x}_1$.
3. Substitute $\mathbf{x}_1$ into $G(\mathbf{x}_1, \mathbf{x}_2) = \mathbf{u}$ to get a solution $\mathbf{x}_2$ given by

$$x_{n+1} = \phi_1^{-1}(u_1 - g_1), \ldots, x_{n+s} = \phi_s^{-1}(u_s - g_s). \qquad (3.2)$$

4. Substitute $\mathbf{x}_1, \mathbf{x}_2$ into the equation $\bar{H}(S_1 \cdot (\mathbf{x}_1, \mathbf{x}_2), \mathbf{x}_3) = 0$ and solve it by Berlekamp's algorithm.

5. Among those solutions, pick a nonzero solution and assign it to $\mathbf{x}_3$. If there is only the zero solution, then let $\mathbf{x}_3 = 0$.
6. Then $\mathbf{x} = (\mathbf{x}_1, \mathbf{x}_2, \mathbf{x}_3)$ is a solution to $F'(\mathbf{x}) = \mathbf{y}$.
7. Finally compute $\mathbf{x} = L_1^{-1}(\mathbf{x}_1, \mathbf{x}_2, \mathbf{x}_3)$ which is then a signature.

From the above signature generation, it is easy to see that for any message, there is always a valid signature. Namely the trapdoor function is a surjective map. This is very important for a signature scheme. In addition, we remark that the MI map F in MI-T-HFE can be replaced by any other trapdoor function.

4   Security Analysis

In this section, we shall analyze the security of MI-T-HFE against current major attacks and discuss the choice of parameters accordingly.

The trapdoor function (3.1)

$$F'(\mathbf{x}_1, \mathbf{x}_2, \mathbf{x}_3) = \acute{F}(\mathbf{x}_1) + S \cdot G(\mathbf{x}_1, \mathbf{x}_2) + T_2 \cdot \bar{H}(T_1 \cdot (\mathbf{x}_1, \mathbf{x}_2), \mathbf{x}_3)$$

of MI-T-HFE is a sum of the following three parts:

1. The inner map is an MI map $F(\mathbf{x}_1)$,
2. The middle map is an extended triangular map $S \cdot G(\mathbf{x}_1, \mathbf{x}_2)$, and
3. The outer map is an HFEv map $T_2 \cdot \bar{H}(T_1 \cdot (\mathbf{x}_1, \mathbf{x}_2), \mathbf{x}_3)$.

From the point of view of perturbation [ZT14], the extended triangular map and the HFEv map in MI-T-HFE are designed to help each other similar to [ZT14]. One reason for this design is that the middle triangular map has an amount of random quadratic terms of the variables $\mathbf{x}_1$ to hide $F(\mathbf{x}_1)$, but its triangular structure makes the additional variables $\mathbf{x}_2$ detectable by high rank attack. The outer map does not have quadratic terms of $\mathbf{x}_1, \mathbf{x}_2$ but has all other quadratic terms of the variables. So the middle map can add random quadratic terms of $\mathbf{x}_1$ to perturb $F(\mathbf{x}_1)$ while the outer map can cover the triangular structure of the middle triangular map if t is big. Further reasons for the design of the trapdoor will become clear in the cryptanalysis below.

We first explain why the design of MI-T-HFE can prevent the simple attack of collecting a large amount of pairs of messages and signatures. In the signature generation, a random value $\mathbf{u} \in \mathbb{F}_q^s$ is assigned to G and $\mathbf{x}_1$ is solved from $F(\mathbf{x}_1) = \mathbf{y} - S \cdot \mathbf{u}$ with $\mathbf{y}$ perturbed by the random value $S \cdot \mathbf{u}$. In addition, notice that $\mathbf{x}_3$ can be zero in the signature generation, but in the signature generation, a nonzero solution to H is preferred and it is of high probability that there is a nonzero solution for a given message by the properties of HFE polynomials. The first feature can randomize $\mathbf{x}_1$ to break relationship between $\mathbf{x}_1$ and $\mathbf{y}$, and the second feature can assure that most $\mathbf{x}_3$ are nonzero so that information of the subspace of vectors $(\mathbf{x}_1, \mathbf{x}_2, 0)$ won't be recovered from the collected pairs of messages and signatures.

In the rest of this section, we will consider rank attacks, differential attack, linearization attack, and attacks to HFE (including MinRank attack and direct attacks).

4.1   Rank Attacks

There are two types of rank attacks, MinRank attack (or called low rank attack) and high rank attack. The MinRank attack tries to find those central polynomials or their linear combinations with the least number r of variables. Its complexity is dominated by $O(q^r)$ and successfully break Triangle-Plus-Minus schemes [GC00]. However, this attack is not applicable to MI-T-HFE in practice, because the least number of variables that the central map has is no less than n which is large enough, noticing that the public map is F' from $\mathbb{F}_q^{n+s+t}$ to $\mathbb{F}_q^n$. The high rank attack, on the contrast, tries to find those central polynomials or their linear combinations with the most number of variables, or equivalently to find those variables which appears the fewest times r in the central map. It has complexity $O(q^r)$ and is a powerful way to break triangular schemes [CSV97, GC00, YC05]. In the case of MI-T-HFE, if the outer map is small, i.e., if t is small, then high rank attack can be applied to find the last variables $\mathbf{x}_3$ first and then find the triangular structure of the second map; namely the three parts of the trapdoor function (3.1) of MI-TT-HFE can be separated. Hence t should be big enough to protect the trapdoor against high rank attack. For example, to have the security level of at least 2^{80}, we should have t such that $q^t \geq 2^{80}$.

4.2   Differential Attack

Although the public map F' of MI-T-HFE is an HFEv map and it has been shown that HFE, HFE- and HFEv are generally secure against differential attack [DST14], the differential attack [FGS05] to Ding's internal perturbation of the MI cryptosystem (IPMI) [Din04] should still be taken into account.

The differential attack to IPMI relies on the two facts: (1) there is a large linear subspace U restricted to which the internal perturbation disappears; (2) a vector $\mathbf{u}$ can be detected if it is in U by checking if the dimension of the kernel of the differential at $\mathbf{u}$ is a specific number.

For MI-T-HFE, we find that the first fact does hold here. Notice that if $\mathbf{x}_3 = 0$, the HFEv polynomial H then automatically disappears. So the linear subspace of vectors $(\mathbf{x}_1, \mathbf{x}_2, 0)$ is an important subspace. If there is no triangular map in the middle, i.e., $s = 0$, then the situation is similar to IPMI and thus the differential attack to IPMI applies. Notice that (extended) triangular maps can resist differential attack and perturbing the MI map by an (extended) triangular perturbation can break the differential invariant. Namely if $s > 0$ then the second fact does not hold anymore, and when s increases, the dimension varies in a bigger range so that the differential attack [FGS05] is no longer applicable here. To resist the differential attack, we guess that s can be just a small number but further careful analysis is needed to estimate it.

4.3   Linearization Attack

The linearization attack is proposed by Patarin [Pat95] to break the MI cryptosystem. The MI cryptosystem and some other schemes may have a large amount of linear equations between $\mathbf{x}$ and $\mathbf{y}$ (or linear on $\mathbf{x}$ but nonlinear on $\mathbf{y}$). From these equations, part of $\mathbf{x}$ may be computed and the rest of $\mathbf{x}$ may be tried one by one. However it is known that linearization attack is not applicable to triangular maps and HFE maps. The trapdoor function (3.1) of MI-TT-HFE is a mixture of an MI map, an extended triangular map and an HFE map which breaks the linear relationship. So if t is big, there would be very few linear equations among $\mathbf{x}_1$ and $\mathbf{y}$ so that linearization attack is resisted. Moreover even if $\mathbf{x}_1$ could be recovered, the rest of the variables $\mathbf{x}_2, \mathbf{x}_3$ are still unknown and the number of them is big enough so that guessing all of them is infeasible.

4.4 Attacks to HFEv

If we lift the trapdoor function (3.1)

$$F'(\mathbf{x}_1, \mathbf{x}_2, \mathbf{x}_3) = F(\mathbf{x}_1) + S \cdot G(\mathbf{x}_1, \mathbf{x}_2) + T_2 \cdot \bar{H}(T_1 \cdot (\mathbf{x}_1, \mathbf{x}_2), \mathbf{x}_3)$$

of MI-T-HFE to the extension field $\mathbb{K}$, it has the following form of an HFEv polynomial

$$\begin{aligned}
H'(V, X) = &\sum a'_{ij} V^{q^i + q^j} + \sum b'_i V^{q^i} \\
&+ \sum_{0 \le i < t} \sum_{1 \le q^j \le D} a_{ij} V^{q^i} X^{q^j} + \sum_{1 \le q^i + q^j \le D} b_{ij} V^{q^i} X^{q^j} + \sum_{1 \le q^j \le D} c_j X^{q^j}.
\end{aligned}$$

Here the vinegar variable V corresponds $(\mathbf{x}_1, \mathbf{x}_2)$ and variable X corresponds to $\mathbf{x}_3$; $F + SG$ corresponds to the sum of the monomials $V^{q^i + q^j}, V^{q^i}$ and $T_2\bar{H}$ corresponds to the sum of the rest monomials.

Attacks applicable to the HFE family are Kipnis-Shamir's attack [KS99] based on the MinRank problem and direct attack [FJ03]. In [DS05a] Ding and Schmidt improve Kipnis-Shamir's attack to cryptanalyze HFEv. They show that Kipnis-Shamir's attack can break HFEv for very small v such as $v = 1$, but as v increases, the complexity increases fast and when v is close to the extension degree of the field $\mathbb{K}$ over $\mathbb{F}_q$, HFEv would be just like a random system of quadratic polynomials.

For direct attack, Ding and Yang provide in [DY13] a solid theoretical estimation on the complexity of direct attack on HFEv and HFEv- by calculating the degree of regularity. Their conclusion is the same as the case of Kipnis-Shamir's attack; namely, direct attack remains feasible for very small v but infeasible for big v. Especially for QUARTZ whose parameters are $(2, 129, 103, 4, 3)$, its degree of regularity is bounded by 9 and its security level is estimated as 2^{92} in [DY13]. Notice that QUARTZ has 4 vinegar variables only.

In the case of MI-T-HFE, the number of vinegar variables is $n+s$ bigger than the extension degree t. So if q^t is big enough, such as $q^t \ge 2^{80}$ and D is around 100, then MI-T-HFE is just like a random system of quadratic polynomials against Kipnis-Shamir's attack, and has high degree of regularity by the formulas in [DY13] so that it is secure against direct attack.

5   A Practical Example and Comparison with QUARTZ

Based on the cryptanalysis in the preceding section, we shall propose a practical parameter set to compare with QUARTZ. It should be mentioned that here we are comparing the essential part of QUARTZ, i.e., the HFEv- scheme with the QUARTZ parameters $(2, 129, 103, 4, 3)$. The full design of QUARTZ [PCG01] applies this essential part a few times iteratively to increase the security but it was later found that this iterative structure does not contribute to the security. We shall propose a parameter set with (almost) identical length of message and same level of security, and compare the key sizes and efficiency.

We suggest the following set of parameters for MI-T-HFE

$$(q, n, s, t, D) = (8, 33, 5, 32, 72).$$

According to the cryptanalysis, the best attack to MI-T-HFE with this set of parameters is the high rank attack, and its complexity is 2^{96}. In other words, MI-T-HFE with parameters $(8, 33, 5, 32, 72)$ has 96-bit security. As a comparison with QUARTZ, its degree of regularity is bounded by 143.5 according to the formulas in [DY13], which is much higher than the bound, 9, for QUARTZ. Based on the degree of regularity, the security level of MI-T-HFE $(8, 33, 5, 32, 72)$ against direct attack should be higher than QUARTZ, which is estimated as 2^{92} in [DY13]. So the overall security of the two schemes are 2^{96} and 2^{92} respectively, which may be regarded as at the same level.

For MI-T-HFE with parameters $(8, 33, 5, 32, 72)$, a message is a vector in $\mathbb{F}_8^{33}$ whose length is 99 bits, and a signature is vector in $\mathbb{F}_8^{70}$ whose length is 210 bits. Its key sizes are calculated as follows. The public map $P : \mathbb{F}_q^{n+s+t} \to \mathbb{F}_q^n$ has n components and each component is a quadratic polynomial with $(n+s+t)(n+s+t+1)/2$ quadratic terms, $n+s+t$ linear terms and 1 constant term. Thus the public key size is

$$\frac{1}{2}n(n+s+t+1)(n+s+t+2)\log_2 q \text{ bits.}$$

With parameters $(8, 33, 5, 32, 72)$, the public key size is 31.6 Kbytes.

The private key consists of several parts. S has ns entries in $\mathbb{F}_q$, T_1, T_2 together have $2nt + st$ entries in $\mathbb{F}_q$, and L_1, L_2 together have $(n+s+t)^2 + n^2$ entries in $\mathbb{F}_q$. G has 3165 coefficients in $\mathbb{F}_q$, and H has 101 coefficients in $\mathbb{K} \cong \mathbb{F}_{q^t}$, equivalently 3232 coefficients in $\mathbb{F}_q$. So the private key size is 5.6 Kbytes.

As comparison, a message of QUARTZ is 100 bits and a signature is 107-bit. Its public key consists of 100 quadratic polynomials each with 107 variables. Thus its public size is 72.3 Kbytes, more than twice that of MI-T-HFE $(8, 33, 5, 32, 72)$. Similarly its private key size is 3.9 Kbytes, a bit smaller than that of MI-T-HFE $(8, 33, 5, 32, 72)$.

We next consider the efficiency of signature generation. In the signature generation of HFEv and the core part of QUARTZ, one first assigns random values to the vinegar variables and then one solve the resulted HFE polynomials; if no solution then try other values of the vinegar variables. This design lowers down the efficiency as one may need to solve HFE polynomials a few times. MI-T-HFE has different design on signature generation: one first solve an MI map to get $\mathbf{x}_1$, then solve a triangular map to get $\mathbf{x}_2$, and finally solve the resulted HFE polynomial only *once*. This is because the resulted HFE equation in MI-T-HFE is of the following form $\sum a_{ij} X^{q^i + q^i} + \sum b_i X^{q^i} = 0$ which always has solutions — a nonzero solution is preferred if there is one. The first two steps are very fast with little computation time, confirmed by computer experiments, as inverting an MI map and a triangular map are both extremely fast. So the main cost for inverting the central map is on inverting the HFE polynomial of MI-T-HFE. Recall that the complexity of inverting an HFE polynomial by Berlekamp's algorithm is $O(nD^2 \log_q D + D^3)$. The value of $nD^2 \log_q D + D^3$ for MI-T-HFE $(8, 33, 5, 32, 72)$ is 1.2×10^6, much smaller than the value 14.2×10^6 for QUARTZ. So it is expected that the complexity of inverting the HFE map of MI-T-HFE is much less than that of HFEv and QUARTZ. We did computer experiments on MAGMA to compare the computation time of inverting their core HFE maps and found that it is on average about 0.42 s for QUARTZ and 0.13 s for MI-T-HFE $(8, 33, 5, 32, 72)$; namely the latter is more than three times faster. Hence we may conclude that MI-T-HFE $(8, 33, 5, 32, 72)$ is about three times faster than the underlying HFEv- of QUARTZ when generating a signature. Full implementation will be conducted to justify this claim in the future.

To summarize, QUARTZ, or its underlying HFEv- scheme with the QUARTZ parameters $(2, 129, 103, 4, 3)$, uses an HFE polynomial with very small number of vinegar variables but relatively higher degree to have a short signature and high enough security level, but the cost is bigger public key size and low efficiency. On the contrary, MI-T-HFE $(8, 33, 5, 32, 72)$ uses a special HFE polynomial with large number of vinegar variables but relatively smaller degree to have smaller public key size, better efficiency and high enough security level, and the only cost is longer signatures. Moreover MI-T-HFE is a definitely surjective scheme but QUARTZ is not.

6   Conclusion

In this paper we have constructed a new multivariate signature scheme, named MI-T-HFE, whose core map is of an HFEv type but has a trapdoor embedded in it. MI-T-HFE has a special HFE polynomial with relatively low degree and a large number of vinegar variables. Unlike the usual HFEv schemes, these vinegar variables are not randomly assigned values but have special structure; namely it is a certain combination of a Matsumoto-Imai map and a kind of extended triangular maps. This trapdoor can also be viewed as a double perturbation of the Matsumoto-Imai cryptosystem by extended triangular maps and HFEv maps. With this trapdoor, MI-T-HFE is a surjective signature scheme, namely there are always valid signatures for any message. The special HFE polynomial of MI-T-HFE and its low degree guarantee its efficiency, while the large amount of vinegar variables backs its security but does not distract efficiency. To be comparable with QUARTZ, we propose a parameter set for MI-T-HFE with the same length of message and same security level as QUARTZ. With the proposed parameters, the public key size of MI-T-HFE is about half of QUARTZ, and signature generation is about three times efficient than the underlying HFEv- scheme with the QUARTZ parameters — thus much more efficient than QUARTZ. Its disadvantage is that its signature length, 210 bits, is about twice that of QUARTZ. Hence we suggest to use MI-T-HFE instead of QUARTZ if longer signatures are accepted.

Acknowledgment. The authors would like to thank the anonymous reviewers for their helpful comments on improving this paper. The first author would like to thank the financial support from the National Natural Science Foundation of China (Grant No. 61572189).

References

[BFP13] Bettale, L., Faugère, J.C., Perret, L.: Cryptanalysis of HFE, multi-HFE and variants for odd and even characteristic. Des. Codes Crypt. **69**(1), 1–52 (2013)

[Cou01] Courtois, N.T.: The security of hidden field equations (HFE). In: Naccache, D. (ed.) CT-RSA 2001. LNCS, vol. 2020, pp. 266–281. Springer, Heidelberg (2001)

[CSV97] Coppersmith, D., Stern, J., Vaudenay, S.: The security of the birational permutation signature schemes. J. Crypt. **10**, 207–221 (1997)

[Din04] Ding, J.: A new variant of the Matsumoto-Imai cryptosystem through perturbation. In: Bao, F., Deng, R., Zhou, J. (eds.) PKC 2004. LNCS, vol. 2947, pp. 305–318. Springer, Heidelberg (2004)

[DS05a] Ding, J., Schmidt, D.: Cryptanalysis of HFEv and internal perturbation of HFE. In: Vaudenay, S. (ed.) PKC 2005. LNCS, vol. 3386, pp. 288–301. Springer, Heidelberg (2005)

[DS05b] Ding, J., Schmidt, D.: Rainbow, a new multivariable polynomial signature scheme. In: Ioannidis, J., Keromytis, A.D., Yung, M. (eds.) ACNS 2005. LNCS, vol. 3531, pp. 164–175. Springer, Heidelberg (2005)

[DST14] Daniels, T., Smith-Tone, D.: Differential properties of the *HFE* cryptosystem. In: Mosca, M. (ed.) PQCrypto 2014. LNCS, vol. 8772, pp. 59–75. Springer, Heidelberg (2014)

[DY09] Ding, J., Yang, B.-Y.: Multivariate public key cryptography. In: Bernstein, D.J., Buchmann, J., Dahmen, E. (eds.) Post-Quantum Cryptography, pp. 193–241. Springer, Berlin (2009)

[DY13] Ding, J., Yang, B.-Y.: Degree of regularity for HFEv and HFEv-. In: Gaborit, P. (ed.) PQCrypto 2013. LNCS, vol. 7932, pp. 52–66. Springer, Heidelberg (2013)

[FGS05] Fouque, P.-A., Granboulan, L., Stern, J.: Differential cryptanalysis for multivariate schemes. In: Cramer, R. (ed.) EUROCRYPT 2005. LNCS, vol. 3494, pp. 341–353. Springer, Heidelberg (2005)

[FJ03] Faugère, J.-C., Joux, A.: Algebraic cryptanalysis of hidden field equation (HFE) cryptosystems using Gröbner bases. In: Boneh, D. (ed.) CRYPTO 2003. LNCS, vol. 2729, pp. 44–60. Springer, Heidelberg (2003)

[GC00] Goubin, L., Courtois, N.T.: Cryptanalysis of the TTM cryptosystem. In: Okamoto, T. (ed.) ASIACRYPT 2000. LNCS, vol. 1976, pp. 44–57. Springer, Heidelberg (2000)

[GJ79] Garey, M.R., Johnson, D.S.: Computers and Intractability: A guide to the Theory of NP-Completeness. W. H. Freeman, New York (1979)

[GJS06] Granboulan, L., Joux, A., Stern, J.: Inverting HFE is quasipolynomial. In: Dwork, C. (ed.) CRYPTO 2006. LNCS, vol. 4117, pp. 345–356. Springer, Heidelberg (2006)

[KPG99] Kipnis, A., Patarin, J., Goubin, L.: Unbalanced oil and vinegar signature schemes. In: Stern, J. (ed.) EUROCRYPT 1999. LNCS, vol. 1592, pp. 206–222. Springer, Heidelberg (1999)

[KS99] Kipnis, A., Shamir, A.: Cryptanalysis of the HFE public key cryptosystem by relinearization. In: Wiener, M. (ed.) CRYPTO 1999. LNCS, vol. 1666, pp. 19–30. Springer, Heidelberg (1999)

[MI88] Matsumoto, T., Imai, H.: Public quadratic polynomial-tuples for efficient signature-verification and message-encryption. In: Günther, C.G. (ed.) EUROCRYPT 1988. LNCS, vol. 330, pp. 419–453. Springer, Heidelberg (1988)

[Pat95] Patarin, J.: Cryptanalysis of the Matsumoto and Imai public key scheme of Eurocrypt '88. In: Coppersmith, D. (ed.) CRYPTO 1995. LNCS, vol. 963, pp. 248–261. Springer, Heidelberg (1995)

[Pat96] Patarin, J.: Hidden fields equations (HFE) and isomorphisms of polynomials (IP): two new families of asymmetric algorithms. In: Maurer, U. (ed.) EUROCRYPT 1996. LNCS, vol. 1070, pp. 33–48. Springer, Heidelberg (1996)

[Pat97] Patarin, J.: The oil and vinegar signature scheme. In: Presented at the Dagstuhl Workshop on Cryptography, September 1997

[PCG99] Patarin, J., Courtois, N., Goubin, L.: FLASH, a fast multivariate signature algorithm. In: Naccache, D. (ed.) CT-RSA 2001. LNCS, vol. 2020, pp. 298–307. Springer, Heidelberg (2001)

[PCG01] Patarin, J., Courtois, N., Goubin, L.: QUARTZ, 128-bit long digital signatures. In: Naccache, D. (ed.) CT-RSA 2001. LNCS, vol. 2020, pp. 282–288. Springer, Heidelberg (2001)

[PG97] Patarin, J., Goubin, L.: Trapdoor one-way permutations and multivariate polynomials. In: Han, Y., Quing, S. (eds.) ICICS 1997. LNCS, vol. 1334, pp. 356–368. Springer, Heidelberg (1997)

[YC05] Yang, B.-Y., Chen, J.-M.: Building secure tame-like multivariate public-key cryptosystems: the new TTS. In: Boyd, C., González Nieto, J.M. (eds.) ACISP 2005. LNCS, vol. 3574, pp. 518–531. Springer, Heidelberg (2005)

[ZT14] Zhang, W., Tan, C.H.: A new perturbed Matsumoto-Imai signature scheme. In: ASIAPKC 2014 Proceedings of the 2nd ACM Workshop on ASIA Public-Key Cryptography, pp. 43–48. ACM, New York (2014)

A New Approach to Efficient Revocable Attribute-Based Anonymous Credentials

David Derler[(✉)], Christian Hanser, and Daniel Slamanig

IAIK, Graz University of Technology, Graz, Austria
{david.derler,christian.hanser,daniel.slamanig}@tugraz.at

Abstract. Recently, a new paradigm to construct very efficient multi-show attribute-based anonymous credential (ABC) systems has been introduced in ASIACRYPT'14. Here, structure-preserving signatures on equivalence classes (SPS-EQ-$\mathcal{R}$), a novel flavor of structure-preserving signatures (SPS), and randomizable polynomial commitments are elegantly combined to yield the first ABC systems with $O(1)$ credential size and $O(1)$ communication bandwidth during issuing and showing. It has, however, been left open to present a full-fledged revocable multi-show attribute-based anonymous credential (RABC) system based on the aforementioned paradigm. As revocation is a highly desired and important feature when deploying ABC systems in a practical setting, this is an interesting challenge.

To this end, we propose an RABC system which builds upon the aforementioned ABC system, preserves its nice asymptotic properties and is in particular entirely practical. Our approach is based on universal accumulators, which nicely fit to the underlying paradigm. Thereby, in contrast to existing accumulator-based revocation approaches, we do not require complex zero-knowledge proofs of knowledge (ZKPKs) to demonstrate the possession of a non-membership witness for the accumulator. This is in part due to the nice rerandomization properties of SPS-EQ-$\mathcal{R}$. Thus, this makes the entire RABC system conceptually simple, efficient and represents a novel direction in credential revocation. We also propose a game-based security model for RABC systems and prove the security of our construction in this model. Finally, to demonstrate the value of our novel approach, we carefully adapt an efficient existing universal accumulator approach (as applied within Microsoft's U-Prove) to our setting and compare the two revocation approaches when used with the same underlying ABC system.

1 Introduction

Credential systems have been envisioned by Chaum [23], with the motivation to develop a concept that allows users to interact anonymously with multiple

The authors have been supported by EU HORIZON 2020 through project PRISMACLOUD (GA No. 644962) and by EU FP7 through project MATTHEW (GA No. 610436). An extended version of this paper is available in the IACR Cryptology ePrint Archive.

J. Groth (Ed.): IMACC 2015, LNCS 9496, pp. 57–74, 2015.
DOI: 10.1007/978-3-319-27239-9_4

organizations online. Thereby, a user can obtain a credential for a pseudonym (nym) from one organization (issuer) and demonstrate possession of the credential to other organizations (verifiers), without revealing his nym. Later on, this idea has been formalized as pseudonym systems in [34] and has, subsequently, been further extended and formalized as anonymous credential (AC) systems in [16]. As privacy in digital interactions has become more and more important over the last decades, various AC systems with different properties and targeting different environments have been proposed [2,4,6,10,11,15–17,19,22,28,29,43]. Today, the most prevalent approaches are IBM's idemix [12] and Microsoft's U-Prove [39]. The former is based on CL signatures [17] supporting an unlimited number of unlinkable showings of a credential (multi-show), where the latter is based on Brands' blind signatures [10] and all showings are linkable (one-show).

While early ACs, such as [16], did not put focus on how credentials should look like, nowadays credentials in ACs are typically viewed as being a collection of users' attributes, e.g., birth date, nationality, sex. In such a setting, users obtain credentials on attributes (issued by some organization). Then, users can prove possession of these credentials anonymously (and in an unlinkable fashion) to any verifier. Thereby, they reveal only (the possession of) some attributes and nothing beyond. Such AC systems are also known as privacy-ABC systems (or simply ABC systems).

Revocation of ABCs. Efficient revocation of credentials is especially important and challenging in practical applications of multi-show ABCs. Unfortunately, this is no trivial task at all. It is clearly not possible to simply blacklist credentials as it can be conveniently done in PKIs. To realize revocable ABCs (RABCs), various different credential revocation mechanisms have been introduced over the years (cf. [31] for an exhaustive discussion). The idea is that a revocation authority (which may be run by the credential issuer) publishes revocation information which allows verifiers to decide whether a credential has been revoked. Ideally, such revocation mechanisms are conceptually simple, scale well and do not add significant additional burden to users and verifiers. However, simple mechanisms are either inflexible or far from practical. Examples are the inclusion of the validity period as attribute into credentials or the re-issuing of all unrevoked credentials triggered by the replacement of the issuer's key material. Obviously, such mechanisms either get insecure due to too long validity periods (and, thus, too long revocation intervals) or require to frequently re-issue a large amount of credentials. More importantly, they do not allow to selectively revoke single credentials in case of loss, theft or fraud.

More sophisticated revocation mechanisms supporting the selective revocation of single credentials are either based on whitelists or blacklists. Whitelist approaches require users to prove that unrevoked credentials are contained in a list. The effort for users (during showings) is typically linear in the number of valid credentials and/or it requires users to download revocation information each time a new credential gets issued. Thus, whitelist approaches do not scale well and cannot be considered practical in general. In contrast, blacklist revocation usually scales far better. The main reason for this is that revocation list updates are only required on revocation (which usually can be considered a rare event in comparison to the issuing of new credentials). Thereby, blacklisting approaches based on verifier-local revocation (VLR) [8] do not require any updates from the users, but require an effort for the verifier that is linear in the number of revoked credentials. Many of the VLR techniques also have the problem of missing the property of backward unlinkability [40], i.e., the revocation of a credential implies the linkability of all past showings (e.g., as it is the case in [29,33]). Furthermore, techniques to add backward unlinkability to VLR either induce a significant additional computational burden on users and verifiers [40] or require frequent updates and computational overhead for verifiers [36]. Another blacklist approach [35] represents blacklists as signatures on ordered credential identifier pairs. This is elegant, since the computational costs for users and verifiers are constant and quite small. Yet, the user and the verifier have to update a significant amount of revocation information on each revocation, as the blacklist has to be recomputed entirely (number of signatures linear in the number of revoked credentials). The remaining and popular choice is to use blacklists based on universal accumulators [1,3,13,18,37]. This approach scales well and requires only constant computational effort for users and verifiers. Although updates of the accumulator and the non-membership witnesses are required on revocation, these are small and often constant in size.

Design Paradigms of Existing (R)ABCs. ABC systems are typically constructed in the following way (with few exceptions [20,21]). A user obtains a signature on (commitments to) attributes using a suitable signature scheme. Then, on a showing, the user randomizes the signature (such that the resulting signature is unlinkable to the issued one) and proves in zero-knowledge the possession of a signature. Thereby, attributes may be selectively revealed and/or relations among attributes may be proven. In one-show ABCs, blind signature schemes are used, and—instead of randomizing the signatures—the same unblinded signature is presented on each showing. A standard way to turn ABCs into RABCs is to add a credential identifier (revocation handle) as an additional never-revealed attribute. Then, for the aforementioned approaches which use explicit ZKPKs, the choice of the revocation mechanisms is somewhat arbitrary. It only has to be guaranteed that the identifier in the credential and the one used for blacklisting (or whitelisting) are identical. Hence, the showing in such an RABC system amounts to providing the ZKPK for the underlying ABC and the ZKPK of the used revocation mechanism plus an additional ZKPK that the identifier in the credential coincides with the identifier used for revocation.

Design Paradigm of the ABC from [30]. The ABC system proposed in [30] is conceptually significantly different from the aforementioned approach. Its main building block are structure-preserving signatures on equivalence classes (SPS-EQ-$\mathcal{R}$). An SPS-EQ-$\mathcal{R}$ signs equivalence classes defined on group element vectors and allows to consistently randomize messages and signatures in the public by changing representatives of the signed class. It is used to sign rerandomizable, constant-size commitments to polynomials. Thereby, the rerandomization of the commitment is compatible with the rerandomization of the SPS-EQ-$\mathcal{R}$.

To perform a showing for a subset of the attributes, the (rerandomized) commitment is partially opened and the rerandomization property of SPS-EQ-$\mathcal{R}$ provides unlinkability, while authenticity is still ensured. Additionally, the approach requires a single, constant-size ZKPK to prevent replays of already conducted interactive showings. Consequently, the so obtained ABC system does not need costly ZKPKs to prove possession of the attributes. In particular, [30] provides the first ABC system with $O(1)$ credential size and $O(1)$ communication bandwidth during both issuing and showing and is thus very efficient. The communication costs of other existing approaches are at least linear in the number of shown/encoded attributes in the ABC system (or constant-size showings can only be achieved for special cases [5,41], e.g., very small attribute domains, at the cost of huge public parameters—linear in the number of all potential values over all attribute domains).

Contribution. The efficiency of the ABC system from [30], e.g., when instantiated with the EUF-CMA secure SPS-EQ-$\mathcal{R}$ scheme from [27], makes it very attractive for practical use. Thus, obtaining an RABC system following the same paradigm is an important step towards highly efficient and practical RABCs. We construct an RABC system based on the ABC system in [30] (which can e.g. be instantiated with the SPS-EQ-$\mathcal{R}$ from [27]), and, thereby, rely on a universal accumulator-based blacklist approach. In contrast to all previous applications of universal accumulators to blacklist revocation [1,3,32,38], we do, however, not require explicit ZKPKs of non-membership witnesses satisfying the accumulator verification equation. We achieve this by rerandomizing the used universal accumulator, which is a novel way of proving possession of a particular non-membership witness.

In order to evaluate our approach, we, in addition, carefully adapt an existing universal accumulator revocation mechanism [1,38] (applied within Microsoft's U-Prove) to the ABC system from [30]. Contrary to our first construction, this revocation mechanism represents a traditional ZKPK approach for demonstrating knowledge of a non-membership witness that satisfies the accumulator verification equation. Thereby, it turns out that regarding the most time critical part, i.e., the showing protocol performed by a (potentially resource constrained [42]) user, our approach outperforms the revocation approach adpoted from U-Prove.

As our revocation mechanisms preserve the asymptotic optimality of the ABC system in [30], our RABC constructions are also the first RABC system with $O(1)$ credential size and $O(1)$ communication costs during issuing as well as showing.

Revocation in ABC systems is typically considered as an add-on and, thus, not considered in the security models of ABCs. To overcome this issue, another contribution of this paper is a comprehensive game-based security model for RABC systems, which explicitly considers backward-unlinkability. We prove our proposed approach secure in this model. Independently to our work, another formal model for ABC systems has been introduced in [14]. It also considers revocation but also additional features such as auditing [15]. However, the model in [14] aims at constructing ABCs by means of a generic composition of numerous building blocks (commitment schemes, NIZKPs, privacy-enhancing attribute-based signatures, revocation schemes and pseudonym schemes), considers only non-interactive protocols (using the notion of tokens) and uses stronger simulation-based security definitions. In particular the stronger security notions add a non-trivial overhead in terms of efficiency to the constructions, which, in turn, makes it less attractive for highly efficient and practical ABC systems.[1]

2 Preliminaries

Definition 1 (Bilinear Map). Let $\mathbb{G}_1 = \langle P \rangle$, $\mathbb{G}_2 = \langle \hat{P} \rangle$ and $\mathbb{G}_T$ be cyclic groups of prime order p, where $\mathbb{G}_1$ and $\mathbb{G}_2$ are additive and $\mathbb{G}_T$ is multiplicative. We call $e \colon \mathbb{G}_1 \times \mathbb{G}_2 \to \mathbb{G}_T$ a *bilinear map* or *pairing* if it is efficiently computable and the following conditions hold:

Bilinearity: $e(aP, b\hat{P}) = e(P, \hat{P})^{ab} = e(bP, a\hat{P}) \quad \forall a, b \in \mathbb{Z}_p$

Non-degeneracy: $e(P, \hat{P}) \neq 1_{\mathbb{G}_T}$, i.e., $e(P, \hat{P})$ generates $\mathbb{G}_T$.

We use lower-case boldface letters for elements in $\mathbb{G}_T$, e.g., $\mathbf{g} = e(P, \hat{P})$.

Definition 2 (Bilinear Group Generator). Let BGGen be an algorithm which takes a security parameter κ and generates a bilinear group $\mathsf{BG} = (p, \mathbb{G}_1, \mathbb{G}_2, \mathbb{G}_T, e, P, \hat{P})$ in the Type-3 bilinear group setting, where the common group order p of the groups $\mathbb{G}_1, \mathbb{G}_2$ and $\mathbb{G}_T$ is a prime of bitlength κ, e is a pairing and P and $\hat{P}$ are generators of $\mathbb{G}_1$ and $\mathbb{G}_2$, respectively.

Definition 3 (Decisional Diffie-Hellman Assumption). The DDH assumption in $\mathbb{G}_i$ states that for all probabilistic polynomial-time (PPT) adversaries $\mathcal{A}$ there is a negligible function $\epsilon(\cdot)$ such that

$$\Pr \left[\begin{matrix} b \xleftarrow{R} \{0,1\}, \ \mathsf{BG} \leftarrow \mathsf{BGGen}(1^\kappa), \ r, s, t \xleftarrow{R} \mathbb{Z}_p \\ b^* \leftarrow \mathcal{A}(\mathsf{BG}, rP_i, sP_i, ((1-b) \cdot t + b \cdot rs)P_i) \end{matrix} : b^* = b \right] - \frac{1}{2} \leq \epsilon(\kappa),$$

where $P_1 = P$ and $P_2 = \hat{P}$ and $i \in \{1, 2\}$.

Definition 4 (Symmetric External Diffie Hellman Assumption). Let BG be a bilinear group. The SXDH assumption states that the DDH assumption holds in $\mathbb{G}_1$ and $\mathbb{G}_2$.

The following assumption [30] is the Type-3 bilinear group counterpart of the strong Diffie-Hellman assumption.

Definition 5 (co-t-Strong Diffie Hellman Assumption). The co-t-SDH$_i^*$ assumption states that for all probabilistic polynomial-time (PPT) adversaries $\mathcal{A}$ there is a negligible function $\epsilon(\cdot)$ such that

$$\Pr \left[\begin{matrix} \alpha \xleftarrow{R} \mathbb{Z}_p, \ \mathsf{BG} \leftarrow \mathsf{BGGen}(1^\kappa), & c \in \mathbb{Z}_p \setminus \{-\alpha\} \\ (c, T_i) \xleftarrow{R} \mathcal{A}(\mathsf{BG}, (\alpha^j P_1)_{j=0}^t, (\alpha^j P_2)_{j=0}^t) : & \wedge \ T_i = \frac{1}{\alpha + c} P_i \end{matrix} \right] \leq \epsilon(\kappa),$$

where $P_1 = P$ and $P_2 = \hat{P}$ and $i \in \{1, 2\}$.

---

[1] We, however, note that the efficiency of our scheme comes at the cost of more complex proofs.

We will use the co-t-SDH$_1^*$ assumption statically, as we will fix t a priori as a system parameter and assume that it is bounded by $\mathsf{poly}(\kappa)$. Then, the security loss which applies when using co-t-SDH$_1^*$ in a non-static way [24] does not apply.

2.1 Universal Accumulators

Cryptographic accumulators [7] represent a finite set $\mathcal{X}$ as a single succinct value $\Pi_{\mathcal{X}}$ and for each $x \in \mathcal{X}$ one can compute a witness ω_x, certifying membership of x in $\mathcal{X}$. Universal accumulators additionally support non-membership witnesses ω_y that certify non-membership of a value $y \notin \mathcal{X}$. Henceforth, we write Π if we do not want to make $\mathcal{X} = \{x_1, \ldots, x_n\}$ explicit. To blacklist credentials, we require a universal accumulator. Subsequently, we restate the accumulator of Au et al. [3] for the Type-3 bilinear group setting and in the model of [25], where we omit the algorithms that are not required in our context, i.e., the dynamic features. The formal model is given in the extended version of this paper.

$\mathsf{Gen}_{\mathsf{Acc}}(\mathsf{BG}, t)$: Given a bilinear group BG and an upper bound t for the number of elements to be accumulated, pick $\lambda \xleftarrow{R} \mathbb{Z}_p^*$, compute $\mathsf{pk}_\Pi \leftarrow ((\lambda^i P)_{i \in [t]}, (\lambda^i \hat{P})_{i \in [t]})$ and return $(\emptyset, \mathsf{pk}_\Pi)$.

$\mathsf{Eval}_{\mathsf{Acc}}(\mathcal{X}, (\emptyset, \mathsf{pk}_\Pi))$: Given a set $\mathcal{X} = \{x_1, \ldots, x_n\}$ and an accumulator public key pk_Π, compute $\pi(X) \leftarrow \prod_{i \in [n]}(X - x_i) = \sum_{i=0}^n a_i \cdot X^i$ and $\Pi_{\mathcal{X}} \leftarrow \sum_{i=0}^n a_i(\lambda^i P)$ and return $\Pi_{\mathcal{X}}$ together with $\mathsf{aux} \leftarrow \mathcal{X}$.

$\mathsf{WitCreate}_{\mathsf{Acc}}(\Pi_{\mathcal{X}}, \mathsf{aux}, y, (\emptyset, \mathsf{pk}_\Pi))$: Given an accumulator $\Pi_{\mathcal{X}}$, some auxiliary information $\mathsf{aux} = \mathcal{X} = \{x_1, \ldots, x_n\}$, a non-member y and an accumulator public key pk_Π, this algorithm checks whether $y \in \mathcal{X}$ and if so returns $\bot$. Otherwise, it computes $\pi(X) \leftarrow \prod_{i \in [n]}(X - x_i)$ and $d \in \mathbb{Z}_p^*$ such that $\pi(X) = g(X)(X - y) + d$ holds. With $g(X) = \sum_{i=0}^{n-1} a_i \cdot X^i$ it computes $\hat{W} \leftarrow \sum_{i=0}^{n-1} a_i(\lambda^i \hat{P})$ and returns $\omega_y \leftarrow (\hat{W}, d)$.

$\mathsf{Verify}_{\mathsf{Acc}}(\Pi, \omega_y, y, \mathsf{pk}_\Pi)$: Given an accumulator Π, a non-membership witness ω_y and some corresponding y, this algorithm parses ω_y as $(\hat{W}, d)$, checks if $d \neq 0$ and $e(\Pi, \hat{P}) = e(\lambda P - yP, \hat{W}) \cdot e(dP, \hat{P})$ holds and if so returns 1 and 0 else.

Scheme 1: Universal accumulator from [3] tailored to non-membership witnesses.

For the Type-3 bilinear setting, in analogy to [3], we can straightforwardly prove the following (where we omit the proof):

Theorem 1. *Scheme 1 is collision-free under the co-t-SDH$_i^*$ assumption, where t is the maximum number of values to be accumulated.*

2.2 Structure-Preserving Signatures on Equivalence Classes

The notion of structure-preserving signature schemes on equivalence classes (SPS-EQ-$\mathcal{R}$) has been introduced in [30]. The authors consider elements of a vector $(M_i)_{i \in [\ell]} \in (\mathbb{G}_1^*)^\ell$ (where $\mathbb{G}_1^* = \mathbb{G}_1 \setminus \{0_{\mathbb{G}_1}\}$, for some prime order group $\mathbb{G}_1$) which share different mutual ratios. These ratios depend on their discrete logarithms and are invariant under the operation $\gamma : \mathbb{Z}_p^* \times (\mathbb{G}_1^*)^\ell \to (\mathbb{G}_1^*)^\ell$ with $(s, (M_i)_{i \in [\ell]}) \mapsto s(M_i)_{i \in [\ell]}$. Thus, one can use this invariance to partition $(\mathbb{G}_1^*)^\ell$ into equivalence classes using the relation $\mathcal{R} = \{(M, N) \in (\mathbb{G}_1^*)^\ell \times (\mathbb{G}_1^*)^\ell : \exists s \in \mathbb{Z}_p^* \text{ such that } N = s \cdot M\} \subseteq (\mathbb{G}_1^*)^{2\ell}$. When signing an equivalence class $[M]_{\mathcal{R}}$ with such a scheme, one actually signs a representative $(M_i)_{i \in [\ell]}$ of class $[M]_{\mathcal{R}}$. The scheme, then, allows to switch to different representatives of the same class and to update corresponding signatures in the public, i.e., without any secret key. The initial instantiation proposed in [30] turned out to only be secure against random-message attacks (cf. [26] and the updated full version of [30]), but together with Fuchsbauer [27] they subsequently presented a scheme that is secure against chosen-message attack (EUF-CMA) in the generic group model.

For our RABC, we need a Type-3 bilinear group setting based, EUF-CMA-secure SPS-EQ-$\mathcal{R}$ that perfectly adapts signatures (formal definitions are provided in the extended version of this paper). The SPS-EQ-$\mathcal{R}$ construction from [27] satisfies all our requirements.

3 An Efficient RABC System

In an RABC system there are different organizations issuing credentials for different users under different pseudonyms.[2] Furthermore, there are revocation authorities which can selectively revoke credentials. Such a system requires that issuings and showings of the same user are unlinkable and is called multi-show RABC system when multiple showings carried out by the same user cannot be linked and one-show RABC system otherwise. A credential cred for user i under pseudonym nym is issued by an organization j for a set $\mathbb{A} = \{(\text{attr}_k, \text{attrV}_k)\}_{k=1}^n$ of attribute labels attr_k and values attrV_k. By $\#\mathbb{A}$ we mean the size of $\mathbb{A}$, which is defined to be the sum of cardinalities of all second components attrV_k of all tuples in $\mathbb{A}$. Moreover, we denote by $\mathbb{A}' \sqsubseteq \mathbb{A}$ a subset of the credential attributes. In particular, for every $k \in [n]$, we have that either $(\text{attr}_k, \text{attrV}_k)$ is missing or $(\text{attr}_k, \text{attrV}_k')$ with $\text{attrV}_k' \subseteq \text{attrV}_k$ is present. A showing with respect to $\mathbb{A}'$ only proves that a valid credential for $\mathbb{A}'$ has been issued, but reveals nothing beyond (selective disclosure). Below, we present our formal RABC model which is based on the ABC model in [30].

Definition 6 (RABC System). A *revocable attribute-based anonymous credential (RABC) system* consists of the following polynomial time algorithms:

Setup: A probabilistic algorithm that takes a security parameter κ and some optional auxiliary information aux (which may fix a universe of attributes and attribute values and other parameters).

---

[2] We stress that in our context pseudonyms are solely used for revocation and not for showing purposes (as e.g., in the model of [14]) and thus one might call ours revocation pseudonyms (but we simply call them pseudonyms henceforth).

RAKeyGen: A probabilistic algorithm that takes input the public parameters pp and outputs a key pair (rsk, rpk) for the revocation authority.

OrgKeyGen: A probabilistic algorithm that takes input the public parameters pp and $j \in \mathbb{N}$ and outputs a key pair $(\mathsf{osk}_j, \mathsf{opk}_j)$ for organization j.

UserKeyGen: A probabilistic algorithm that takes input the public parameters pp and $i \in \mathbb{N}$ and outputs a key pair $(\mathsf{usk}_i, \mathsf{upk}_i)$ for user i.

(Obtain, Issue): These (probabilistic) algorithms are run by user i and organization j, who interact during execution. Obtain takes input the public parameters pp, the user's secret key usk_i, an organization's public key opk_j, a pseudonym nym and an attribute set $\mathbb{A}$. Issue takes input the public parameters pp, the public key of the revocation authority rpk, the user's public key upk_i, an organization's secret key osk_j, a pseudonym nym and an attribute set $\mathbb{A}$. At the end, Obtain outputs a credential $\mathsf{cred}_{\mathsf{nym}}$ for $\mathbb{A}$ for user i with respect to nym.

(Show, Verify): These (probabilistic) algorithms are run by user i and a verifier, who interact during execution. Show takes input public parameters pp, the public revocation key rpk, the user's secret key usk_i, the organization's public key opk_j, a credential $\mathsf{cred}_{\mathsf{nym}}$ for the attribute set $\mathbb{A}$, a second set $\mathbb{A}' \sqsubseteq \mathbb{A}$ and some information $\mathbb{R}_S^{\mathsf{nym}}$ to prove that $\mathsf{cred}_{\mathsf{nym}}$ has not been revoked. Verify takes input pp, rpk, opk_j, a set $\mathbb{A}'$ and some revocation information $\mathbb{R}_V$. At the end, Verify outputs 1 or 0 indicating whether the credential showing was accepted or not.

Revoke: This (probabilistic) algorithm takes input the public parameters pp, the revocation key pair (rsk, rpk) and two disjoint lists NYM and RNYM holding valid and revoked pseudonyms, respectively. It outputs the revocation information $\mathbb{R} = (\mathbb{R}_V, \mathbb{R}_S)$. $\mathbb{R}_V$ is needed for verifying the revocation status and $\mathbb{R}_S$ is a list holding the revocation information per nym.

3.1   Security Model for RABCs

The subsequent security model is adapted from [30]. We note that we consider only a single organization (identified by $j = 1$) in our model (since all organizations have independent signing keys, the extension is straightforward). Basically, an RABC system needs to be *correct*, *unforgeable* and *anonymous*. To provide formal definitions of these properties we introduce several global variables and oracles. To keep track of all, honest and corrupt users as well as users, whose secret keys and credentials have leaked, we introduce the sets U, HU, CU and KU, respectively. Furthermore, we introduce the sets N and RN for keeping track of all pseudonyms and all revoked pseudonyms, respectively. We use the variables RI and NYM_{LoR} (initially set to $\bot$) to store the globally maintained revocation information $\mathbb{R}$ and the pseudonyms used in the $\mathcal{O}^{LoR}$ oracle. All these sets as well as RI and NYM_{RoR} are maintained by the environment and are available to the adversary for read access. We use the lists UPK, USK, CRED and ATTR to track issued user keys, credentials and corresponding attributes (per pseudonym). These lists are only accessible to the environment. We introduce the subsequent oracles and assume the public parameters pp to be implicitly available to them:

$\mathcal{O}^{HU+}(i)$: It takes input a user identity i. If $i \in U$ return $\bot$. Otherwise, it creates a new user i by running $(USK[i], UPK[i]) \leftarrow UserKeyGen(pp, i)$, adding i to U and to HU and returning $UPK[i]$.

$\mathcal{O}^{CU+}(pk, i)$: It takes input a user public key pk and a user i. If $i \notin U$, $i \in CU$, or $NYM_{LoR} \cap N[i] \neq \emptyset$ return $\bot$. Otherwise, it adds user i to the set of corrupted users CU, removes i from HU, and sets $UPK[i] \leftarrow pk$.

$\mathcal{O}^{KU+}(i)$: It takes input a user i. If $i \notin U$, $i \in KU$, or $NYM_{LoR} \cap N[i] \neq \emptyset$ return $\bot$. Otherwise, it reveals the credentials and the secret key of user i by returning $USK[i]$ and the credentials $CRED[nym]$ for all $nym \in N[i]$. Finally, it adds i to KU.

$\mathcal{O}^{RN+}(rsk, rpk, REV)$: It takes input the revocation secret key rsk, the revocation public key rpk and a list REV of pseudonyms to be revoked. If $REV \cap RN \neq \emptyset$ or $REV \not\subseteq N$ return $\bot$. Otherwise, set $RN \leftarrow RN \cup REV$ and $RI \leftarrow Revoke(pp, rsk, rpk, N \setminus RN, RN)$.

$\mathcal{O}^{U_l O_o}(osk, opk, rsk, rpk, i, nym, \mathbb{A})$: It takes input the organization key pair (osk, opk), the revocation key pair (rsk, rpk), a user i, a pseudonym nym and a set of attributes $\mathbb{A}$. If $i \notin HU$ or $nym \in N$ return $\bot$. Otherwise, it issues a credential cred on $\mathbb{A}$ and nym for an honest user $i \in HU$. Here, the oracle plays the role of the user as well as the organization. It runs

$$(cred, \emptyset) \leftarrow (Obtain(pp, USK[i], opk, nym, \mathbb{A}), Issue(pp, rpk, UPK[i], osk, nym, \mathbb{A})).$$

Finally, it sets $(CRED[nym], ATTR[nym]) \leftarrow (cred, \mathbb{A})$, appends nym to $N[i]$ and runs $RI \leftarrow Revoke(pp, rsk, rpk, N \setminus RN, RN)$. The caller does not get any output.

$\mathcal{O}^{U_l}(osk, opk, rsk, rpk, i, nym, \mathbb{A})$: It takes input the organization key pair (osk, opk), the revocation key pair (rsk, rpk), a user i, a pseudonym nym and a set of attributes $\mathbb{A}$. If $i \notin HU$ or $nym \in N$ return $\bot$. Otherwise, it plays the role of an honest user who gets issued a credential for $\mathbb{A}$ and nym. It runs

$$(cred, \emptyset) \leftarrow (Obtain(pp, USK[i], opk, nym, \mathbb{A}), Issue(pp, rpk, UPK[i], osk, nym, \mathbb{A})),$$

where Obtain is run on behalf of honest user i and Issue is executed by the caller (the dishonest organization). Finally, it sets $(CRED[nym], ATTR[nym]) \leftarrow (cred, \mathbb{A})$, appends nym to $N[i]$ and runs $RI \leftarrow Revoke(pp, rsk, rpk, N \setminus RN, RN)$.

$\mathcal{O}^{O_o}(osk, opk, rsk, rpk, i, nym, usk_i, \mathbb{A})$: It takes input the organization key pair (osk, opk), the revocation key pair (rsk, rpk), a user i, a pseudonym nym, a user secret key usk_i and a set of attributes $\mathbb{A}$. If $i \notin CU$ or $nym \in N$ return $\bot$. Otherwise, it plays the role of the organization when interacting with a dishonest user, i.e., a corrupted user whose public key has been replaced (thus, the corresponding secret key usk_i is not stored in USK). It runs

$$(cred, \emptyset) \leftarrow (Obtain(pp, usk_i, opk, nym, \mathbb{A}), Issue(pp, rpk, UPK[i], osk, nym, \mathbb{A})),$$

where Obtain is executed by the caller and sets $(CRED[nym], ATTR[nym]) \leftarrow (cred, \mathbb{A})$, appends nym to $N[i]$ and runs $RI \leftarrow Revoke(pp, rsk, rpk, N \setminus RN, RN)$.

$\mathcal{O}^{U_v}(opk, rpk, nym, \mathbb{A}', \mathbb{R}_V)$: It takes input the organization public key opk, the public revocation key rpk, a user i, a pseudonym nym, a set of attributes

A' certified to the user i_{nym} (that is the index such that $\text{nym} \in N[i_{\text{nym}}]$) and the revocation information $\mathbb{R}_V$. If $\text{nym} \notin N$, $i_{\text{nym}} \notin HU$, $A' \not\sqsubseteq \text{ATTR}[\text{nym}]$ or $\text{nym} \in RN$ return $\bot$. Otherwise, it plays the role of an honest user i_{nym} and runs

$$(\emptyset, b) \leftarrow \big(\text{Show}(\text{pp}, \text{rpk}, \text{USK}[i_{\text{nym}}], \text{opk}, \text{CRED}[\text{nym}], \text{ATTR}[\text{nym}],$$
$$A', \text{RI}[2][\text{nym}]), \text{Verify}(\text{pp}, \text{rpk}, \text{opk}, A', \mathbb{R}_V)\big),$$

where Verify is executed by the caller (the dishonest verifier).
$\mathcal{O}^{LoR}(\text{osk}, \text{opk}, \text{rsk}, \text{rpk}, b, \text{nym}_0, \text{nym}_1, A', \mathbb{R}_V)$: It takes input the organization and revocation key pairs (osk, opk) and (rsk, rpk), a bit b, two pseudonyms nym_0 and nym_1 and a set of attributes A'. It returns $\bot$ if for $j \in \{0, 1\}$

$$\text{nym}_j \notin N \ \lor \ i_{\text{nym}_j} \notin HU \ \lor \ i_{\text{nym}_j} \in KU \ \lor \ A' \not\sqsubseteq \text{ATTR}[\text{nym}_j] \ \lor \ \text{nym}_j \in RN,$$

where i_{nym_j} is such that $\text{nym}_j \in N[i_{\text{nym}_j}]$. Else, it adds nym_0 and nym_1 to NYM_{LoR} and interacts with the adversary during an execution of the (Show, Verify) protocol for the credential with the pseudonym nym_b and attributes A'.

Now, we are ready to introduce an exact definition of a *secure* RABC system:

Definition 7 (Correctness). An RABC system is *correct*, if

$\forall \kappa > 0, \ \forall \text{aux}, \ \forall A, \ \forall A' \sqsubseteq A \ \forall j, \ \forall i,$
$\forall \text{NYM}, \text{RNYM} \subseteq N : \text{NYM} \cap \text{RNYM} = \emptyset, \ \forall \text{nym} \in \text{NYM},$
$\forall \text{pp} \leftarrow \text{Setup}(1^\kappa, \text{aux}), \ \forall (\text{rsk}, \text{rpk}) \leftarrow \text{RAKeyGen}(\text{pp}) :$
$(\text{osk}_j, \text{opk}_j) \leftarrow \text{OrgKeyGen}(\text{pp}, j), \ (\text{usk}_i, \text{upk}_i) \leftarrow \text{UserKeyGen}(\text{pp}, i),$
$(\text{cred}, \emptyset) \leftarrow (\text{Obtain}(\text{pp}, \text{usk}_i, \text{opk}_j, \text{nym}, A), \text{Issue}(\text{pp}, \text{upk}_i, \text{osk}_j, \text{nym}, A)),$
$(\mathbb{R}_S, \mathbb{R}_V) \leftarrow \text{Revoke}(\text{pp}, (\text{rsk}, \text{rpk}), \text{NYM}, \text{RNYM})$ it holds that
$(\emptyset, 1) \leftarrow (\text{Show}(\text{pp}, \text{usk}_i, \text{opk}_j, \text{cred}, A, A', \mathbb{R}_S[\text{nym}]), \text{Verify}(\text{pp}, \text{opk}_j, A', \mathbb{R}_V)).$

Definition 8 (Unforgeability). We call an RABC system *unforgeable*, if for all PPT-adversaries $\mathcal{A}$ there is a negligible function $\epsilon(\cdot)$ such that

$$\Pr\left[\begin{array}{l} \text{pp} \leftarrow \text{Setup}(1^\kappa, \text{aux}), (\text{rsk}, \text{rpk}) \leftarrow \text{RAKeyGen}(\text{pp}), \\ (\text{osk}, \text{opk}) \leftarrow \text{OrgKeyGen}(\text{pp}, 1), \mathcal{O} \leftarrow \{\mathcal{O}^{HU+}(\cdot), \mathcal{O}^{CU+}(\cdot, \cdot), \\ \mathcal{O}^{KU+}(\cdot), \mathcal{O}^{RN+}(\text{rsk}, \text{rpk}, \cdot), \mathcal{O}^{U_I O_O}(\text{osk}, \text{opk}, \text{rsk}, \text{rpk}, \cdot, \cdot, \cdot), \\ \mathcal{O}^{U_V}(\text{opk}, \text{rpk}, \cdot, \cdot, \text{RI}[0]), \mathcal{O}^{O_O}(\text{osk}, \text{opk}, \text{rsk}, \text{rpk}, \cdot, \cdot, \cdot, \cdot)\}, \\ (\mathcal{A}'^{*}, \text{state}) \leftarrow \mathcal{A}^{\mathcal{O}}(\text{pp}, \text{opk}, \text{rpk}), \\ (\emptyset, b^*) \leftarrow (\mathcal{A}(\text{state}), \text{Verify}(\text{pp}, \text{opk}, \text{rpk}, A'^{*}, \text{RI}[1])) : \\ \qquad\qquad b^* = 1 \ \land \ (\text{nym}^* = \bot \ \lor \ (\text{nym}^* \neq \bot \ \land \\ (A'^{*} \not\sqsubseteq \text{ATTR}[\text{nym}^*] \ \lor \ (i^*_{\text{nym}^*} \in HU \setminus KU \ \lor \ \text{nym}^* \in RN))) \end{array}\right] \leq \epsilon(\kappa),$$

where the credential shown by $\mathcal{A}$ in the second phase corresponds to pseudonym nym^* and to user $i^*_{\text{nym}^*}$ (that is the index such that $\text{nym}^* \in N[i^*_{\text{nym}^*}]$). Thereby, $\bot$ indicates that no such index nym^* exists.

The winning conditions in the unforgeability game are chosen following the subsequent rationale. The first condition ($\text{nym}^* = \bot$) captures showings of credentials, which have never been issued (existential forgeries). The second condition ($\text{nym}^* \neq \bot \wedge \mathbb{A}'^* \not\sqsubseteq \text{ATTR}[\text{nym}^*]$) captures showings with respect to existing credentials, but invalid attribute sets. The third condition ($\text{nym}^* \neq \bot \wedge i^*_{\text{nym}^*} \in \text{HU} \setminus \text{KU}$) covers showings with respect to honest users, whose credentials and respective secrets the adversary does not know. This essentially boils down to replayed showings. Finally, the last condition ($\text{nym}^* \neq \bot \wedge \text{nym}^* \in \text{RN}$) covers that showings cannot be performed with respect to revoked pseudonyms.

Definition 9 (Anonymity). We call an RABC system *anonymous*, if for all PPT-adversaries $\mathcal{A}$ there is a negligible function $\epsilon(\cdot)$ such that

$$\Pr\left[\begin{array}{l} \text{pp} \leftarrow \text{Setup}(1^\kappa, \text{aux}),\ b \overset{R}{\leftarrow} \{0,1\}, \\ (\text{osk}, \text{opk}) \leftarrow \text{OrgKeyGen}(\text{pp}, 1), \\ (\text{rsk}, \text{rpk}) \leftarrow \text{RAKeyGen}(\text{pp}), \\ \mathcal{O} \leftarrow \{\mathcal{O}^{\text{HU+}}(\cdot), \mathcal{O}^{\text{CU+}}(\cdot, \cdot), \mathcal{O}^{\text{KU+}}(\cdot), \mathcal{O}^{\text{RN+}}(\text{rsk}, \text{rpk}, \cdot),\ : b^* = b \\ \quad \mathcal{O}^{U_\text{I}}(\text{osk}, \text{opk}, \text{rsk}, \text{rpk}, \cdot, \cdot, \cdot), \mathcal{O}^{U_\text{V}}(\text{opk}, \text{rpk}, \cdot, \cdot, \text{RI}[0]), \\ \quad \mathcal{O}^{LoR}(\text{osk}, \text{opk}, \text{rsk}, \text{rpk}, b, \cdot, \cdot, \cdot, \text{RI}[0])\}, \\ b^* \leftarrow \mathcal{A}^{\mathcal{O}}(\text{pp}, \text{osk}, \text{opk}, \text{rsk}, \text{rpk}) \end{array}\right] - \frac{1}{2} \leq \epsilon(\kappa).$$

Observe, that the pseudonyms contained in NYM_{LoR} can later be revoked using the $\mathcal{O}^{\text{RN+}}$ oracle. This explicitly requires that even if pseudonyms get revoked and the adversary has access to all previous showing transcripts, users still remain anonymous (backward unlinkability).

4 Construction of the RABC System

We first recall the intuition behind the ABC system in [30]. Then, we present the intuition behind our construction and finally we present our RABC system.

4.1 Intuition of the ABC System

The ABC construction in [30] requires an EUF-CMA secure SPS-EQ-$\mathcal{R}$ scheme with perfect adaption of signatures and DDH holding on the message space (subsumed as class-hiding property in [30]; e.g., the Scheme in [27]). It further requires randomizable polynomial commitments with factor openings (PolyCommitFO, cf. [30]) and one single, constant-size ZKPK to prevent replays of previously shown credentials. Below, we recall how the building blocks are combined.

In [30], a credential cred_i for user i is a vector of two group elements (C_1, P) together with a signature of the organization under the SPS-EQ-$\mathcal{R}$ scheme, where C_1 is a polynomial commitment to a polynomial that encodes the attribute set $\mathbb{A}$ of the credential. The encoding of the attribute set $\mathbb{A} = \{(\text{attr}_k, \text{attrV}_k)\}_{k=1}^n$ to a polynomial in $\mathbb{Z}_p[X]$ is defined by the following encoding function, where $H : \{0,1\}^* \to \mathbb{Z}_p^*$ is a collision-resistant hash function:

$$\mathsf{enc} : \mathbb{A} \mapsto \prod_{k=1}^{n} \prod_{M \in \mathtt{attrV_k}} \left(X - H(\mathtt{attr}_k \| M) \right).$$

Additionally, C_1 includes the private key r_i corresponding to the public key $R_i = r_i P$ of user i.

On a showing for some attribute set $\mathbb{A}' \sqsubseteq \mathbb{A}$, a credential owner proceeds as follows. To achieve unlinkability, the user randomizes the credential using a random scalar ρ. This is simply done by changing the representative of (C_1, P) with signature σ to the representative $\rho(C_1, P)$ and signature σ' (using $\mathsf{ChgRep}_{\mathcal{R}}$ of SPS-EQ-$\mathcal{R}$). Then, a user provides the randomized credential together with a selective opening of the polynomial commitment ρC_1 with respect to the encoding of the revealed attributes $\mathsf{enc}(\mathbb{A}')$. This so called factor opening includes a consistently randomized witness (by using ρ), attesting that $\mathbb{A}' \sqsubseteq \mathbb{A}$ while hiding the unrevealed attribute set $\overline{\mathbb{A}'}$.[3] Thereby, the rerandomization of $\mathsf{PolyCommitFO}$ is compatible with the rerandomization of the SPS-EQ-$\mathcal{R}$ scheme. Additionally, the user provides a ZKPK (denoted PoK) to demonstrate knowledge of ρ in ρP with respect to P to guarantee freshness, i.e., to prevent replaying of past showings.

Now, to verify a credential, the verifier starts by checking the signature σ' on the obtained credential $(\rho C_1, \rho P)$ (using the organization's SPS-EQ-$\mathcal{R}$ public key). Then, it verifies whether the factor opening to $\mathsf{enc}(\mathbb{A}')$ is correct with respect to the randomized polynomial commitment ρC_1 (via $\mathsf{VerifyFactor_{PC}}$ [30]). In particular, it checks whether the polynomial that encodes $\mathbb{A}'$ is indeed a factor of the polynomial committed to in ρC_1 by using the witness to $\overline{\mathbb{A}'}$ and without learning anything about $\overline{\mathbb{A}'}$. By construction this also guarantees that the prover knows the respective secret key (without revealing it). Furthermore, the verifier only accepts if PoK holds to guarantee that the showing is fresh (and no replay).

Example: To illustrate the attribute sets, we restate a short example from [30]. Suppose that we are given a user with the following attribute set: $\mathbb{A} = \{(\mathtt{age}, \{> 16, > 18\}), (\mathtt{drivinglicense}, \{\#, car\})\}$, where $\#$ indicates an attribute value that proves the possession of an attribute without revealing any concrete value. A showing could involve the attributes $\mathbb{A}' = \{(\mathtt{age}, \{> 18\}), (\mathtt{drivinglicense}, \{\#\})\}$ and its hidden complement $\overline{\mathbb{A}'} = \{(\mathtt{age}, \{> 16\}), (\mathtt{drivinglicense}, \{car\})\}$.

4.2 Incorporating Blacklist Revocation

To enable revocation, we need to augment the credentials in the ABC construction of [30] to include a unique nym. Recall that in our context pseudonyms are more or less credential identifiers that are never being revealed during showings and solely used for revocation purposes. In a nutshell, the revocation authority holds a list of revoked nyms $\mathtt{RNYM} = \{\mathsf{nym}_i\}_{i \in [n]}$ and unrevoked nyms $\mathtt{NYM} = \{\mathsf{nym}_i\}_{i \in [m]}$, respectively. It publishes an accumulator Π, which represents the

---

[3] Such a witness is basically a consistently randomized commitment (by using ρ) to $\overline{\mathbb{A}'}$.

list of revoked pseudonyms RNYM. Additionally, the revocation authority maintains a public list WIT of non-membership witnesses $\{\omega_{\mathsf{nym}_i}\}_{i \in [m]}$ for unrevoked users. An unrevoked user then demonstrates that the nym encoded in the credential has not been blacklisted, i.e., nym is not contained in the accumulator, during a showing. We assume that two dummy nyms are initially inserted into the accumulator so that the accumulator Π as well as witnesses ω_{nym_i} match the form, which is required for the respective algorithms to work. We emphasize that, in contrast to existing accumulator-based approaches, we avoid to prove in zero-knowledge the possession of such a non-membership witness which satisfies the accumulator verification relation. Furthermore, we note that one could also allow the users to update their witnesses on their own by using the dynamic features of the accumulator construction in [3].

4.3 Our Construction

Our revocation mechanism is based on the observation that the accumulator in Scheme 1 is compatible with the rerandomizations of the credentials (due to similarities between Scheme 1 and PolyCommitFO in [30]). In particular, we extend the original credential by two values C_2 and C_3, resulting in a credential cred $= ((C_1, C_2, C_3, P), \sigma)$. We choose the second credential component C_2 to be $C_2 = u_i(\lambda P - \mathsf{nym} \cdot P)$ (which can directly be used in the Verify$_{\mathsf{Acc}}$ algorithm). Here, u_i is an additional user secret key that is required for anonymity (similar to the secret r_i in C_1) and corresponds to $U_i = u_i P$ in the augmented public key (R_i, U_i). Furthermore, for technical reasons, we include a third credential component $C_3 = u_i Q$, where Q (as in the original scheme) is a random element in $\mathbb{G}_1$ with unknown discrete logarithm. During showings, rerandomized versions of the credential will be presented, which is due to the nature of the credential scheme in [30]. To preserve the correctness of the accumulator verification relation, the prover must present consistently rerandomized versions of the accumulator Π (and of the non-membership witnesses as well). Apparently, the prover must be restricted to present only honestly rerandomized versions thereof.[4]

Scheme 2 presents our RABC system, where we require t, t' to be bounded by poly(κ). If a check does not yield 1 or a PoK is invalid, the respective algorithm terminates with a failure and the algorithm Verify accepts only if Verify$_{\mathcal{R}}$, VerifyFactor$_{\mathsf{PC}}$, Verify$_{\mathsf{Acc}}$ return 1. Note that in Scheme 2, we use a slightly modified version of the algorithm Verify$_{\mathsf{Acc}}$, which directly takes $\mathbf{d} = e(dP, \hat{P})$ instead of a scalar d as part of the witness (as done in Scheme 1). This version uses the verification relation $e(\Pi, \hat{P}) = e(\lambda P - yP, \hat{W}) \cdot \mathbf{d}$. Also note that the prover can compute the commitment of the $\mathbf{d}'$-part of the proof using a pairing, which is typically faster than a corresponding exponentiation in $\mathbb{G}_T$ in state-of-the-art pairing implementations. In addition to PoK on the discrete logarithm of $\mathbf{d}'$,

---

[4] To ensure the authenticity of the rerandomized revocation information, we require users to prove knowledge of the randomizer used for randomizing the original accumulator and for proof-technical reasons we require the user to prove knowledge of $\log_Q C_3$.

we must also check whether $\mathbf{d}' \neq 1$ to ensure the correct form of the presented witness $(\hat{W}', \mathbf{d}')$ (recall that $d \neq 0$ is required). Furthermore, the accumulator Π needs to be available in an authentic fashion. Finally, we note that the first move in the showing protocol can be combined with the first move of PoK. Thus, a showing consists of a total of three moves.

4.4 Security of the RABC System

Theorem 2. *The RABC system in Scheme 2 is correct.*

The correctness of Scheme 2 follows from inspection.

Theorem 3. *If* PolyCommitFO *is factor-sound,* $\{(H_s, s)\}_{s \in S}$ *is a collision-resistant hash function family, the underlying SPS-EQ-$\mathcal{R}$ is* EUF-CMA *secure and perfectly adapts signatures,* Acc *is collision-free and the DDH assumption holds in* $\mathbb{G}_1$, *then Scheme 2 is unforgeable.*

We prove Theorem 3 in the extended version of this paper. Now, for anonymity of Scheme 2 we introduce two plausible assumptions in the Type-3 bilinear group setting.

Definition 10. Let BG *be a bilinear group with* $\log_2 p = \kappa$. *Then, for every* PPT *adversary* $\mathcal{A}$ *there is a negligible function* $\epsilon(\cdot)$ *such that*

$$\Pr\left[\begin{matrix} b \xleftarrow{R} \{0,1\}, \ r,s,t,u,v \xleftarrow{R} \mathbb{Z}_p, b^* \leftarrow \mathcal{A}(\mathsf{BG}, rP, r\hat{P}, sP, s\hat{P}, \\ tP, ru\hat{P}, stuP, \mathbf{g}^{(1-b)\cdot v + b \cdot ut}) \end{matrix} : b^* = b \right] - \frac{1}{2} \leq \epsilon(\kappa).$$

We emphasize that the assumption in Definition 10 can easily be justified in the uber-assumption framework [9], i.e., by setting $\mathsf{R} = \langle 1, r, s, t, stu \rangle, \mathsf{S} = \langle 1, r, s, ru \rangle, \mathsf{T} = \langle 1 \rangle, f = ut$. The subsequent assumption is closely related to the assumption in Definition 10, but does not fit the uber-assumption framework due to the decision-part being in $\mathbb{G}_2$. Consequently, we analyze the assumption in the generic group model.

Definition 11. Let BG *be a bilinear group with* $\log_2 p = \kappa$. *Then, for every* PPT *adversary* $\mathcal{A}$ *there is a negligible function* $\epsilon(\cdot)$ *such that*

$$\Pr\left[\begin{matrix} b \xleftarrow{R} \{0,1\}, \ r,s,t,u,v \xleftarrow{R} \mathbb{Z}_p, b^* \leftarrow \mathcal{A}(\mathsf{BG}, rP, r\hat{P}, sP, s\hat{P}, \\ tP, stuP, ((1-b) \cdot v + b \cdot ru)\hat{P}) \end{matrix} : b^* = b \right] - \frac{1}{2} \leq \epsilon(\kappa).$$

Proposition 1. *The assumption in Definition 11 holds in generic Type-3 bilinear groups and reaches the optimal, quadratic simulation error bound.*

The proof of the above proposition is given in the extended version of this paper.

Theorem 4. *If the underlying SPS-EQ-$\mathcal{R}$ perfectly adapts signatures, DDH in* $\mathbb{G}_1$ *and the assumptions in Definitions 10 and 11 hold, then Scheme 2 is anonymous.*

We prove Theorem 4 in the extended version of this paper.

Setup: Given $(1^\kappa, \mathsf{aux})$, parse $\mathsf{aux} \leftarrow (t, t')$, run $\mathsf{pp}' = (\mathsf{BG}, (\alpha^i P)_{i \in [t]}, (\alpha^i \hat{P})_{i \in [t]}) \leftarrow \mathsf{Setup}_{\mathsf{PC}}(1^\kappa, t)$ and $\mathsf{pp}'' = ((\lambda^i P)_{i \in [t']}, (\lambda^i \hat{P})_{i \in [t']}) \leftarrow \mathsf{Gen}_{\mathsf{Acc}}(\mathsf{BG}, t')$. Then, let $\mathbf{g} \leftarrow e(P, \hat{P})$ and $H_s : \{0,1\}^* \to \mathbb{Z}_p^*$ be a collision-resistant keyed hash function used inside $\mathsf{enc}(\cdot)$, drawn uniformly at random from a family of collision-resistant keyed hash functions $\{(H_s, s)\}_{s \in S}$. Finally, choose $Q \xleftarrow{R} \mathbb{G}_1$ and output $\mathsf{pp} \leftarrow (H_s, \mathsf{enc}, Q, \mathbf{g}, \mathsf{pp}', \mathsf{pp}'')$.

RAKeyGen: Given pp return $(\mathsf{rsk}, \mathsf{rpk}) \leftarrow (\emptyset, \mathsf{pp}'')$.

OrgKeyGen: Given pp and $j \in \mathbb{N}$, return $(\mathsf{osk}_j, \mathsf{opk}_j) \leftarrow \mathsf{KeyGen}_{\mathcal{R}}(1^\kappa, \ell = 4)$.

UserKeyGen: Given pp and $i \in \mathbb{N}$, pick $r_i, u_i \xleftarrow{R} \mathbb{Z}_p^*$, compute $(R_i, U_i) \leftarrow (r_i P, u_i P)$ and return $(\mathsf{usk}_i, \mathsf{upk}_i) \leftarrow ((r_i, u_i), (R_i, U_i))$.

(Obtain, Issue): Obtain and Issue interact in the following way:

$\mathsf{Issue}(\mathsf{pp}, \mathsf{rpk}, \mathsf{upk}_i, \mathsf{osk}_j, \mathsf{nym}, \mathbb{A})$		$\mathsf{Obtain}(\mathsf{pp}, \mathsf{usk}_i, \mathsf{opk}_j, \mathsf{nym}, \mathbb{A})$
$\dfrac{e(C_1, \hat{P})}{e(R_i, \mathsf{enc}(\mathbb{A})(\alpha)\hat{P})}$	$\xleftarrow{C_1, C_2, C_3}$	$(C_1, C_2, C_3) \leftarrow (r_i \mathsf{enc}(\mathbb{A})(\alpha)P,$
$e(C_2, \hat{P}) = e(U_i, \lambda\hat{P} - \mathsf{nym} \cdot \hat{P})$	$\xleftrightarrow{\mathsf{PoK}}$	$u_i(\lambda P - \mathsf{nym} \cdot P), u_i Q)$
$\sigma \leftarrow \mathsf{Sign}_{\mathcal{R}}((C_1, C_2, C_3, P), \mathsf{osk}_j)$	$\xrightarrow{\sigma}$	$\mathsf{Verify}_{\mathcal{R}}((C_1, C_2, C_3, P), \sigma, \mathsf{opk}_j) = 1$
		$\mathsf{cred}_{\mathsf{nym}} \leftarrow ((C_1, C_2, C_3, P), \sigma)$

where PoK is: $\mathsf{PoK}\{(\psi) : C_3 = \psi Q \ \wedge \ U_i = \psi P\}$.

(Show, Verify): Show and Verify interact in the following way, where $\mathbb{R}_V = \Pi \leftarrow \mathbb{R}[1]$ and $\mathbb{R}_S^{\mathsf{nym}} = (\Pi, (\hat{W}, d)) \leftarrow (\mathbb{R}[1], \mathbb{R}[2][\mathsf{nym}])$:

$\mathsf{Verify}(\mathsf{pp}, \mathsf{rpk}, \mathsf{opk}_j, \mathbb{A}', \mathbb{R}_V)$		$\mathsf{Show}(\mathsf{pp}, \mathsf{rpk}, \mathsf{usk}_i, \mathsf{opk}_j, \mathsf{cred}_{\mathsf{nym}}, \mathbb{A}, \mathbb{A}', \mathbb{R}_S^{\mathsf{nym}})$
		$\rho, \nu \xleftarrow{R} \mathbb{Z}_p^*$
		$(\hat{W}', \mathbf{d}') \leftarrow (\nu\hat{W}, e(\rho\nu u_i dP, \hat{P}))$
		$\Pi' \leftarrow \rho\nu u_i \Pi$
		$\mathsf{cred} \leftarrow \mathsf{ChgRep}_{\mathcal{R}}(\mathsf{cred}_{\mathsf{nym}}, \rho, \mathsf{opk}_j)$
$\Big[\mathsf{Verify}_{\mathcal{R}}(\mathsf{cred}, \mathsf{opk}_j) \ \wedge$	$\xleftarrow{\mathsf{cred}, C_{\overline{\mathbb{A}'}}, \Pi', \hat{W}', \mathbf{d}'}$	$C_{\overline{\mathbb{A}'}} \leftarrow (\rho \cdot r_i) \cdot \mathsf{enc}(\overline{\mathbb{A}'})(\alpha)P$
$\mathbf{d}' \neq 1_{\mathbb{G}_T} \wedge \mathsf{VerifyFactor}_{\mathsf{PC}}(\mathsf{pp}',$		
$C_1, \mathsf{enc}(\mathbb{A}'), C_{\overline{\mathbb{A}'}}) \wedge \mathsf{Verify}_{\mathsf{Acc}}(\Pi',$		
$(\hat{W}', \mathbf{d}'), C_2, \mathsf{pp}'')\Big] = 1$	$\xleftrightarrow{\mathsf{PoK}}$	

where $\mathsf{cred} = ((C_1, C_2, C_3, C_4), \sigma)$ and PoK is: $\mathsf{PoK}\{(\gamma, \delta, \eta, \zeta, \psi) : Q = \eta P \ \vee \ (C_3 = \psi Q \ \wedge \ C_4 = \gamma P \ \wedge \ \mathbf{d}' = \mathbf{g}^\delta \ \wedge \ \Pi' = \zeta\Pi)\}$.

Revoke: Given pp, $(\mathsf{rsk}, \mathsf{rpk})$, NYM and RNYM, this algorithm computes $\Pi \leftarrow \mathsf{Eval}_{\mathsf{Acc}}(\mathsf{RNYM}, (\emptyset, \mathsf{pp}''))$. Then, for all $\mathsf{nym} \in \mathsf{NYM}$ it computes $(W'_{\mathsf{nym}}, d_{\mathsf{nym}}) \leftarrow \mathsf{WitCreate}_{\mathsf{Acc}}(\Pi, \mathsf{RNYM}, \mathsf{nym}, (\emptyset, \mathsf{pp}''))$, sets $\mathsf{WIT}[\mathsf{nym}] \leftarrow (W'_{\mathsf{nym}}, d_{\mathsf{nym}})$ and returns $\mathbb{R} \leftarrow (\Pi, \mathsf{WIT})$.

Scheme 2: Our multi-show RABC system.

5 Discussion

The presented revocation mechanism for the RABC system uses similar building blocks as the original ABC system. In particular, it does not use a complex ZKPK for demonstrating the knowledge of a non-membership witness, which satisfies the verification relation of the accumulator. It only requires a simple ZKPK of the dicrete logarithms in $\mathbf{d}', \Pi', C_3$ (and C_4 which is already required in the original ABC system from [30]) for technical reasons. Consequently, this concept yields a new direction for revocation in ABC systems.

In the exented version of this paper, we carefully adapt an existing universal accumulator revocation mechanism [1,38] (applied within Microsoft's U-Prove) to the ABC system from [30] and prove it secure in the model proposed in this paper. Due to space limitations, this part is not included in this version. We note, however, that due to the high number of zero-knowledge proofs of knowledge in [1,38], the approach presented here is the more efficient choice (see the full version for a detailed comparison).

Acknowledgements. We would like to thank the anonymous reviewers for their valuable comments.

References

1. Acar, T., Chow, S.S.M., Nguyen, L.: Accumulators and U-Prove revocation. In: Sadeghi, A.-R. (ed.) FC 2013. LNCS, vol. 7859, pp. 189–196. Springer, Heidelberg (2013)
2. Akagi, N., Manabe, Y., Okamoto, T.: An efficient anonymous credential system. In: Tsudik, G. (ed.) FC 2008. LNCS, vol. 5143, pp. 272–286. Springer, Heidelberg (2008)
3. Au, M.H., Tsang, P.P., Susilo, W., Mu, Y.: Dynamic universal accumulators for DDH groups and their application to attribute-based anonymous credential systems. In: Fischlin, M. (ed.) CT-RSA 2009. LNCS, vol. 5473, pp. 295–308. Springer, Heidelberg (2009)
4. Baldimtsi, F., Lysyanskaya, A.: Anonymous credentials light. In: ACM CCS. ACM (2013)
5. Begum, N., Nakanishi, T., Funabiki, N.: Efficient proofs for CNF formulas on attributes in pairing-based anonymous credential system. In: Kwon, T., Lee, M.-K., Kwon, D. (eds.) ICISC 2012. LNCS, vol. 7839, pp. 495–509. Springer, Heidelberg (2013)
6. Belenkiy, M., Camenisch, J., Chase, M., Kohlweiss, M., Lysyanskaya, A., Shacham, H.: Randomizable proofs and delegatable anonymous credentials. In: Halevi, S. (ed.) CRYPTO 2009. LNCS, vol. 5677, pp. 108–125. Springer, Heidelberg (2009)
7. Benaloh, J.C., de Mare, M.: One-way accumulators: a decentralized alternative to digital signatures. In: Helleseth, T. (ed.) EUROCRYPT 1993. LNCS, vol. 765, pp. 274–285. Springer, Heidelberg (1994)
8. Boneh, D., Shacham, H.: Group signatures with verifier-local revocation. In: ACM CCS (2004)

9. Boyen, X.: The uber-assumption family – a unified complexity framework for bilinear groups. In: Galbraith, S.D., Paterson, K.G. (eds.) Pairing 2008. LNCS, vol. 5209, pp. 39–56. Springer, Heidelberg (2008)

10. Brands, S.: Rethinking public-key Infrastructures and Digital Certificates: Building in Privacy. MIT Press, Cambridge (2000)

11. Camenisch, J., Dubovitskaya, M., Haralambiev, K., Kohlweiss, M.: Composable and modular anonymous credentials: definitions and practical constructions. IACR Cryptology ePrint Archive

12. Camenisch, J., Herreweghen, E.V.: Design and implementation of the idemix anonymous credential system. In: ACM CCS. ACM (2002)

13. Camenisch, J., Kohlweiss, M., Soriente, C.: An accumulator based on bilinear maps and efficient revocation for anonymous credentials. In: Jarecki, S., Tsudik, G. (eds.) PKC 2009. LNCS, vol. 5443, pp. 481–500. Springer, Heidelberg (2009)

14. Camenisch, J., Krenn, S., Lehmann, A., Mikkelsen, G.L., Neven, G., Pedersen, M.O.: Formal treatment of privacy-enhancing credential systems (2015)

15. Camenisch, J., Lehmann, A., Neven, G., Rial, A.: Privacy-preserving auditing for attribute-based credentials. In: Kutyłowski, M., Vaidya, J. (eds.) ICAIS 2014, Part II. LNCS, vol. 8713, pp. 109–127. Springer, Heidelberg (2014)

16. Camenisch, J.L., Lysyanskaya, A.: An efficient system for non-transferable anonymous credentials with optional anonymity revocation. In: Pfitzmann, B. (ed.) EUROCRYPT 2001. LNCS, vol. 2045, pp. 93–118. Springer, Heidelberg (2001)

17. Camenisch, J.L., Lysyanskaya, A.: A signature scheme with efficient protocols. In: Cimato, S., Galdi, C., Persiano, G. (eds.) SCN 2002. LNCS, vol. 2576, pp. 268–289. Springer, Heidelberg (2003)

18. Camenisch, J.L., Lysyanskaya, A.: Dynamic accumulators and application to efficient revocation of anonymous credentials. In: Yung, M. (ed.) CRYPTO 2002. LNCS, vol. 2442, p. 61. Springer, Heidelberg (2002)

19. Camenisch, J.L., Lysyanskaya, A.: Signature schemes and anonymous credentials from bilinear maps. In: Franklin, M. (ed.) CRYPTO 2004. LNCS, vol. 3152, pp. 56–72. Springer, Heidelberg (2004)

20. Canard, S., Lescuyer, R.: Anonymous credentials from (indexed) aggregate signatures. In: DIM. ACM (2011)

21. Canard, S., Lescuyer, R.: Protecting privacy by sanitizing personal data: a new approach to anonymous credentials. In: ASIA CCS. ACM (2013)

22. Chase, M., Meiklejohn, S., Zaverucha, G.M.: Algebraic MACs and keyed-verification anonymous credentials. In: ACM CCS. ACM (2014)

23. Chaum, D.: Security without identification: transaction systems to make big brother obsolete. Commun. ACM 28(10), 1030–1044 (1985)

24. Cheon, J.H.: Security analysis of the strong diffie-hellman problem. In: Vaudenay, S. (ed.) EUROCRYPT 2006. LNCS, vol. 4004, pp. 1–11. Springer, Heidelberg (2006)

25. Derler, D., Hanser, C., Slamanig, D.: Revisiting cryptographic accumulators, additional properties and relations to other primitives. In: Nyberg, K. (ed.) CT-RSA 2015. LNCS, vol. 9048, pp. 127–144. Springer, Heidelberg (2015)

26. Fuchsbauer, G.: Breaking existential unforgeability of a signature scheme from Asiacrypt 2014. IACR Cryptology ePrint Archive (2014)

27. Fuchsbauer, G., Hanser, C., Slamanig, D.: EUF-CMA-Secure structure-preserving signatures on equivalence classes. IACR Cryptology ePrint Archive (2014)

28. Garman, C., Green, M., Miers, I.: Decentralized anonymous credentials. In: NDSS (2014)

29. Hajny, J., Malina, L.: Unlinkable attribute-based credentials with practical revocation on smart-cards. In: Mangard, S. (ed.) CARDIS 2012. LNCS, vol. 7771, pp. 62–76. Springer, Heidelberg (2013)

30. Hanser, C., Slamanig, D.: Structure-preserving signatures on equivalence classes and their application to anonymous credentials. In: Sarkar, P., Iwata, T. (eds.) ASIACRYPT 2014. LNCS, vol. 8873, pp. 491–511. Springer, Heidelberg (2014)

31. Lapon, J., Kohlweiss, M., De Decker, B., Naessens, V.: Analysis of revocation strategies for anonymous idemix credentials. In: De Decker, B., Lapon, J., Naessens, V., Uhl, A. (eds.) CMS 2011. LNCS, vol. 7025, pp. 3–17. Springer, Heidelberg (2011)

32. Li, J., Li, N., Xue, R.: Universal accumulators with efficient nonmembership proofs. In: Katz, J., Yung, M. (eds.) ACNS 2007. LNCS, vol. 4521, pp. 253–269. Springer, Heidelberg (2007)

33. Lueks, W., Alpár, G., Hoepman, J.H., Vullers, P.: Fast revocation of attribute-based credentials for both users and verifiers. In: Federrath, H., Gollmann, D. (eds.) SEC 2015. IFIP AICT, vol. 455, pp. 463–478. Springer, Heidelberg (2015)

34. Lysyanskaya, A., Rivest, R.L., Sahai, A., Wolf, S.: Pseudonym systems (extended abstract). In: Heys, H.M., Adams, C.M. (eds.) SAC 1999. LNCS, vol. 1758, pp. 184–199. Springer, Heidelberg (2000)

35. Nakanishi, T., Fujii, H., Hira, Y., Funabiki, N.: Revocable group signature schemes with constant costs for signing and verifying. In: Jarecki, S., Tsudik, G. (eds.) PKC 2009. LNCS, vol. 5443, pp. 463–480. Springer, Heidelberg (2009)

36. Nakanishi, T., Funabiki, N.: Verifier-local revocation group signature schemes with backward unlinkability from bilinear maps. In: Roy, B. (ed.) ASIACRYPT 2005. LNCS, vol. 3788, pp. 533–548. Springer, Heidelberg (2005)

37. Nguyen, L.: Accumulators from bilinear pairings and applications. In: Menezes, A. (ed.) CT-RSA 2005. LNCS, vol. 3376, pp. 275–292. Springer, Heidelberg (2005)

38. Nguyen, L., Paquin, C.: U-prove designated-verifier accumulator revocation extension. Technical report, Microsoft Research (2014)

39. Paquin, C., Zaverucha, G.: U-prove cryptographic specification v1.1, revision 3. Technical report, Microsoft Corporation (2013)

40. Song, D.X.: Practical forward secure group signature schemes. In: ACM CCS. ACM (2001)

41. Sudarsono, A., Nakanishi, T., Funabiki, N.: Efficient proofs of attributes in pairing-based anonymous credential system. In: Fischer-Hübner, S., Hopper, N. (eds.) PETS 2011. LNCS, vol. 6794, pp. 246–263. Springer, Heidelberg (2011)

42. Unterluggauer, T., Wenger, E.: Efficient pairings and ECC for embedded systems. In: Batina, L., Robshaw, M. (eds.) CHES 2014. LNCS, vol. 8731, pp. 298–315. Springer, Heidelberg (2014)

43. Verheul, E.R.: Self-blindable credential certificates from the weil pairing. In: Boyd, C. (ed.) ASIACRYPT 2001. LNCS, vol. 2248, pp. 533–551. Springer, Heidelberg (2001)

Symmetric Cryptography

Tweak-Length Extension
for Tweakable Blockciphers

Kazuhiko Minematsu[1]([⊠]) and Tetsu Iwata[2]

[1] NEC Corporation, Kawasaki, Japan
k-minematsu@ah.jp.nec.com
[2] Nagoya University, Nagoya, Japan
iwata@cse.nagoya-u.ac.jp

Abstract. Tweakable blockcipher (TBC) is an extension of standard blockcipher introduced by Liskov, Rivest and Wagner in 2002. TBC is a versatile building block for efficient symmetric-key cryptographic functions, such as authenticated encryption.

In this paper we study the problem of extending tweak of a given TBC of fixed-length tweak, which is a variant of popular problem of converting a blockcipher into a TBC, i.e., blockcipher mode of operation. The problem is particularly important for known dedicated TBCs since they have relatively short tweak. We propose a simple and efficient solution, called XTX, for this problem. XTX converts a TBC of fixed-length tweak into another TBC of arbitrarily long tweak, by extending the scheme of Liskov, Rivest and Wagner that converts a blockcipher into a TBC. Given a TBC of n-bit block and m-bit tweak, XTX provides $(n + m)/2$-bit security while conventional methods provide $n/2$ or $m/2$-bit security. We also show that XTX is even useful when combined with some blockcipher modes for building TBC having security beyond the birthday bound.

Keywords: Tweakable blockcipher · Tweak extension · Mode of operation · LRW

1 Introduction

Tweakable Blockcipher. Tweakable blockcipher (TBC) is an extension of standard blockcipher introduced by Liskov, Rivest and Wagner in 2002 [17]. An encryption of TBC takes a parameter called tweak, in addition to key and plaintext. Tweak is public and can be arbitrarily chosen. Due to its versatility, TBC is getting more and more popularity. Known constructions of TBCs are generally classified into two categories: dedicated design and blockcipher mode of operation, i.e. using a blockcipher as a black box.

For the first category, Hasty pudding cipher [30] and Mercy [10] are examples of early designs. More recently Skein hash function uses a dedicated TBC called Threefish [2]. Jean, Nikolić and Peyrin [13] developed several dedicated TBCs as components of their proposals to CAESAR [1], a competition of authenticated encryption (AE).

© Springer International Publishing Switzerland 2015
J. Groth (Ed.): IMACC 2015, LNCS 9496, pp. 77–93, 2015.
DOI: 10.1007/978-3-319-27239-9_5

For the second category, Liskov et al. [17] provided two blockcipher modes to build TBC, and their second mode is known as LRW[1]. Rogaway [28] refined LRW and proposed XE and XEX modes, and Minematsu [22] proposed a generalization of LRW and XEX. These schemes have provable security up to so-called "birthday bound", i.e. they can be broken if adversary performs around $2^{n/2}$ queries, where n is the block length of TBC. The first scheme to break this barrier was shown by Minematsu [23], though it was limited to short tweak and rekeyed for every tweak. Landecker, Shrimpton and Terashima [16] showed that a chain of two LRWs has security beyond the birthday bound, which is the first scheme with this property which does not use rekeying. Lampe and Seurin [15] extended the work of [16] for longer chains. Tweakable variants of Even-Mansour cipher [11] are studied by Cogliati, Lampe and Seurin [8] and Mennink [21]. A concrete example is seen in a CAESAR proposal, called Minalpher [29].

Tweak Extension. In this paper, we study tweak extension of a given TBC. More formally, let $\widetilde{E}$ be a TBC of n-bit block and m-bit tweak, and we want to arbitrarily extend m-bit tweak of $\widetilde{E}$ keeping n-bit block. Here m is considered to be fixed. At first sight the problem looks trivial since most of previous studies in the second category already cover the case of arbitrarily long tweak when combined with a universal hash (UH) function of variable-length input, and a TBC with any fixed tweak is also a blockcipher. Coron et al. (Theorem 6, [9]) pointed out another simple solution by applying a UH function H to tweak and then use the hash value $H(T)$ as the tweak of $\widetilde{E}$. However, the problem is security. For TBC $\widetilde{E}$ of n-bit block and m-bit tweak, applying LRW or XEX to (fixed-tweak) $\widetilde{E}$ the security is up to $O(2^{n/2})$ queries. Coron et al.'s solution is also secure up to $O(2^{m/2})$ queries. We would get a better security bound by using the chained LRW [15,16], but it would significantly increase the computation cost from $\widetilde{E}$.

In this paper we provide an alternative solution, called XTX, which can be explained as an intuitive yet non-trivial combination of LRW and Coron et al.'s method mentioned above, applicable to any black-box TBC. Specifically, XTX converts a TBC $\widetilde{E}$ of n-bit block and m-bit tweak into another TBC of n-bit block and t-bit tweak for any $t > m$, using H which is a (variant of) UH function of t-bit input and $(n+m)$-bit output. See Fig. 1 for XTX. We proved the security bound of $q^2\epsilon$ where ϵ denotes the bias of UH function. This implies security up to $O(2^{(n+m)/2})$ queries if ϵ is ideally small. As well as LRW, XTX needs one calls of $\widetilde{E}$ and H, hence the computation cost is only slightly increased, and H is called only once for multiple encryptions sharing the same tweak, by caching the output of H.

We observe that tweak length of existing dedicated TBCs are relatively short, at least not much longer than the block length. For instance, KIASU-BC [13] has 128-bit block and 64-bit tweak, and Threefish has 256 or 512 or 1024-bit block

---

[1] The two schemes shown by [17] are also called LRW1 and LRW2, and we refer to LRW2 throughout this paper.

with 128-bit tweak for all block sizes. One apparent reason for having fixed, short tweak is that it is enough for their primary applications, however, if tweak can be effectively extended it would expand their application areas. For another reason, we think the complexity of security analysis for dedicated TBC is expected to be dependent on the size of tweak space, since we have to make sure that for each tweak the instance should behave as an independently-keyed blockcipher. The TWEAKEY framework of Jean et al. [13] provided a systematic way to build TBCs by incorporating tweak in the key schedule, and it shows that building efficient TBCs from scratch is far from trivial, in particular when we want to have long tweaks. Our XTX can efficiently extend tweak of dedicated TBCs with reasonably small security degradation in terms of the maximum number of allowable queries. In addition, XTX is even useful when applied to some modes of operations when the baseline TBC from a (non-tweakable) blockcipher has beyond-birthday-bound security. We summarize these results in Sect. 4.

Applications of Tweak Extension. We remark that a TBC with long tweak is useful. In general, a tweak of a TBC can be used to contain various additional information associated with plaintext block, hence it would be desirable to make tweak substantially longer than the block length (say 128 bits). For concrete examples, a large-block TBC of Shrimpton and Terashima [31] used a TBC with variable-length tweak, which was instantiated by a combination of techniques from [9,16]. Hirose, Sasaki and Yasuda [12] presented an AE scheme using TBC with tweak something longer than the unit block.

2   Preliminaries

Notation. Let $\{0,1\}^*$ be the set of all finite bit strings. For an integer $\ell \geq 0$, let $\{0,1\}^\ell$ be the set of all bit strings of ℓ bits. For $X \in \{0,1\}^*$, $|X|$ is its length in bits, and for $\ell \geq 1$, $|X|_\ell = \lceil |X|/\ell \rceil$ is the length in ℓ-bit blocks. When ℓ denotes the length of a binary string we also write ℓ_n to mean $\lceil \ell/n \rceil$. A sequence of a zeros is denoted by 0^a. For set $\mathcal{S} \subseteq \{0,1\}^n$ and $x \in \{0,1\}^n$, $\mathcal{S} \oplus x$ denotes the set $\{s \oplus x : s \in \mathcal{S}\}$. If random variable X is uniformly distributed over $\mathcal{X}$ we write $X \in_U \mathcal{X}$.

Cryptographic Functions. For any keyed function we assume that its first argument denotes the key. For keyed function $F : \mathcal{K} \times \mathcal{X} \rightarrow \mathcal{Y}$, we write $F_K(x)$ to denote $F(K, x)$ for the evaluation of input $x \in \mathcal{X}$ with key $K \in_U \mathcal{K}$.

A blockcipher $E : \mathcal{K} \times \mathcal{M} \rightarrow \mathcal{M}$ is a keyed permutation over the message space $\mathcal{M}$. We write encryption of M using K as $C = E_K(M)$ and its inverse as $M = E_K^{-1}(C)$. Similarly, a tweakable blockcipher (TBC) is a family of n-bit blockcipher indexed by tweak $T \in \mathcal{T}$. It is written as $\widetilde{E} : \mathcal{K} \times \mathcal{T} \times \mathcal{M} \rightarrow \mathcal{M}$. If $\mathcal{M} = \{0,1\}^n$ and $\mathcal{T} = \{0,1\}^t$, we say $\widetilde{E}$ is an (n,t)-bit TBC. An encryption of message M with tweak T is written as $\widetilde{E}_K^T(M)$, and if we have $C = \widetilde{E}_K^T(M)$

then $M = \widetilde{E}_K^{T,-1}(C)$ holds for any (T, M). Let $\mathrm{Perm}(n)$ be the set of all n-bit permutations. For a finite set $\mathcal{X}$, let $\mathrm{Perm}^{\mathcal{X}}(n)$ be the set of all functions $: \mathcal{X} \times \{0,1\}^n \to \{0,1\}^n$ such that, for any $f \in \mathrm{Perm}^{\mathcal{X}}(n)$ and $x \in \mathcal{X}$, $f(x,*)$ is a permutation. An n-bit uniform random permutation (URP) is a keyed permutation with uniform key distribution over $\mathrm{Perm}(n)$ (where a key directly represents a permutation). Note that implementation of n-bit URP is impractical when n is a block size of conventional blockciphers (say, 64 or 128). We also define an n-bit tweakable URP (TURP) with tweak space $\mathcal{T}$ as a keyed tweakable permutation with uniform key distribution over $\mathrm{Perm}^{\mathcal{T}}(n)$.

Let $\mathcal{A}$ be the adversary trying to distinguish two oracles, $\mathcal{O}_1$ and $\mathcal{O}_2$, by possibly adaptive queries (which we call chosen-plaintext attack, CPA for short). We denote the event that the final binary decision of $\mathcal{A}$ after querying oracle $\mathcal{O}$ is 1 by $\mathcal{A}^{\mathcal{O}} \Rightarrow 1$. We write

$$\mathrm{Adv}^{\mathrm{cpa}}_{\mathcal{O}_1, \mathcal{O}_2}(\mathcal{A}) \stackrel{\mathrm{def}}{=} \Pr[\mathcal{A}^{\mathcal{O}_1} \Rightarrow 1] - \Pr[\mathcal{A}^{\mathcal{O}_2} \Rightarrow 1], \qquad (1)$$

where the probabilities are defined over the internal randomness of $\mathcal{O}_i$ and $\mathcal{A}$. In particular if $\mathcal{O}_1 = E_K$ and $\mathcal{O}_2 = G_{K'}$ for two keyed permutations, E_K and $G_{K'}$, we assume $\mathcal{A}$ performs a chosen-ciphertext attack (CCA), i.e., has encryption and decryption queries and define

$$\mathrm{Adv}^{\mathrm{cca}}_{E,G}(\mathcal{A}) \stackrel{\mathrm{def}}{=} \Pr[\mathcal{A}^{(E_K, E_K^{-1})} \Rightarrow 1] - \Pr[\mathcal{A}^{(G_{K'}, G_{K'}^{-1})} \Rightarrow 1], \qquad (2)$$

where $\mathcal{A}^{(E_K, E_K^{-1})}$ denotes that $\mathcal{A}$ can choose one of E_K or E_K^{-1} for each query. In the same manner we define $\mathrm{Adv}^{\mathrm{cca}}_{\widetilde{E}_K, \widetilde{G}_{K'}}(\mathcal{A})$ for two keyed tweakable permutations, where tweaks in queries are arbitrarily chosen.

For n-bit blockcipher E_K and (n, t)-bit TBC $\widetilde{E}_K$, we define SPRP (for strong pseudorandom permutation) and TSPRP (for tweakable SPRP) advantages for $\mathcal{A}$ as

$$\mathrm{Adv}^{\mathrm{sprp}}_E(\mathcal{A}) \stackrel{\mathrm{def}}{=} \mathrm{Adv}^{\mathrm{cca}}_{E,\mathsf{P}}(\mathcal{A}), \text{ and } \mathrm{Adv}^{\mathrm{tsprp}}_{\widetilde{E}}(\mathcal{A}) \stackrel{\mathrm{def}}{=} \mathrm{Adv}^{\mathrm{cca}}_{\widetilde{E},\widetilde{\mathsf{P}}}(\mathcal{A}), \qquad (3)$$

where P is n-bit URP and $\widetilde{\mathsf{P}}$ is (n, t)-bit TURP.

If $\mathcal{A}$ is information-theoretic, it is only limited in the numbers and lengths of queries. If $\mathcal{A}$ is computational, it also has a limitation on computation time in some fixed model, which is required to define computationally-secure objects, e.g. pseudorandom function (PRF). In this paper most security proofs are information-theoretic, i.e. the target schemes are built upon URP or TURP. When their components are substituted with conventional blockcipher or TBC, a computational security bound is obtained using a standard technique [4].

2.1 Universal Hash Function and Polynomial Hash Function

We will need a class of non-cryptographic functions called universal hash function [7] defined as follows.

Definition 1. *For function $H : \mathcal{K} \times \mathcal{X} \to \mathcal{Y}$ being keyed by $K \in_U \mathcal{K}$, we say it is ϵ-almost uniform (ϵ-AU) if*

$$\max_{x \neq x'} \Pr_K[H_K(x) = H_K(x')] \leq \epsilon \tag{4}$$

holds. Moreover if $\mathcal{Y} = \{0,1\}^n$ for some n, we say it is ϵ-almost XOR uniform (ϵ-AXU) if

$$\max_{x \neq x', \Delta \in \{0,1\}^n} \Pr_K[H_K(x) \oplus H_K(x') = \Delta] \leq \epsilon \tag{5}$$

holds.

From the definition if H is ϵ-AXU then it is also ϵ-AU.

Next we introduce polynomial hash function as a popular class of AU and AXU functions. Let $\mathsf{Poly}[a] : \mathcal{L} \times \{0,1\}^* \to \{0,1\}^a$ for key space $\mathcal{L} = \mathrm{GF}(2^a)$ be the polynomial hash function defined over $\mathrm{GF}(2^a)$. Formally, we have

$$\mathsf{Poly}[a]_L(X) = \sum_{i=1,\ldots,|X|_a} L^{|X|_a - i + 1} \cdot X[i], \tag{6}$$

where multiplications and additions are over $\mathrm{GF}(2^a)$, and $(X[1], \ldots, X[|X|_a])$ denotes an a-bit partition of $X \in \{0,1\}^*$ with a mapping between $\{0,1\}^a$ and $\mathrm{GF}(2^a)$ and a padding for partial message. Here, padding must have the property that the original message is uniquely recovered from the padded message. For example we can pad the non-empty sequence with $v = 100 \ldots 0$ so that $|X\|v|$ is a multiple of a. Moreover, we write $\mathsf{Poly}[a,b] : \mathcal{L} \times \{0,1\}^* \to \{0,1\}^a \times \{0,1\}^b$ for $\mathcal{L} = \mathrm{GF}(2^a) \times \mathrm{GF}(2^b)$ to denote the function $\mathsf{Poly}[a,b]_{(L_1,L_2)}(X) = (\mathsf{Poly}[a]_{L_1}(X), \mathsf{Poly}[b]_{L_2}(X))$, where L_1 and L_2 are independent. Further extensions, such as $\mathsf{Poly}[a,b,c]$, are similarly defined.

If we limit the input space of Poly to $\{0,1\}^\ell$ for some predetermined ℓ, we have the following.

Proposition 1. *A polynomial hash function $\mathsf{Poly}[n] : \mathcal{L} \times \{0,1\}^\ell \to \{0,1\}^n$ is ϵ-AXU with $\epsilon = \ell_n / 2^n$. Moreover, $\mathsf{Poly}[n_1, n_2, \ldots, n_c] : \mathcal{L} \times \{0,1\}^\ell \to \mathcal{Y}$ for $\mathcal{L} = GF(2^{n_1}) \times \cdots \times GF(2^{n_c})$ and $\mathcal{Y} = \{0,1\}^{n_1} \times \cdots \times \{0,1\}^{n_c}$ is ϵ-AXU for $\epsilon = \prod_i (\ell_{n_i} / 2^{n_i})$.*

Polynomial hash function can work over inputs of different lengths if combined with appropriate encoding. However, for simplicity this paper mainly discusses the case where Poly has a fixed input length, and in this respect we treat tweak length (in block or bit) appeared in the security bound as a constant, which is usually denoted by ℓ. Recall that we use ℓ_n to denote $\lceil \ell/n \rceil$.

3 Main Construction

3.1 Previous Schemes

We start with a description of one of the most popular TBC schemes based on blockcipher. It is the second construction of Liskov et al. [17] and is called LRW.

Using Blockcipher $E : \mathcal{K} \times \mathcal{M} \to \mathcal{M}$ with $\mathcal{M} = \{0,1\}^n$ and a keyed function $H : \mathcal{L} \times \mathcal{T} \to \mathcal{M}$, LRW is described as

$$\text{LRW}_{K,L}^T(M) = H_L(T) \oplus E_K(M \oplus H_L(T)), \tag{7}$$

where $T \in \mathcal{T}$ is a tweak and $K \in_U \mathcal{K}$ and $L \in_U \mathcal{L}$ are independent keys. Let $\text{LRW}_{\mathsf{P},L}$ denote LRW using n-bit URP, P, as a blockcipher and H with independent key $L \in_U \mathcal{L}$. Its TSPRP-advantage is bounded as[2]

$$\text{Adv}_{\text{LRW}_{\mathsf{P},L}}^{\text{tsprp}}(\mathcal{A}) \leq \epsilon \cdot q^2, \tag{8}$$

for any CCA-adversary $\mathcal{A}$ using q queries, if H is ϵ-AXU. Since $\epsilon \geq 1/2^n$ this implies provable security up to the birthday bound. The bound is tight in that there is an attack matching the bound. Rogaway's XEX [28] and Minematsu's scheme [22] reduce the two keys of LRW to one blockcipher key.

For tweak extension of given (n, m)-bit TBC, $\widetilde{E}$, we have two previous solutions: the first one is to use LRW with blockcipher instantiated by $\widetilde{E}$ taking a fixed tweak. This has security bound of (8), hence $n/2$-bit security when H is (e.g.) $\mathsf{Poly}[n]$. The second one, proposed by Coron et al. [9] as mentioned earlier, is to use $H : \mathcal{L} \times \mathcal{T} \to \mathcal{V}$ and combine $\widetilde{E}$ and H as $C = \widetilde{E}_K^V(M)$ for $V = H_L(T)$. If H is ϵ-AU, this clearly has security bound of $O(\epsilon q^2)$ which implies $m/2$-bit security at best. Then, what will happen if we use both solutions all together? In the next section we show that in fact this combination gives a better result.

3.2 XTX

We describe our proposal. Let $\widetilde{E} : \mathcal{K} \times \mathcal{V} \times \mathcal{M} \to \mathcal{M}$ be a TBC of message space $\mathcal{M} = \{0,1\}^n$ and tweak space $\mathcal{V} = \{0,1\}^m$. Let $\mathcal{T}$ be another (larger) tweak space. Let $H : \mathcal{L} \times \mathcal{T} \to \mathcal{M} \times \mathcal{V}$ be a function keyed by $L \in_U \mathcal{L}$. We define $\text{XTX} : (\mathcal{K} \times \mathcal{L}) \times \mathcal{T} \times \mathcal{M} \to \mathcal{M}$ be a TBC of message space $\mathcal{M}$ and tweak space $\mathcal{T}$ and key space $(\mathcal{K} \times \mathcal{L})$, using $\widetilde{E}$ and H, such that

$$\text{XTX}_{K,L}^T(M) = \widetilde{E}_K^V(M \oplus W) \oplus W, \text{ where } (W, V) = H_L(T). \tag{9}$$

Figure 1 shows the scheme. For security we need that H is a variant of ϵ-AXU function defined as follows.

Definition 2. *Let H be a keyed function $H : \mathcal{L} \times \mathcal{T} \to \{0,1\}^n \times \{0,1\}^m$. We say H is (n, m, ϵ)-partial AXU $((n, m, \epsilon)$-pAXU$)$ if it satisfies*

$$\max_{\substack{x,x' \in \mathcal{T}, x \neq x' \\ \Delta \in \{0,1\}^n}} \Pr_L[H_L(x) \oplus H_L(x') = (\Delta, 0^m)] \leq \epsilon. \tag{10}$$

Clearly an ϵ-AXU function of $(n + m)$-bit output is also (n, m, ϵ)-pAXU.

A change of a tweak in XTX affects to both block and tweak of inner TBC, whereas in LRW or XEX it affects only to input block, and in Coron et al.'s

---

[2] Originally proved by [17] with a slightly larger constant, then improved by [22].

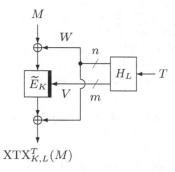

Fig. 1. XTX. A thick black line in $\widetilde{E}$ denotes tweak input.

method it affects only to inner tweak. XTX's structure is somewhat similar to a construction of (n, n)-bit TBC presented by Mennink [20], though [20] was based on ideal-cipher model, while XTX works on standard model and has no limitation on the outer tweak length. As well as LRW, XTX needs two keys. However it can be easily reduced to one by reserving one tweak bit for key generation. For instance, when $\mathcal{L} = \{0, 1\}^n$, we let $L = \widetilde{E}^{(0^m)}(0^n)$ and use $(n, m - 1)$-bit TBC defined as $\widetilde{E}^{(*\|1)}(*)$.

3.3   Security

We prove the security of XTX when underlying TBC is perfect, i.e. a TURP.

Theorem 1. *Let* $XTX_{\widetilde{P}, L}$ *be XTX using* (n, m)*-bit tweakable URP,* $\widetilde{P}$*, and* $H :$ $\mathcal{L} \times \mathcal{T} \to \{0, 1\}^n \times \{0, 1\}^m$ *for tweak space* $\mathcal{T}$ *with independent key* $L \in_U \mathcal{L}$. *Then we have*

$$\mathrm{Adv}^{\mathrm{tsprp}}_{XTX_{\widetilde{P}, L}}(\mathcal{A}) \leq \epsilon \cdot q^2, \tag{11}$$

for any CCA-adversary $\mathcal{A}$ *using* q *queries if* H *is* (n, m, ϵ)*-pAXU.*

In particular, when $\mathcal{T} = \{0, 1\}^\ell$ for some ℓ and H is $\mathsf{Poly}[n + m]$, we have $\mathrm{Adv}^{\mathrm{tsprp}}_{XTX_{\widetilde{P}, L}}(\mathcal{A}) \leq \ell_{n+m} q^2 / 2^{n+m}$ from Theorem 1 and Proposition 1.

3.4   Proof of Theorem 1

Overview. Following the proof of LRW [22], our proof is based on the method developed by Maurer [18][3], though other methods such as game-playing proof [6] or Coefficient-H technique [24] can be used as well. Basically, the proof is an

---

[3] In some special cases the result obtained by the method of [18] cannot be converted into computational counterparts [19,25]. However the proof presented here does not have such difficulty. A bug in a theorem of [18] was pointed out by Jetchev, Özen and Stam [14], however we did not use it.

extension of LRW proofs [17,22], which shows that the advantage is bounded by the probability of "bad" event, defined as a non-trivial input collision in the underlying blockcipher of LRW. Intuitively, the security bound of XTX is obtained by extending this observation, and we can set "bad" event as non-trivial, simultaneous collisions of input *and* tweak in the underlying TBC.

Proof. We start with basic explanations on Maurer's method. They are mostly the same as those of [18], with minor notational changes. Consider the game that an adversary tries to distinguish two keyed functions, F and G, with queries. The game we consider is information-theoretic, that is, adversary has no computational limitation and F and G have no computational assumption, say, they are URF or URP. There may be some conditions of valid adversaries, e.g., no repeating queries etc. Let α_i denote an event defined at time i, i.e., when adversary performs ith query and receives a response from oracle. Let $\overline{\alpha_i}$ be the negation of α_i. We assume α_i is monotone, i.e., α_i never occurs if $\overline{\alpha_{i-1}}$ occurs. For instance, α_i is monotone if it indicates that all i outputs are distinct. An infinite sequence of monotone events $\alpha = \alpha_0\alpha_1\ldots$ is called a *monotone event sequence* (MES). Here, α_0 denotes some tautological event. Note that $\alpha \wedge \beta = (\alpha_0 \wedge \beta_0)(\alpha_1 \wedge \beta_1)\ldots$ is a MES if $\alpha = \alpha_0\alpha_1\ldots$ and $\beta = \beta_0\beta_1\ldots$ are both MESs. Here we may abbreviate $\alpha \wedge \beta$ as $\alpha\beta$. For any sequence of random variables, $X_1, X_2, \ldots$, let X^i denote $(X_1, \ldots, X_i)$. We use $\text{dist}(X^i)$ to denote that $X_1, X_2, \ldots, X_i$ are distinct. We also write $\text{dist}((X, Y)^i)$ to denote that $(X_1, Y_1), \ldots, (X_i, Y_i)$ are distinct. Let MESs α and β be defined for two keyed functions, $F : \mathcal{K} \times \mathcal{X} \to \mathcal{Y}$ and $G : \mathcal{K}' \times \mathcal{X} \to \mathcal{Y}$, respectively. For simplicity, we omit the description of keys in this explanation. Let $X_i \in \mathcal{X}$ and $Y_i \in \mathcal{Y}$ be the ith input and output. Let P^F be the probability space defined by F. For example, $P^F_{Y_i|X^iY^{i-1}}(y^i, x^i)$ means $\Pr[Y_i = y_i | X^i = x^i, Y^{i-1} = y^{i-1}]$ where $Y_j = F(X_j)$ for $j \geq 1$. If $P^F_{Y_i|X^iY^{i-1}}(y^i, x^i) = P^G_{Y_i|X^iY^{i-1}}(y^i, x^i)$ for all possible (y^i, x^i), i.e. all assignments for which probabilities are defined, then we write $P^F_{Y_i|X^iY^{i-1}} = P^G_{Y_i|X^iY^{i-1}}$. Inequalities such as $P^F_{Y_i|X^iY^{i-1}} \leq P^G_{Y_i|X^iY^{i-1}}$ are similarly defined. Using MES $\alpha = \alpha_0\alpha_1,\ldots$ and $\beta = \beta_0\beta_1,\ldots$ defined for F and G, we define the following notations, which will be used in our proof.

Definition 3. *We write $F^\alpha \equiv G^\beta$ if $P^F_{Y_i\alpha_i|X^iY^{i-1}\alpha_{i-1}} = P^G_{Y_i\beta_i|X^iY^{i-1}\beta_{i-1}}$ holds for all $i \geq 1$, which means $P^F_{Y_i\alpha_i|X^iY^{i-1}\alpha_{i-1}}(y^i, x^i) = P^G_{Y_i\beta_i|X^iY^{i-1}\beta_{i-1}}(y^i, x^i)$ holds for all possible (y^i, x^i) such that both $P^F_{\alpha_{i-1}|X^{i-1}Y^{i-1}}(y^{i-1}, x^{i-1})$ and $P^G_{\beta_{i-1}|X^{i-1}Y^{i-1}}(y^{i-1}, x^{i-1})$ are positive.*

Definition 4. *We write $F|\alpha \equiv G|\beta$ if $P^F_{Y_i|X^iY^{i-1}\alpha_i} = P^G_{Y_i|X^iY^{i-1}\beta_i}$ holds for all $i \geq 1$.*

In general if $F^\alpha \equiv G^\beta$, then $F|\alpha \equiv G|\beta$ holds, but not vice versa.

Definition 5. *We define $\nu(F, \overline{\alpha_q})$ as the maximal probability of $\overline{\alpha_q}$ for any adversary using q queries to F, considered as valid in the definition of game, which we assume clear in the context.*

Theorem 2 (*Theorem* 1 (i) *of* [18]). *If* $F^\alpha \equiv G^\beta$ *or* $F|\alpha \equiv G$ *holds, we have* $\mathrm{Adv}^{\mathrm{cpa}}_{F,G}(\mathcal{A}) \leq \nu(F, \overline{\alpha_q})$ *for any adversary using* q *queries.*

We also use the following two lemmas of [18].

Lemma 1 (*Lemma* 1 (iv) *of* [18]). *Let MESs* α *and* β *be defined for* F *and* G. *Moreover, let* X_i *and* Y_i *denote the* ith *input and output of* F *(or* G), *respectively. Assume* $F|\alpha \equiv G|\beta$. *If* $P^F_{\alpha_i|X^iY^{i-1}\alpha_{i-1}} \leq P^G_{\beta_i|X^iY^{i-1}\beta_{i-1}}$ *for* $i \geq 1$, *which means* $P^F_{\alpha_i|X^iY^{i-1}\alpha_{i-1}}(x^i, y^{i-1}) \leq P^G_{\beta_i|X^iY^{i-1}\beta_{i-1}}(x^i, y^{i-1})$ *holds for all* (x^i, y^{i-1}) *such that* $P^F_{\alpha_{i-1}|X^{i-1}Y^{i-1}}(x^{i-1}, y^{i-1})$ *and* $P^G_{\beta_{i-1}|X^{i-1}Y^{i-1}}(x^{i-1}, y^{i-1})$ *are positive. Then there exists an MES* γ *defined for* G *such that* $F^\alpha \equiv G^{\beta\gamma}$.

Lemma 2 (*Lemma* 6 (iii) *of* [18]). $\nu(F, \overline{\alpha_q \wedge \beta_q}) \leq \nu(F, \overline{\alpha_q}) + \nu(F, \overline{\beta_q})$.

Analysis of XTX. We abbreviate $\mathrm{XTX}_{\widetilde{\mathsf{P}}, L}$ to XTX_1. We define XTX_2 be TURP with tweak space $\mathcal{T}$. What is needed is the indistinguishability of XTX_1 and XTX_2 for CCA adversary.

We write the adversary's query as $\mathbf{X}_i = (X_i, T_i, B_i) \in \{0,1\}^n \times \mathcal{T} \times \{0,1\}$. Here $B_i = 0$ ($B_i = 1$) indicates that ith query is an encryption (a decryption) query. Let $Y_i \in \{0,1\}^n$ be the corresponding response and we write $H_L(T_i) = (W_i, V_i)$ following (9). We also assume XTX_2 has computation of $H_L(T_i) = (W_i, V_i)$ as dummy, using independent and uniform sampling of L. In XTX_2, W_i and V_i are not used in the computation of Y_i. We write the set of scripts for all $i = 1, \ldots, j$th queries as $Z^j = (\mathbf{X}_1, \ldots, \mathbf{X}_j, Y_1, \ldots, Y_j)$. We may use M_i to denote X_i when $B_i = 0$ or Y_i when $B_i = 1$, and use C_i to denote Y_i when $B_i = 0$ or X_i when $B_i = 1$. We say Z^j is valid if $T_i = T_j$ and $M_i \neq M_j$ ($C_i \neq C_j$) then $C_i \neq C_j$ ($M_i \neq M_j$) holds. We note that a transcript which is not valid is one that cannot be obtained from a TBC.

We define $S_i = M_i \oplus W_i$ and $U_i = C_i \oplus W_i$ for both XTX_1 and XTX_2. They correspond to the input and output of $\widetilde{\mathsf{P}}$ in XTX_1, and dummy variables in XTX_2. MESs are defined as $\alpha_q = \mathrm{dist}((S,V)^q)$ and $\beta_q = \mathrm{dist}((U,V)^q)$. We observe that in XTX_1, α_q and β_q are equivalent, however not equivalent in XTX_2. Let us define $D(V_q) \overset{\text{def}}{=} \{1 \leq i < q : V_i = V_q\}$ and for n-bit variable $A \in \{X, Y, W, S, U\}$ define $A[D(V_q)] \overset{\text{def}}{=} \{A_i : i \in D(V_q)\}$. Here $A[D(V_q)]^c = \{0,1\}^n \backslash A[D(V_q)]$. Figure 2 shows XTX_1 with the labels mentioned above.

We investigate the distribution $P^G_{Y_q|Z^{q-1}\mathbf{X}_q\alpha_q\beta_q}$ for $G \in \{\mathrm{XTX}_1, \mathrm{XTX}_2\}$. We have

$$P^G_{Y_q|Z^{q-1}\mathbf{X}_q\alpha_q\beta_q} = \sum_L P^G_{Y_q|Z^{q-1}\mathbf{X}_q\alpha_q\beta_q L} \cdot P^G_{L|Z^{q-1}\mathbf{X}_q\alpha_q\beta_q}, \tag{12}$$

where the summation is taken for all values of $L = l$. We first focus on the term $P^G_{L|Z^{q-1}\mathbf{X}_q\alpha_q\beta_q}$. Let us assume $B_q = 0$. For both $G = \mathrm{XTX}_1$ and XTX_2, L is uniform over all values consistent with the conditional clause (note that

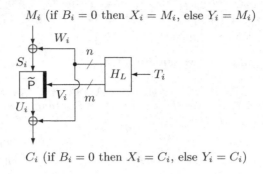

M_i (if $B_i = 0$ then $X_i = M_i$, else $Y_i = M_i$)

C_i (if $B_i = 0$ then $X_i = C_i$, else $Y_i = C_i$)

Fig. 2. XTX$_1$ with labels used in the proof of Theorem 1.

L defines W^q, V^q, S^q and U^{q-1}, thus α_q and β_{q-1} are deterministic events given L). Hence we have

$$P^{\mathrm{XTX}_1}_{L|Z^{q-1}\mathbf{X}_q\alpha_q\beta_q} = P^{\mathrm{XTX}_2}_{L|Z^{q-1}\mathbf{X}_q\alpha_q\beta_q}. \tag{13}$$

For $P^G_{Y_q|Z^{q-1}\mathbf{X}_q\alpha_q\beta_q L}$, if $B_q = 0$ and $G = \mathrm{XTX}_1$, U_q is uniform over $\mathcal{U} \overset{\mathrm{def}}{=} U[D(V_q)]^c$, thus $Y_q = C_q = U_q \oplus W_q$ is uniform over $\mathcal{U} \oplus W_q$. If $B_q = 0$ and $G = \mathrm{XTX}_2$ and there is no conditional clause $\alpha_q\beta_q$, $Y_q(= C_q)$ is uniform over $\mathcal{C} \overset{\mathrm{def}}{=} C[D(T_q)]^c$. Here U_q is uniform over

$$\{C_i \oplus W_q : i \in D(T_q)\}^c = \{C_i \oplus W_i : i \in D(T_q)\}^c = \{U_i : i \in D(T_q)\}^c. \tag{14}$$

With condition $\alpha_q\beta_q$ (here only β_q is relevant since α_q is deterministic given L), U_i for $i \in D(V_q)$ (but $T_i \neq T_q$) is further removed from possible values for U_q, hence U_q is uniform over $\mathcal{U} = U[D(V_q)]^c$ and Y_q is uniform over $\mathcal{U} \oplus W_q$. Therefore C_q's distributions are identical for both XTX_1 and XTX_2. The same analysis holds for the case $B_q = 1$, and we have

$$P^{\mathrm{XTX}_1}_{Y_q|Z^{q-1}\mathbf{X}_q\alpha_q\beta_q L} = P^{\mathrm{XTX}_2}_{Y_q|Z^{q-1}\mathbf{X}_q\alpha_q\beta_q L}. \tag{15}$$

Thus Y_q's distributions are identical for both XTX_1 and XTX_2 if conditioned by $\alpha_q\beta_q$ and $L = l$ for any l. Therefore from (13) and (15) we have

$$P^{\mathrm{XTX}_1}_{Y_q|Z^{q-1}\mathbf{X}_q\alpha_q\beta_q} = P^{\mathrm{XTX}_2}_{Y_q|Z^{q-1}\mathbf{X}_q\alpha_q\beta_q}, \text{ that is, } \mathrm{XTX}_1|\alpha\beta \equiv \mathrm{XTX}_2|\alpha\beta. \tag{16}$$

Let us assume $B_q = 0$, and we focus on $p(G) = P^G_{\alpha_q\beta_q|Z^{q-1}\mathbf{X}_q\alpha_{q-1}\beta_{q-1}L}$. Note that the conditional clause uniquely determines whether α_q holds or not. If α_q does not hold, $p(G) = 0$ for both $G = \mathrm{XTX}_1$ or XTX_2. If α_q holds, $p(\mathrm{XTX}_1) = 1$ as $\beta_q \equiv \alpha_q$ in XTX_1, however $p(\mathrm{XTX}_2) < 1$ since β_q depends on U_q which is not determined by the conditional clause. This shows that

$$P^{\mathrm{XTX}_2}_{\alpha_q\beta_q|Z^{q-1}\mathbf{X}_q\alpha_{q-1}\beta_{q-1}L} \leq P^{\mathrm{XTX}_1}_{\alpha_q\beta_q|Z^{q-1}\mathbf{X}_q\alpha_{q-1}\beta_{q-1}L}. \tag{17}$$

Moreover using similar argument as (12), we have

$$P_{L|Z^{q-1}\mathbf{X}_q\alpha_{q-1}\beta_{q-1}}^{\mathrm{XTX}_1} = P_{L|Z^{q-1}\mathbf{X}_q\alpha_{q-1}\beta_{q-1}}^{\mathrm{XTX}_2}. \tag{18}$$

Thus, from (17) and (18), we have

$$P_{\alpha_q\beta_q|Z^{q-1}\mathbf{X}_q\alpha_{q-1}\beta_{q-1}}^{\mathrm{XTX}_2} \le P_{\alpha_q\beta_q|Z^{q-1}\mathbf{X}_q\alpha_{q-1}\beta_{q-1}}^{\mathrm{XTX}_1}. \tag{19}$$

From (16) and (19) and Lemma 1, we observe that $\mathrm{XTX}_1^{\alpha\beta\gamma} \equiv \mathrm{XTX}_2^{\alpha\beta}$ holds true for some MES γ. With this equivalence, Theorem 2 and Lemma 2, we have

$$\mathbf{Adv}_{\mathrm{XTX}_1,\mathrm{XTX}_2}^{\mathrm{cca}}(\mathcal{A}) \le \nu(\mathrm{XTX}_2, \overline{\alpha_q \wedge \beta_q}) \le \nu(\mathrm{XTX}_2, \overline{\alpha_q}) + \nu(\mathrm{XTX}_2, \overline{\beta_q}) \tag{20}$$

for any CCA adversary $\mathcal{A}$ using q queries.

Let $\mathrm{XTX}_2[\widetilde{p}]$ be XTX_2 using a fixed tweakable permutation $\widetilde{p} \in \mathrm{Perm}^{\mathcal{T}}(n)$. We observe that the last two terms of (20) are bounded as

$$\nu(\mathrm{XTX}_2, \overline{\alpha_q}) \le \max_{\widetilde{p} \in \mathrm{Perm}^{\mathcal{T}}(n)} \nu(\mathrm{XTX}_2[\widetilde{p}], \overline{\alpha_q}) \tag{21}$$

$$\nu(\mathrm{XTX}_2, \overline{\beta_q}) \le \max_{\widetilde{p} \in \mathrm{Perm}^{\mathcal{T}}(n)} \nu(\mathrm{XTX}_2[\widetilde{p}], \overline{\beta_q}). \tag{22}$$

As $\widetilde{p}$ is fixed, the adversary can evaluate it without oracle access, hence the right hand side terms of (21) are obtained by considering the maximum of possible and valid (M^q, T^q, C^q). For fixed (M^q, T^q, C^q), the probabilities of $\overline{\alpha_q}$ and $\overline{\beta_q}$ are determined by W^q and V^q. Thus, for any $\widetilde{p}$ we have

$$\nu(\mathrm{XTX}_2[\widetilde{p}], \overline{\alpha_q})$$

$$\le \max_{\substack{(M^q,T^q,C^q) \\ \mathrm{valid}}} \Pr_{\substack{(W^q,V^q) \\ (W_i,V_i)=H_L(T_i)}} [^\exists i,j, \text{ s.t. } (W_i \oplus W_j = M_i \oplus M_j) \wedge (V_i = V_j)] \tag{23}$$

$$\le \binom{q}{2} \cdot \epsilon, \text{ and}$$

$$\nu(\mathrm{XTX}_2[\widetilde{p}], \overline{\beta_q})$$

$$\le \max_{\substack{(M^q,T^q,C^q) \\ \mathrm{valid}}} \Pr_{\substack{(W^q,V^q) \\ (W_i,V_i)=H_L(T_i)}} [^\exists i,j, \text{ s.t. } (W_i \oplus W_j = C_i \oplus C_j) \wedge (V_i = V_j)] \tag{24}$$

$$\le \binom{q}{2} \cdot \epsilon,$$

since H is (n, m, ϵ)-pAXU. From (20), (23) and (24), we conclude the proof. $\square$

Tightness of Our Bound. We note that the bound is tight in the sense that we have an attack with about $q = O(2^{(n+m)/2})$ queries. The attack is simple, and let $M = 0^n$. The adversary makes q encryption queries $(M, T_1), \ldots, (M, T_q)$, where $T_1, \ldots, T_q$ are distinct tweaks. With a high probability, we have i and j such that $C_i = C_j$, where C_i is the ciphertext for (M, T_i) and C_j is that for (M, T_j). Now, the adversary can make two more encryption queries (M', T_j)

and (M', T_j) for any $M' \neq M$, and see if the corresponding ciphertexts collide, in which case, with a high probability, the oracle is the tweakable blockcipher.

The attack works since in the ideal case, there exit i and j such that $C_i = C_j$ with a non-negligible probability, but we have the collision between ciphertexts of (M', T_i) and (M', T_j) with only a negligible probability.

4   Applications

Suppose we have an (n, m)-bit TBC $\widetilde{E}$ and want to extend tweak by applying XTX. We first remark if $\widetilde{E}$ is obtained by LRW this is almost pointless because $\widetilde{E}$ itself has only security up to the birthday bound. In this case a simple solution would be to extend the input domain of UH function used in LRW. However if $\widetilde{E}$ is a dedicated TBC, or a mode of operation having security beyond the birthday bound, application of XTX to $\widetilde{E}$ can have practical merits.

4.1   Dedicated TBC

Let us assume $\widetilde{E}$ is an (n, m)-bit dedicated TBC. As mentioned, using $\widetilde{E}$ with fixed tweak then applying LRW with some UH function only provides $n/2$-bit security, and Coron's method only provides $m/2$-bit security, while XTX provides $(n+m)/2$-bit security. For example, KIASU-BC [13] is a $(128, 64)$-bit TBC based on AES. By combining XTX using H as Poly[192] or Poly[64, 64, 64] we obtain a TBC of longer tweak with 96-bit security with respect to the number of queries, while previous methods provide 64 or 32-bit security. Similarly, a $(256, 128)$-bit TBC version of Threefish can be conveted into a TBC of longer tweak having 192-bit security, using XTX with H being Poly[384] or Poly[128, 128, 128].

We remark that the use of Poly$[m, m, m]$ for $m = n/3$ instead of Poly$[n + m]$ can reduce the implementation size and gain efficiency. For example Aoki and Yasuda [3] proposed to use Poly$[n/2, n/2]$ instead of Poly$[n]$ used in GCM authenticated encryption. A drawback is that it will increase the advantage with respect to tweak length, from linear to cubic in our case (though we assumed it as a constant in Sect. 2.1). Therefore, the use of a polynomial hash function with a small field is not desirable if the impact of such increase is not negligible. In addition we have to be careful with the existence of weak keys in polynomial hash function pointed out by Procter and Cid [27].

4.2   Rekeying Construction

Minematsu's rekeying construction for TBC [23] is described as follows. Using a blockcipher $E : \mathcal{K} \times \mathcal{M} \to \mathcal{M}$ with $\mathcal{K} = \mathcal{M} = \{0, 1\}^n$, [23] builds a (n, m)-bit TBC for $m < n$ such that

$$\mathsf{Min}_K^T(M) = E_{K_T'}(M) \text{ where } K_T' = E_K(T \| 0^{n-m}). \tag{25}$$

The security bound of this construction is as follows. For any $\mathcal{A}$ using q queries with τ time, we have another adversary $\mathcal{B}$ using q queries with $\tau' = \tau + O(q)$

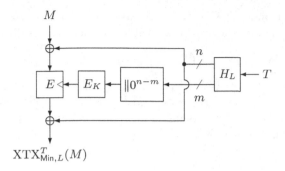

$$\mathrm{XTX}^T_{\mathsf{Min},L}(M)$$

Fig. 3. XTX applied to Minematsu's TBC.

time such that $\mathrm{Adv}^{\mathrm{tsprp}}_{\mathsf{Min}}(\mathcal{A}) \le (\eta+1)\mathrm{Adv}^{\mathrm{sprp}}_{E}(\mathcal{B}) + \frac{\eta^2}{2^{n+1}}$, where $\eta = \min\{q, 2^m\}$.
As analyzed by [23] this can provide a TBC with beyond-birthday security when
$m < n/2$. In particular [23] suggested $m = n/3$ which provides security against
$2^{n-m} = 2^{2n/3}$ queries. Despite the simple construction, one big shortcoming
is its short tweak length, as mentioned by (e.g.) [15,16,20]. This is, however,
recovered if (25) is combined with XTX. Let $\mathrm{XTX}_{\mathsf{Min},L}$ be XTX with internal
TBC being Min having n-bit block, m-bit tweak using n-bit blockcipher E. Here
we assume that tweak space of $\mathrm{XTX}_{\mathsf{Min},L}$ is $\mathcal{T} = \{0,1\}^\ell$, and underlying H :
$\mathcal{L} \times \mathcal{T} \to \{0,1\}^{n+m}$ is $\mathsf{Poly}[n+m]$. Then for any adversary $\mathcal{A}$ using q queries
and τ time, from (25) and Proposition 1 and Theorem 1, we have

$$\mathrm{Adv}^{\mathrm{tsprp}}_{\mathrm{XTX}_{\mathsf{Min},L}}(\mathcal{A}) \le (\eta+1)\mathrm{Adv}^{\mathrm{sprp}}_{E}(\mathcal{B}) + \frac{\eta^2}{2^{n+1}} + \frac{\ell_{n+m}q^2}{2^{n+m}}, \qquad (26)$$

for some adversary $\mathcal{B}$ using q queries with $\tau + O(q)$ time, where $\eta = \min\{q, 2^m\}$
as above. For choosing m, we can assume that $\mathrm{Adv}^{\mathrm{sprp}}_{E}(\mathcal{B})$ is at least $q/2^n$ for
adversary $\mathcal{B}$ using q queries and τ time when q is about τ (since E has n-bit
key and $\mathcal{B}$ can perform exhaustive key search, as observed by Bellare et al. [5]).
Ignoring ℓ_{n+m} and substituting η with 2^m in the bound, the first and last terms
are about $q/2^{n-m}$ and $q^2/2^{n+m}$. Then $m = n/3$ is a reasonable choice which
makes these terms $(q/2^{2n/3})^i$ for $i = 1$ and $i = 2$. This shows that we can
extend tweak keeping the original security of rekeying construction. The resulting
scheme is shown in Fig. 3, where a triangle in E denotes key input, and H_L
denotes $\mathsf{Poly}[n+m]$. Still, we need rekeying for each tweak and this can be
another drawback for performance.

4.3   Chained LRW

A provably-secure TBC construction which does not rely on rekeying construction [23] was first proposed by Landecker et al. [16]. It is an independently-
keyed chain of LRW, and is called[4] CLRW2. Assuming LRW shown as (7) using

---

[4] The name CLRW2 means it is a chain of the second construction of [17], which we
simply call LRW.

$H : \mathcal{L} \times \mathcal{T} \to \mathcal{M}$ and $E : \mathcal{K} \times \mathcal{M} \to \mathcal{M}$ for $\mathcal{M} = \{0,1\}^n$ and $\mathcal{T} = \{0,1\}^\ell$ as underlying components, they proposed the construction described as

$$\text{CLRW2}_{K_1,K_2,L_1,L_2}^T(M) = \text{LRW}_{K_2,L_2}^T(\text{LRW}_{K_1,L_1}^T(M)). \tag{27}$$

The authors proved[5] that its TSPRP-advantage is $O(q^3\epsilon^2)$ when H is ϵ-AXU. More formally, TSPRP-advantage is at most $8q^3\hat{\epsilon}^2/(1 - q^3\hat{\epsilon}^2)$ where $\hat{\epsilon}$ is defined as $\max\{\epsilon, 1/(2^n - 2q)\}$. Thus, assuming $\hat{\epsilon} = \epsilon$ and the denominator being larger than $1/2$, the bound is at most

$$16q^3\epsilon^2. \tag{28}$$

If H is $\mathsf{Poly}[n]$, we have $\epsilon = \ell_n/2^n$, then the bound is $16q^3\ell_n^2/2^{2n}$. In this case CLRW2 needs $2\ell_n$ $\text{GF}(2^n)$ multiplications for each ℓ-bit tweak.

A natural extension of CLRW2, i.e. a longer chain more than two, was proposed by Lampe and Seurin [15]. The construction for r chains is simply described as r-$\text{CLRW}_{K_1,\ldots,K_r,L_1,\ldots,L_r}^T(M)$ in the same manner to (27), where 2-CLRW is equivalent to CLRW2. If r blockciphers are independent URPs, they proved that r-CLRW for any even r has TSPRP-advantage of

$$c_r q^{\frac{(r+2)}{4}} \epsilon^{\frac{r}{4}}, \text{ where } c_r = \frac{4\sqrt{2}}{\sqrt{r+2}} \cdot 2^{\frac{r}{4}}, \tag{29}$$

when the underlying $H : \mathcal{L} \times \mathcal{T} \to \mathcal{M}$ is ϵ-AXU. If $\mathcal{T}$ is $\{0,1\}^\ell$ and H is $\mathsf{Poly}[n]$, the bound is

$$\frac{c_r q^{\frac{(r+2)}{4}} \ell_n^{\frac{r}{4}}}{2^{\frac{nr}{4}}}. \tag{30}$$

Let r-CLRW(m) be the r-CLRW with n-bit blockcipher E_K and m-bit tweak (for some fixed $m > 0$) processed by independently-keyed r instances of $\mathsf{Poly}[n]$. We note that r-CLRW(ℓ) needs $r\ell_n$ multiplications over $\text{GF}(2^n)$.

r-CLRW Combined with XTX. Let $r > 2$ be an even integer. We apply XTX with H being $\mathsf{Poly}[n,n]$ using two keys in $\text{GF}(2^n)$ to r-CLRW(n), to build an (n,ℓ)-bit TBC. The resulting scheme uses $r + 2\ell_n$ $\text{GF}(2^n)$ multiplications, hence uses fewer multiplications than r-CLRW(ℓ) if $\ell_n > 1$ and $r \geq 4$. From Theorem 1, Proposition 1 and (30), TSPRP-advantage of the resulting scheme is

$$\frac{c_r q^{\frac{(r+2)}{4}}}{2^{\frac{nr}{4}}} + \frac{\ell_n^2 q^2}{2^{2n}}. \tag{31}$$

This provides the same level of security as (30), unless ℓ_n is huge.

In case $r = 2$ the above combination gives no efficiency improvement. Still, by combining CLRW2(m) for some $m < n$ with XTX a slight improvement is possible. This is because CLRW2 needs two n-bit UH functions and the product

---

[5] Originally the constant was 6, however an error in the proof was pointed out by Procter [26]. He fixed the proof with an increased constant, 8.

of their biases is multiplied by q^3, while the bias of $(n + m)$-bit UH function in XTX is multiplied by q^2. For example, assuming n is divisible by 3, we set $m = n/3$, and consider $\mathrm{CLRW2}(m)$ using two $\mathsf{Poly}[n]$ (with padding of tweak), combined with XTX using $\mathsf{Poly}[n, m]$ to process ℓ-bit tweak. This requires ℓ_n $\mathrm{GF}(2^n)$ multiplications and $3\ell_n$ $\mathrm{GF}(2^m)$ multiplications. It is not straightforward to compare the complexity of one multiplication over $\mathrm{GF}(2^n)$ and three multiplications over $\mathrm{GF}(2^m)$, however, in most cases the latter is considered to be lighter than the former, though the gain will be depending on whether the underlying computing platform operates well over m-bit words. If this is the case our scheme will have a better complexity than the plain use of CLRW2.

As a more concrete example, let us consider CLRW2 using two instances of $\mathsf{Poly}[m, m, m]$ with $m = n/3$. For ℓ-bit tweak, this CLRW2 requires $2 \cdot 3 \cdot 3\ell_n = 18\ell_n$ multiplications over $\mathrm{GF}(2^m)$ and its TSPRP-advantage is, based on (28), at most

$$16 \cdot q^3 \left(\frac{(3\ell_n)^3}{(2^{\frac{n}{3}})^3} \right)^2 = \frac{11664 \cdot q^3 \cdot \ell_n^6}{2^{2n}}. \tag{32}$$

If we combine this instance of $\mathrm{CLRW2}(m)$ with XTX using $\mathsf{Poly}[m, m, m, m]$, then the advantage is at most

$$\frac{16 \cdot 729 q^3}{2^{2n}} + q^2 \frac{(3\ell_n)^4}{(2^{\frac{n}{3}})^4} = \frac{11664 q^3}{2^{2n}} + \frac{81 \ell_n^4 q^2}{2^{\frac{4n}{3}}}. \tag{33}$$

As shown by (32) and (33), for a moderate tweak length both bounds indicate the security against about $2^{2n/3}$ queries, while the combined scheme uses fewer $\mathrm{GF}(2^m)$ multiplications, i.e. $6 + 4 \cdot 3\ell_n = 6 + 12\ell_n$.

5 Conclusion

In this paper, we have studied the problem of tweak extension for a tweakable blockcipher having fixed-length tweak. We proposed XTX as an effective solution to this problem, by extending the work of Liskov et al. XTX uses one call of a given tweakable blockcipher, $\widetilde{E}$, and a variant of universal hash function, H, for processing global tweak. When $\widetilde{E}$ has n-bit block and m-bit tweak, XTX provides $(n + m)/2$-bit security, which is better than the conventional methods known as Liskov et al.'s LRW or Corol et al.'s solution. The proposed method is useful in extending tweak of dedicated tweakable blockciphers, which typically have relatively short, fixed-length tweak. Moreover, XTX is even useful when applied to some blockcipher modes for tweakable blockcipher which have beyond-birthday-bound security. A natural open problem here is to find tweak extension schemes that have better security bounds than that of XTX.

Acknowledgments. The authors would like to thank the anonymous reviewers for highly constructive and insightful comments. The work by Tetsu Iwata was supported in part by JSPS KAKENHI, Grant-in-Aid for Scientific Research (B), Grant Number 26280045.

References

1. CAESAR: Competition for authenticated encryption: security, applicability, and robustness. http://competitions.cr.yp.to/caesar.html/
2. Skein Hash Function: SHA-3 submission (2008). http://www.skein-hash.info/
3. Aoki, K., Yasuda, K.: The security and performance of "GCM" when short multiplications are used instead. In: Kutyłowski, M., Yung, M. (eds.) Inscrypt 2012. LNCS, vol. 7763, pp. 225–245. Springer, Heidelberg (2013)
4. Bellare, M., Desai, A., Jokipii, E., Rogaway, P.: A concrete security treatment of symmetric encryption. In: 38th Annual Symposium on Foundations of Computer Science, FOCS 1997, 19–22 Oct 1997, pp. 394–403. IEEE Computer Society, Miami Beach (1997)
5. Bellare, M., Krovetz, T., Rogaway, P.: Luby-Rackoff backwards: increasing security by making block ciphers non-invertible. In: Nyberg, K. (ed.) EUROCRYPT 1998. LNCS, vol. 1403, pp. 266–280. Springer, Heidelberg (1998)
6. Bellare, M., Rogaway, P.: The security of triple encryption and a framework for code-based game-playing proofs. In: Vaudenay, S. (ed.) EUROCRYPT 2006. LNCS, vol. 4004, pp. 409–426. Springer, Heidelberg (2006)
7. Carter, L., Wegman, M.N.: Universal classes of hash functions. J. Comput. Syst. Sci. 18(2), 143–154 (1979)
8. Cogliati, B., Lampe, R., Seurin, Y.: Tweaking Even-Mansour ciphers. In: Gennaro, R., Robshaw, M. (eds.) CRYPTO 2015, Part I. LNCS, vol. 9215, pp. 189–208. Springer, Heidelberg (2015). Full version in Cryptology ePrint Archive, Report 2015/539. http://eprint.iacr.org/
9. Coron, J.-S., Dodis, Y., Mandal, A., Seurin, Y.: A domain extender for the ideal cipher. In: Micciancio, D. (ed.) TCC 2010. LNCS, vol. 5978, pp. 273–289. Springer, Heidelberg (2010)
10. Crowley, P.: Mercy: a fast large block cipher for disk sector encryption. In: Schneier, B. (ed.) FSE 2000. LNCS, vol. 1978, pp. 49–63. Springer, Heidelberg (2001)
11. Even, S., Mansour, Y.: A construction of a cipher from a single pseudorandom permutation. J. Cryptol. 10(3), 151–162 (1997)
12. Hirose, S., Sasaki, Y., Yasuda, K.: IV-FV authenticated encryption and triplet-robust decryption. In: Early Symetric Crypto, ESC 2015 (2015)
13. Jean, J., Nikolic, I., Peyrin, T.: Tweaks and keys for block ciphers: the TWEAKEY framework. In: Sarkar, P., Iwata, T. (eds.) ASIACRYPT 2014. LNCS, vol. 8874, pp. 274–288. Springer, Heidelberg (2014)
14. Jetchev, D., Özen, O., Stam, M.: Understanding adaptivity: random systems revisited. In: Wang, X., Sako, K. (eds.) ASIACRYPT 2012. LNCS, vol. 7658, pp. 313–330. Springer, Heidelberg (2012)
15. Lampe, R., Seurin, Y.: Tweakable blockciphers with asymptotically optimal security. In: Moriai, S. (ed.) FSE 2013. LNCS, vol. 8424, pp. 133–152. Springer, Heidelberg (2014)
16. Landecker, W., Shrimpton, T., Terashima, R.S.: Tweakable blockciphers with beyond birthday-bound security. In: Safavi-Naini, R., Canetti, R. (eds.) CRYPTO 2012. LNCS, vol. 7417, pp. 14–30. Springer, Heidelberg (2012)
17. Liskov, M., Rivest, R.L., Wagner, D.: Tweakable block ciphers. In: Yung, M. (ed.) CRYPTO 2002. LNCS, vol. 2442, pp. 31–46. Springer, Heidelberg (2002)
18. Maurer, U.M.: Indistinguishability of random systems. In: Knudsen, L.R. (ed.) EUROCRYPT 2002. LNCS, vol. 2332, pp. 110–132. Springer, Heidelberg (2002)

19. Maurer, U.M., Pietrzak, K.: Composition of random systems: when two weak make one strong. In: Naor, M. (ed.) TCC 2004. LNCS, vol. 2951, pp. 410–427. Springer, Heidelberg (2004)

20. Mennink, B.: Optimally secure tweakable blockciphers. In: Leander, G. (ed.) FSE 2015. LNCS, vol. 9054, pp. 428–448. Springer, Heidelberg (2015)

21. Mennink, B.: XPX: generalized tweakable even-mansour with improved security guarantees. IACR Cryptology ePrint Archive 2015, 476 (2015)

22. Minematsu, K.: Improved security analysis of XEX and LRW modes. In: Biham, E., Youssef, A.M. (eds.) SAC 2006. LNCS, vol. 4356, pp. 96–113. Springer, Heidelberg (2007)

23. Minematsu, K.: Beyond-birthday-bound security based on tweakable block cipher. In: Dunkelman, O. (ed.) FSE 2009. LNCS, vol. 5665, pp. 308–326. Springer, Heidelberg (2009)

24. Patarin, J.: The "coefficients H" technique. In: Avanzi, R.M., Keliher, L., Sica, F. (eds.) SAC 2008. LNCS, vol. 5381, pp. 328–345. Springer, Heidelberg (2009)

25. Pietrzak, K.: Composition does not imply adaptive security. In: Shoup, V. (ed.) CRYPTO 2005. LNCS, vol. 3621, pp. 55–65. Springer, Heidelberg (2005)

26. Procter, G.: A note on the CLRW2 tweakable block cipher construction. IACR Cryptology ePrint Archive 2014, 111 (2014)

27. Procter, G., Cid, C.: On weak keys and forgery attacks against polynomial-based MAC schemes. In: Moriai, S. (ed.) FSE 2013. LNCS, vol. 8424, pp. 287–304. Springer, Heidelberg (2014)

28. Rogaway, P.: Efficient instantiations of tweakable blockciphers and refinements to modes OCB and PMAC. In: Lee, P.J. (ed.) ASIACRYPT 2004. LNCS, vol. 3329, pp. 16–31. Springer, Heidelberg (2004)

29. Sasaki, Y., Todo, Y., Aoki, K., Naito, Y., Sugawara, T., Murakami, Y., Matsui, M., Hirose, S.: Minalpher. A submission to CAESAR

30. Schroeppel, R.: Hasty pudding cipher. AES submission (1998). http://www.cs.arizona.edu/rcs/hpc/

31. Shrimpton, T., Terashima, R.S.: A modular framework for building variable-input-length tweakable ciphers. In: Sako, K., Sarkar, P. (eds.) ASIACRYPT 2013, Part I. LNCS, vol. 8269, pp. 405–423. Springer, Heidelberg (2013)

Rogue Decryption Failures: Reconciling AE Robustness Notions

Guy Barwell, Daniel Page, and Martijn Stam$^{(\boxtimes)}$

Department of Computer Science, University of Bristol, Bristol BS8 1UB, UK
{guy.barwell,daniel.page,martijn.stam}@bristol.ac.uk

Abstract. An authenticated encryption scheme is deemed secure (AE) if ciphertexts both look like random bitstrings and are unforgeable. AE is a much stronger notion than the traditional IND–CCA. One shortcoming of AE as commonly understood is its idealized, all-or-nothing decryption: if decryption fails, it will always provide the *same single* error message *and nothing more*. Reality often turns out differently: encode-then-encipher schemes often output decrypted ciphertext before verification has taken place whereas pad-then-MAC-then-encrypt schemes are prone to distinguishable verification failures due to the subtle interaction between padding and the MAC-then-encrypt concept. Three recent papers provided what appeared independent and radically different definitions to model this type of decryption leakage.

We reconcile these three works by providing a reference model of security for authenticated encryption in the face of decryption leakage from invalid queries. Having tracked the development of AE security games, we provide a single expressive framework allowing us to compare and contrast the previous notions. We find that at their core, the notions are essentially equivalent, with their key differences stemming from definitional choices independent of the desire to capture real world behaviour.

Keywords: Provable security · Authenticated encryption · Multiple errors · Unverified plaintext · Robustness

1 Introduction

Nowadays, authenticated encryption (AE) is understood to mean that ciphertexts both look like random bitstrings (IND\$–CPA) and are unforgeable (INT–CTXT). Moreover, the customary syntax of AE considers encryption deterministic and stateless, instead accepting a nonce (number-used-once) and associated data to ensure that repeated encryption of the same message does not lead to repeated ciphertexts. Preferably security degrades gracefully if nonces are repeated. AE thus defined is more flexible and considerably stronger than the traditional notion of IND–CCA symmetric encryption.

The CAESAR competition [4] served as a catalyst to strengthen the security models used in AE even further. One particular shortcoming is the traditional reliance on an idealised, all-or-nothing decryption: if decryption fails, it

© Springer International Publishing Switzerland 2015
J. Groth (Ed.): IMACC 2015, LNCS 9496, pp. 94–111, 2015.
DOI: 10.1007/978-3-319-27239-9_6

will only ever provide a single error message. For various reasons, this is not a realistic assumption. Especially MAC-then-encrypt schemes (or rather, decrypt-then-verify) are prone to real-world security flaws, on the one hand due to distinguishable verification failures and on the other due to the need to output (or at least, store) decrypted ciphertext before verification has taken place.

Three recent works improve the "robustness" of AE schemes by considering how well their security guarantees hold up under incorrect usage or when implemented non-ideally. Boldyreva et al. [6] investigated the effect of multiple decryption errors for both probabilistic and stateful encryption (BDPS). Later, Andreeva et al. [2] moved to a nonce-based setting, introducing a framework to capture the release of unverified plaintexts (RUP). Concurrently, Hoang et al. [11] coined an alternative notion, robust authenticated encryption (RAE), which they claim is radically different from RUP.

On the surface, these papers take very different approaches, with quite different goals in mind. BDPS concentrates on decryption errors, and does not consider nonce-based encryption. RUP extends AE by syntactically adding explicit, fixed-size tags and considering separate verification and decryption algorithms. It models the leakage of candidate plaintexts, with an eye on the online or nonce-abuse settings. In contrast, RAE considers schemes with variable, user-specified stretch as authentication mechanism, and decryption is given a much richer syntax, extending semantics for ciphertexts not generated by the encryption algorithm. This raises the natural questions how these models relate to each other and how well each captures real-world decryption leakage.

Our Contribution. Inspired by the above works, we provide a framework taking in the best of all worlds, where our key goal is to reconcile RUP and RAE with BDPS, both notationally and conceptually. Our framework allows us to draw parallels and highlight where the works agree or differ, while ensuring any goals described can be easily interpreted and compared to the scenarios they model.

Our framework revolves around a broad reference game that models adversarial access, and demonstrate that classic reductions still hold. This allows us to define "subtle Authenticated Encryption" (SAE) as the strongest security goal relevant to (deterministic) decryption leakage. The term *subtle* highlights that security in the real-world is very much dependent of the subtleties of *how* decryption is implemented. As illustration, in the full version we describe a natural yet insecure implementation of AEZ [11], refuting its robustness. Finally, we compare results from the three noted papers within our framework, using SAE as a reference point. After clarifying some (misconceived) terminology, we find that for schemes *with fixed stretch* the notions essentially collapse.

The fundamental difference between the models is philosophical: Is authenticated encryption primarily a primitive like a blockcipher, whose security should be measured with reference to the ideal object of the given syntax and where the authentication level might be set to zero; or is it a means to authenticate and encrypt where security should be measured against a—possibly unobtainable—ideal?

2 Security Games for the Real World

2.1 Standard Syntax of Authenticated Encryption

Current understanding of authenticated encryption is the culmination of many years of work (see the full version [3] for an overview). Modern AEAD schemes take a number of standard inputs and produce a single output. The corresponding spaces are named after the elements they represent: the *key* space K, the *message* space M, the *nonce* space N, the *associated data* space A, and finally the *ciphertext* space C. Each of these spaces is a subset of $\{0,1\}^*$ and we make no assertions over the sizes of these spaces.

An *authenticated encryption* (AE) scheme is a pair of deterministic algorithms $\Pi = (\mathcal{E}, \mathcal{D})$ (encrypt and decrypt) satisfying

$$\mathcal{E}: \mathsf{K} \times \mathsf{N} \times \mathsf{A} \times \mathsf{M} \to \mathsf{C}$$
$$\mathcal{D}: \mathsf{K} \times \mathsf{N} \times \mathsf{A} \times \mathsf{C} \to \mathsf{M} \cup \{\bot\}.$$

We use subscripts for keys, superscripts for public information (nonce and associated data) and put content data in parentheses.

To be *correct*, decryption must be a left inverse of encryption: if $C = \mathcal{E}_k^{N,A}(M)$ then $\mathcal{D}_k^{N,A}(C) = M$. Conversely, a scheme is *tidy* if decryption is a right inverse: if $\mathcal{D}_k^{N,A}(C) = M \neq \bot$ then $\mathcal{E}_k^{N,A}(M) = C$. Together then, correctness and tidiness imply encryption and decryption are inverses. For schemes that are both correct and tidy, $\mathcal{E}_k$ uniquely determines $\mathcal{D}_k$, which implies that security can be regarded as a property of $\mathcal{E}_k$ only [13].

The *stretch* measures the amount of ciphertext expansion (or redundancy). We require that the stretch $\tau(M) = |\mathcal{E}_k^{N,A}(M)| - |M|$, depends only on the length of the message, and so $\tau(M) = \tau(|M|)$ (for all k, N, A, and M). We call such schemes τ-*length-regular*, extending the accepted term *length–regular* to describe *how* the length is regulated. To minimise ciphertext expansion, most modern schemes set τ to be constant. We restrict ourselves to length-regular schemes: those whose stretch depends only on the length of the message, meaning $\tau(M) = \tau(|M|)$.

One might deviate from the syntax above. On the one hand, RUP uses an equivalent formulation with explicit tag space in addition to the ciphertext space (see Sect. 3.2). On the other hand, RAE uses an explicit input of the encryption indicating what size of tag is desired. In Sect. 3.3 we discuss the implication of user-defined tag-sized explicitly, and our rationale for omitting it from our framework.

2.2 Syntax of Subtle Authenticated Encryption

Just as a plan seldom survives contact with the enemy, so it goes with authenticated encryption: several provably secure schemes have fallen when implemented in practice. Especially for the decryption of invalid ciphertext it is challenging to ensure an adversary really only learns the invalidity of the ciphertext, and not some additional information. Additional information that has been considered in the past (and we will encounter again shortly) are multiple error symbols and unverified plaintext. Both can be classified as *leakage*, leading to our new notion of a *subtle authenticated encryption* scheme.

A subtle AE (SAE) scheme is a triple of deterministic algorithms $\Pi = (\mathcal{E}, \mathcal{D}, \Lambda)$, where Λ corresponds to leakage from the decryption function. We restrict ourselves to leakage functions that are deterministic functions on their inputs, and only provide leakage to invalid decryption queries. Thus the leakage function looks like

$$\Lambda : \mathsf{K} \times \mathsf{N} \times \mathsf{A} \times \mathsf{C} \to \{\top\} \cup \mathsf{L}$$

where the *leakage space* L can be any non-empty set not containing $\top$, and the distinguished symbol $\top$ refers to a message that is valid. So, for any (N, A, C), either $\mathcal{D}_k^{N,A}(C) = \bot$ or $\Lambda_k^{N,A}(C) = \top$, but not both: a message is either valid (and so decryption returns the plaintext but there is no leakage) or is invalid (and so decrypts to $\bot$ and leakage is available). The generality of L caters for any type of leakage, including schemes with multiple errors [6], those which output candidate plaintexts [2] or those which return arbitrary strings when presented invalid ciphertexts [11].

Explicitly separating Λ from $\mathcal{D}$ emphasises that leakage is a property of the decryption *implementation*, rather than of the decryption *function*. Consequently, security (for correct and tidy schemes) becomes a property of both the encryption function and the decryption implementation's leakage. A scheme may be proven secure for some leakage model Λ, but such a result is only meaningful as long as Λ accurately reflects the actual leakage as observed in practice. Even minor optimizations of the same decryption function can change the associated implementation so much that the scheme goes from being provably secure under some robust security definition to trivially insecure (we show how AEZ is affected in the full version). From this perspective, security becomes a subtle rather than robust affair, hence the name *subtle authenticated encryption*, a term inspired by the *SubtleCrypto* interface of the WebCryptoApi [17].

Comparison with the Traditional Model. There is a canonical mapping from any SAE scheme $(\mathcal{E}, \mathcal{D}, \Lambda)$ to a more traditional one $(\mathcal{E}, \mathcal{D})$ simply by removing access to the leakage oracle: correctness and tidiness of the subtle scheme clearly imply correctness and tidiness of the traditional one. Note that many distinct SAE schemes map to the same traditional form, implying that the canonical mapping induces an equivalence relation on SAE schemes. One could turn a traditional $(\mathcal{E}, \mathcal{D})$ scheme into a subtle form by inverting the above canonical map, for which the obvious preimage is setting $\Lambda^{N,A}(C) = \bot$ if $\mathcal{D}_k^{N,A}(C) = \bot$ and otherwise $\top$, again preserving correctness and tidiness. This corresponds to the SAE scheme whose implementation does not leak at all, so we expect our security notion to match the traditional one in this case (and it does).

Contrast with Leakage Resilience. Our separation into $\mathcal{D}$ and Λ is possible because decryption is deterministic, and its inputs (i.e. N, A, C) may be provided

to Λ_k. Within the leakage resilience community [10], leakage is generally charac-
terised as an auxiliary output from the original algorithm (often supported by an
auxiliary input to control the type of leakage); moreover one would expect both
encryption and decryption to leak. This integrated perspective reflects the real
world more closely (as leakage results from running some algorithm) and is more
expressive. For example, if the decryption routine were probabilistic, the leakage
may require access to the internal randomness, or if the scheme is stateful it may
require the correct state variables. Some of these issues could be overcome by
(for example) assuming the adversary always calls $\mathcal{D}_k$ directly before calling Λ_k,
and that Λ_k has access to the previous internal state (from which it can deduce
the operation of $\mathcal{D}_k$ if required), however ultimately which syntax works best
depends on the context.

In the context of capturing subtle implementation differences for modern
authenticated encryption (where decryption is stateless and deterministic) we
feel separating leakage and decryption is a useful abstraction. Though our work
could be recast into a form more closely aligned with the leakage resilience
literature, the notation would become more cumbersome, for instance when an
adversary can only observe the leakage.

2.3 Authentication and Encryption Security Games

In most modern AE definitions, an adversary is given access to a pair of oracles claiming to implement encryption and decryption. They are either real, and
act as claimed, or ideal, returning the appropriate number of random bits for
encryptions and rejecting all decryption attempts. To win the game, the adversary must decide which version it is interacting with. Certain queries would lead
to trivial wins, for example asking for the decryption of a message output by the
encryption oracle. These queries are forbidden (or their output suppressed).

This contrasts with the original definition of AE as IND–CPA plus INT–
CTXT, where in both constituent games an adversary only has access to a single,
real encryption oracle (and no decryption oracle); moreover, in the IND–CPA
only a single challenge ciphertext is present and for INT–CTXT only a single
ciphertext needs to be forged.

At first sight the two definitions may appear quite different, yet they are
known to be equivalent. Where does this difference stem from and should one
prefer one over the other?

We argue that both definitions can be cast as simplifications of a single
reference game. This reference game is itself a distinguishing game where an
adversary has access to two sets of oracles: one set of oracles will be used to
capture the *goal* of the adversary, whereas the other matches the *powers* of the
adversary. For instance, to capture AE an adversary has access to *four* oracles:
the two oracles from the modern definition (implementing either the real or ideal
scenario) *and* the two oracles from the traditional IND–CCA definition (namely
true encryption and decryption oracles).

Four oracles may seem overly complicated, but we posit that our approach
using a reference game has several advantages:

1. *Generality:* Hybrid arguments and composition results—the techniques implicitly underlying the standard definition—do not always hold when enriching the security model to take into account real-world phenomena such as key dependent messages or leakage (e.g. [9]). In these cases, one typically undoes certain simplifications; relying on our reference game instead is more transparent.
2. *Granularity:* Because adversarial goal and power are clearly separated, one can immediately identify a natural lattice of security notions and argue about possible equivalences *depending on the context*.
3. *Intuition:* The simplified games are less intuitive when considering real-life scenarios. For instance, even if an adversary knows it has seen a number of true plaintext–ciphertext pairs, for any set of fresh purported plaintext–ciphertext pair it should be clueless as to its validity. This statement follows directly from our reference game, yet for the simplified games one would need a hybrid argument.
4. *Tightness:* In real world scenarios, obtaining challenge ciphertexts versus known ciphertexts might carry different costs, which can be more easily reflected in our reference game (as the queries go to different oracles). A security analysis directly in our game is potentially more tight than one in a simplified game (whose results subsequently need to be ported to the more fine-grained real-world setting).

Security Games. We refer to games in the form GOAL–POWER, clearly separating the adversary's objective from its resources. The complete lists of powers and goals are presented in Table 1, and described. Security of scheme Π in game XXX against an adversary $\mathcal{A}$ is written as an advantage $\mathsf{Adv}_{\Pi}^{\mathrm{XXX}}[\Pi](\mathcal{A})$ and captures the adversary's ability to distinguish between two worlds. In both worlds the adversary has oracle access that depends on the scheme Π (initiated using some random and secret key $k \leftarrow_{\$} \mathsf{K}$); the oracles corresponding to the goal differ between the worlds, whereas the oracles corresponding to the power will be identical. The notation $\Delta_{\mathcal{O}_a,\mathcal{O}_b}^{\mathcal{O}_1,\mathcal{O}_2}$, short-hand for the advantage in distinguishing between $(\mathcal{O}_1,\mathcal{O}_2)$ and $(\mathcal{O}_a,\mathcal{O}_b)$, is used to make the oracles explicit. A scheme is XXX *secure* if $\mathsf{Adv}_{\Pi}^{\mathrm{XXX}}$ is sufficiently small for all reasonably resourced adversaries.

Goals. The goal oracles Enc and Dec either implement the true scheme or an idealised version. In each case they return ξ if their inputs are not elements of the appropriate spaces. If $b = 0$, we are in the *real world*, where Enc and Dec implement $\mathcal{E}_k$ and $\mathcal{D}_k$ respectively, and if $b = 1$ we are in the *ideal world*, where they implement $\$$ and $\perp$.

The oracle $\perp$ matches the syntax of $\mathcal{D}_k$ but returns $\perp$ in response to any queries. The oracle $\$$ is a random function: for each nonce-associated data pair, it samples an element $\$_{N,A}$ uniformly at random from the set of all τ-length-regular functions $f : \mathsf{M} \to \mathsf{C}$. When queried, $\$(N, A, M) := \$_{N,A}(M)$. When the adversary is forbidden from repeating queries, this corresponds to uniformly sampling $|M| + \tau(|M|)$ random bits.

Table 1. A compact table of goals and powers. The challenge oracles Enc and Dec specify the adversary's goal: they either implement honest encryption and decryption or their idealised versions, where Enc samples responses randomly and Dec returns $\perp$ to all queries. The honest oracles $\mathcal{E}_k$ and $\mathcal{D}_k$ capture the adversary's power. Each game corresponds to a 5-bit bitstring $b_1b_2b_3b_4b_5$, with for example CTI–CPA (equivalent to INT–CTXT) being 01100, and IND–sCCA (i.e. IND-CCA with decryption leakage) as 10111.

Oracles	Type	Challenge			Honest			Leakage	
	Role	Enc	Dec		$\mathcal{E}_k$	$\mathcal{D}_k$		Λ_k	
	Bit	1	2		3	4		5	
	Names	0	0	n/a	0	0	PAS	0	No leakage
		1	0	IND	1	0	CPA	1	Leakage (s)
		0	1	CTI	0	1	CDA		
		1	1	AE	1	1	CCA		

The *goal* is defined based on which oracles an adversary is given access to. We code this access using 2-bit strings, where the first bit is set in the presence of an Enc oracle, leaving the second bit for Dec. This leads to three possible goals (it does not make sense to have no challenge oracle): indistinguishability (IND, 10), authenticated encryption (AE, 11), and ciphertext integrity (CTI, 01).

Ideal Versus Attainable. Our ideal encryption oracle responds random bitstrings for fresh calls. This corresponds to security as one would expect it to hold; it can be considered as a computational analogue of Shannon's notion of perfect security where the uncertainty of a ciphertext given a message should be maximal. Similarly, the ideal decryption oracle is unforgiven, implying (traditional) integrity of ciphertexts.

Consequently, for some classes of constructions the advantage cannot be small. For instance, for online schemes it will be easy to distinguish by looking at prefixed and for schemes without sufficient stretch, randomly choosing a ciphertext can be used to forge.

One could bypass these impossibilities by adapting the ideal oracles accordingly [1,12]). Hoang et al. [11] suggest to use attainable security as benchmark; one can see the resulting security notions as (ever more complicated) extensions of the pseudorandom permutation notion typical for blockcipher security. This immediately reveals that to some extent, this choice is one of abstraction boundaries. When purely studying how to transform one primitive to another, it makes sense to used the ideal primitive as benchmark (as that will be the best attainable). Yet, we prefer a security definition that is both robust and meaningful: When non-experts use the primitive in larger designs, there should be as few implementation and configuration pitfalls as possible plus a small adversarial advantage should imply security as intuitively understood.

Powers. Traditionally, the adversary's *powers* describe what access they are given to honest encryption and decryption oracles, with which to learn about the scheme. Again, we identify these with 2-bit strings, listed in Table 1. The standard notions are a *passive* attack (PAS, 00) a *chosen plaintext* attack (CPA, 10), and a *chosen ciphertext* attack (CCA, 11). Access to only a decryption oracle is known as (DEM) CCA in the KEM–DEM setting (e.g. [7,8]), we will refer to it as a *chosen decryption* attack (CDA, 01). Unless *overall* encryption access is restricted as in the DEM scenario, the CDA scenario is of limited relevance (see Sect. 2.5).

Leakage Oracle. We add a third honest oracle implementing Λ_k, that models how schemes behave when subject to imperfect decryption implementations. Again, we use a bit to indicate whether a game provides an adversary access or not. If not, the standard notions arise, but presence leads to a range of new notions, which we will call their "subtle" variant. The name is chosen to emphasise that security critically depends on implementation subtleties.

As an example, power 101 stands for "subtle Chosen Plaintext Attack", or sCPA in short (note the "s" prefix). The power 001 corresponds to an adversary who cannot make decryption queries, yet it can observe leakage from them. This seeming contradiction makes sense when recalling that Λ_k only gives out information when queried with invalid ciphertexts. For instance, an adversary might learn how long it takes for ciphertexts to be rejected, but not what plaintexts correspond to valid ciphertexts. Given the implied validity checking capability and following the literature, we will refer to this power as a *chosen verification* attack (CVA) instead of a subtle passive attack.

2.4   Restrictions on the Adversary

With these lists in place, we consider what domain separations are required to prevent trivial wins. That is, we ask in what cases must the adversary be forbidden from taking the output of one oracle and using it as input to a second. The domain separation required for inputs to Λ_k is the same as $\mathcal{D}_k$, although we do not place any restrictions on the output of Λ_k: any seemingly trivial wins that occur from this are weaknesses of the scheme and demonstrate such a Λ_k cannot be secure. In the reference game, the adversary may make any queries he wishes that are not prohibited. In the effective game, he does not make superfluous queries either.

Trivial Wins. Any messages repeated between the two encryption oracles will distinguish the Enc oracle. Similarly, attempting to decrypt the output of Enc will allow the adversary to immediate determine whether Enc is random, since he will receive the initial plaintext if not. Attempting to decrypt the output of the honest encryption oracle $\mathcal{E}_k$ will also trivially identify whether Dec is real or idealised. Since the scheme is assumed to be tidy, we have that for any $C \in \mathsf{C}$, $\mathcal{E}_k(\mathcal{D}_k(C)) = C$. So, any output from the honest decryption oracle $\mathcal{D}_k$ cannot be passed to the challenge encryption oracle Enc, since this would trivially distinguish the schemes.

Superfluous Queries. A superfluous query is one to which the adversary need not make. Sending the output of $\mathcal{E}_k$ to $\mathcal{D}_k$ is superfluous since by correctness the answer is already known. Similarly, tidiness implies the opposite: output from $\mathcal{D}_k$ need not be sent to $\mathcal{E}_k$. As soon as Dec outputs something other than $\perp$, the adversary can distinguish it as the real case, and so might as well terminate, meaning no outputs from Dec need ever be queried to the encryption oracles. Finally, though not displayed in the diagram, assuming the game is deterministic and stateless (such as in the nonce or IV–based settings) it is superfluous to repeat queries or make any that return $\frac{1}{2}$, since neither yields useful information.

Nonces. If the adversary is *nonce-respecting* if he does not query (N, A, M') to either Enc or $\mathcal{E}_k$ if he has already queried either of them with (N, A, M) for some M. Note that we do not require the adversary be nonce-respecting, leaving this choice to specific security notion: relations between games are independent of strategies the adversary may or may not use, such as being nonce-respecting or nonce-abusing. That said, this behaviour can be enforced by the security game suppressing all such queries and returning $\frac{1}{2}$, making such queries superfluous.

2.5 Effective Games

Since there are 32 possible games and countless probabilistic adversaries, it would be prudent to begin by removing those which are directly equivalent. We give these in terms of the corresponding bitstring, where **x**, **y** and **z** signify bits that may (but need not) be set. We write X $\implies$ Y to signify that security in game X implies security in game Y, meaning that for any adversary $\mathcal{A}$ against game Y there is an adversary $\mathcal{B}$ against game X who uses similar resources and wins with similar probability.

Proposition 1 lists three (classes of) implications, which allows us to reduce the 32 games to only 4 interesting ones in Corollary 1. The proof for Proposition 1 can be found in the full version.

Proposition 1. *We may assume the adversary is deterministic and makes no superfluous or prohibited queries. Against such an adversary, several games are trivially related:*

1. *Adding extra oracles never makes the adversary weaker.*
2. *$x1y0z \iff x1y1z$: a decryption oracle does not help if a Dec challenge oracle is present.*
3. *$1x0yz \iff 1x1yz$: an encryption oracle does not help if a Enc challenge oracle is present.*

However, no further (generic) reductions are possible.

Corollary 1. *The effective games are just $1100x$, $1000x$, $1001x$ and $0110x$ (where x signifies a bit that might or might not be set). These correspond to AE–PAS, IND–PAS, IND–CDA, CTI–CPA and their subtle variants.*

2.6 Error Simulatability

We now define ERR (for *Error Simulatability*) to be the goal of distinguishing Λ_k from Λ_l, where $l \leftarrow_\$ \mathsf{K}$ is drawn independently of k. As always, this can be paired with any set of powers, leading to (for example) ERR–CCA:

$$\mathsf{Adv}_{\Pi}^{\mathrm{ERR-CCA}} := \Delta_{\varepsilon_k, \mathcal{D}_k, \Lambda_t}^{\varepsilon_k, \mathcal{D}_k, \Lambda_k}.$$

Initially this may appear unnecessarily specific: why should a definition of simulatability be given that restricts the simulator so tightly? As the following lemma shows, if there exists any good simulator, then Λ_l is one. Choosing this as our reference definition means security is completely described by $(\mathcal{E}, \mathcal{D}, \Lambda)$, rather than also requiring a description of the simulator. Obviously proof authors are welcome to use any simulator they wish, but a reference definition should be no more complex than absolutely necessary. After providing the lemma in question, we give some initial observations. Both results are proven in the full version.

Lemma 1. *If there exists a good simulator, Λ_l is one. That is, if there exists some stateful simulator S such that $\Delta_{\varepsilon_k, \mathcal{D}_k, \mathsf{S}}^{\varepsilon_k, \mathcal{D}_k, \Lambda_k}$ is small, then so is ERR–CCA. The inverse also holds.*

Lemma 2. *We observe that $\mathsf{Adv}_{\Pi}^{\mathrm{ERR-PAS}} = 0$. Also,* CTI–CPA + ERR–CCA $\iff$ CTI-sCPA | ERR–CPA.

2.7 Subtle Authenticated Encryption (SAE)

We define *Subtle Authenticated Encryption* (SAE) as a more succinct name for AE-sCCA, the strongest goal describable within this framework (i.e. 11111). The name, inspired by WebCryptoAPI [17], highlights the importance of the subtleties in implementations when applying such results. Thus, a secure SAE scheme is a triple $(\mathcal{E}, \mathcal{D}, \Lambda)$ along with appropriate spaces such that the AE-sCCA advantage is sufficiently small. So, the adversary has access to challenge encryption and decryption oracles, as well as honest encryption and decryption oracles, and certain amounts of leakage from the decryption function, and can make any query that does not leak to a trivial win. This characterisation clearly describes the situation from the real-world perspective.

From the designers point of view, due to reductions described in Proposition 1 it suffices to demonstrate that the scheme is AE-sPAS (i.e. AE-CVA, 11001) against an adversary who does not make useless queries or those that lead to trivial wins. Clearly there are various ways of doing this. Looking ahead somewhat, we will provide description of SAE in terms of the RUP definitions, as well as comparing it with RAE. The most intuitive method for proving a scheme SAE secure is likely to be through the following decomposition.

Theorem 1. *The SAE goal can be trivially decomposed:*

$$SAE \iff AE + ERR\text{-}CCA \iff IND\text{-}CPA + CTI\text{-}CPA + ERR\text{-}CCA.$$

Table 2. Notions from BDPS and RUP that directly translate into our framework.

Our notion	IND–CPA	CTI–CPA	CTI–sCPA	IND–sCCA	IND–CVA
Simplified bitstring	10000	01100	01101	10011	10001
BDPS notion	IND\$-CPA	INT–CTXT*	INT–CTXT	IND\$–CCA	IND\$–CVA
Reference (in [6])	Definition 5	Definition 7	Definition 7	Definition 5	Definition 5
Direct translation	10000	01110	01111	10011	10001
RUP notion	IND–CPA	INT–CTXT	INT–RUP		
Reference (in [2])	Definition 1	Definition 4	Definition 8		
Direct translation	10000	01100	01111		

3   Comparison of Recent AE Notions

Three recent papers introduced strengthened AE notions to capture distinguishable decryption failures [6], releasing unverified plaintext [2], and "robust" authenticated encryption [11]. In every case the encryption oracle can be cast as

$$E : K \times N \times A \times M \to C$$

but their authors make slightly different definitional choices depending on which aspect of the implementation they had in mind when developing the notion. The main differences are how decryption and its leakage are defined, when a ciphertext is considered valid, and what security to aim for. In the remainder of this section we will show how each of these three notions can be cast into our framework. With the appropriate modifications, it turns out that each of these three notions are essentially equivalent to our more general notion. As an obvious corollary, the three existing notions turn out to be not quite that radically different.

3.1   Distinguishable Decryption Failures (BDPS, [6])

Several provably secure IND-CCA secure schemes have succumbed to practical attacks as a result of different decryption failures being distinguishable, both in the public key and symmetric settings [5,18]. Boldyreva et al. [6] initiated a systematic study of the effects of symmetric schemes with multiple decryption errors. They emphasised probabilistic and stateful schemes, omitting a more modern nonce-based treatment. Below we describe the nonce-based analogues of their syntax and security notions.

A nonce-based, multi-error AE scheme a la BDPS, is a pair (E_k, D_k),

$$E : K \times N \times A \times M \to C$$
$$D : K \times N \times A \times C \to M \cup L$$

satisfying the classical definition of correctness. The idea is that if decryption fails, it may output any error symbol from L. (BDPS stipulate finite L, but this restriction appears superfluous and we omit it.)

To cast a (nonce-based) BDPS scheme into our SAE syntax, we observe the obvious (invertible) mapping from a scheme (E_k, D_k) by setting $\mathcal{E}_k = E_k$, $\mathcal{D}_k(C) = D_k(C)$ whenever $D_k(C) \in M$ or otherwise $\perp$, and $\Lambda_k(C) = D_k(C)$ whenever $D_k(C) \in L$ or otherwise $\top$.

Notions. BDPS define a number of notions, including both IND and IND$ concepts. Once adapted to a nonce-based setting, several of their notions directly translate into our framework, as listed in Table 2. Additionally, BDPS define *error invariance* [6, Definition 8], which (roughly) says that it should be hard for an adversary with access to honest encryption and decryption oracles to achieve any leakage other than a particular value. This notion, INV–ERR, implies an adversary cannot learn anything from decryption leakage and can be thought of as a special case of ERR–CCA since the simulator need just return this common value. However, error invariance is strictly stronger than leakage simulatability.

The strongest goal defined by BDPS is IND$–CCA3 [6, Definition 19], which incorporates multiple errors to the classical authenticated encryption notion. It is characterised by two oracles: an adversary has to distinguish between (E_k, D_k) (real) and $(\$, \perp)$ (ideal), where the error $\perp$ is a parameter of the notion. Thus despite the desire to capture multiple errors, in the ideal case the adversary is still only presented with a single error symbol. This curious artefact results from using INV–ERR rather than ERR–CCA to characterise "acceptable" leakage. Unfortunately, it leads to a reference definition that does not model the real-world problem satisfactorily, for instance it fails to capture the release of unverified plaintext.

Implications and Separations. BDPS provide several implications and separations between their notions. Although originally stated and proven for probabilistic and stateful schemes, the results easily carry over to a nonce-based setting. Using our naming convention, BDPS show that IND–CVA + CTI–sCPA $\implies$ IND–sCCA, yet IND–CVA + CTI–CPA $\not\implies$ IND–sCCA. This immediately implies a separation between CTI–sCPA and CTI–CPA. They also prove that AE and INV–ERR jointly are equivalent to their IND$–CCA3 notion (Theorem 20), which itself implies IND–CVA and CTI–sCPA.

Since INV-ERR implies ERR–CCA, this means IND$–CCA3 implies SAE. Moreover, the separation between ERR–CCA and INV–ERR carries over when comparing IND$–CCA3 and SAE. For completeness, we give the following theorem, which is a direct result of combination of Theorems 1 and 20 of BDPS (after incorporating nonces) with the observation that INV–ERR is more restrictive than ERR–CCA.

Theorem 2. *The IND$–CCA3 notion of BDPS is stronger than SAE solely in its requirement of simulatable errors. That is,*

$$\text{IND\$–CCA3} \iff \text{AE + INV–ERR} \implies \text{AE + ERR–CCA} \iff \text{SAE}.$$

3.2 Releasing Unverified Plaintext (RUP, [2])

Andreeva et al. [2] set out to model decryption more accurately for schemes that calculate a candidate plaintext before confirming its validity. In practice, such a candidate plaintext often becomes available (including to an adversary), even if validation fails. Examples include all schemes that need to decrypt or decipher before integrity can be checked (covering MAC–then–Encrypt, MAC–and–Encrypt, and encode–then–encipher) as well as schemes sporting online decryption (for instance single-pass CBC–then–MAC decryption). Andreeva et al. provide a large number of new definitions, covering security under decryption-leakage for both confidentiality and integrity.

Differences Between Frameworks. The RUP framework includes an explicit tag T, however the tag and ciphertext terms are always used together. This allows us to consider the ciphertext as (C, T) instead, which can be injectively mapped into C, e.g. by $C' := C\|T$ if the stretch is fixed. Following their motivating scenario, the RUP paper models decryption using a decryption oracle D and a verification oracle V satisfying

$$\mathsf{D} : \mathsf{K} \times \mathsf{N} \times \mathsf{A} \times \mathsf{C} \to \mathsf{M} = \mathsf{L}$$
$$\mathsf{V} : \mathsf{K} \times \mathsf{N} \times \mathsf{A} \times \mathsf{C} \to \{\top, \bot\}.$$

When called with a valid ciphertext, D_k returns the plaintext, and V_k returns $\top$. Conversely, when called with an invalid ciphertext, D_k will return some leakage information (nominally, the eponymous "unverified plaintext") and V_k will return $\bot$.

By changing perspective, we can cast a RUP scheme into the SAE framework: let $\mathcal{D}_k(C) = \mathsf{D}_k(C)$ if $\mathsf{V}_k(C) = \top$ (otherwise $\mathcal{D}_k(C) = \bot$) and $\Lambda_k(C) = \mathsf{D}_k(C)$ whenever $\mathsf{V}_k(C) = \bot$ (and otherwise $\Lambda_k(C) = \top$). Then, $(\mathcal{E}, \mathcal{D}, \Lambda)$ is an SAE scheme (where $\mathcal{E} = \mathsf{E}$), with leakage space $\mathsf{L} = \mathsf{M}$.

Notions. Andreeva et al. refer to the classic "encryption-only" notions of confidentiality and integrity under their customary names IND–CPA and INT–CTXT (our CTI–CPA). When decryption comes into play, a large number of new notions is suggested, typically defined in terms of adversarial access to their D_k and V_k oracles.

For integrity, dubbed INT–RUP for integrity under release of unverified plaintext, the adversary is given full access to all three honest oracles ($\mathcal{E}_k$, D_k, and V_k), and challenged to make a forgery. INT–RUP directly translates into our framework, where it corresponds to CTI–sCCA (itself equivalent to CTI–sCPA). This makes Andreeva et al.'s INT–RUP equivalent to BDPS's INT–CTXT notion.

For the myriad of RUP's confidentiality notions, an adversary is—for whatever reason—not provided with a verification oracle. This makes translation into our syntax cumbersome as any direct method would implicitly provide access to V_k functionality.

Implications and Separations. Andreeva et al. provide a number of implications and separations involving their new notions. They show that PA2 and DI are equivalent (Theorems 8 and 9), and imply PA1 (Theorem 1). Moreover, when combined with IND–CPA, PA2 provides a meaningful increase in security (Theorems 2 and 3), whereas PA1 does not (Theorems 4 and 5). Finally, they provide an alternative proof that CTI–sCPA is strictly stronger than CTI–CPA (Theorem 10).

Comments and Comparisons. The RUP model restricts any decryption leakage to the message space. This is unnecessarily restrictive: it does not directly cover multiple decryption errors; moreover a scheme may conceivably leak some internal variable (say a buffer) that is not in the message space.

The verification oracle for most of the RUP confidentiality notions is missing. For instance, the RUP version of IND–CCA security only gives an adversary access to the leakage, which raises the question whether RUP's IND–CCA security implies classical IND–CCA once the leakage is ignored. If the scheme is tidy, the RUP decryption and encryption oracle together suffice to implement the verification oracle. For a tuple (N, A, C), request $M \leftarrow D_k^{N,A}(C)$ and "accept" if and only if $C = C' \leftarrow E_k^{N,A}(M)$. Unfortunately, the domain separation in place for RUP's IND–CCA prohibits this sequence of queries. As a result, it is unclear whether RUP's IND–CCA implies standard IND–CCA or not, even though the former is defined as part of a framework of stronger notions.

Authenticated Encryption Definition. Andreeva et al. suggest that an authenticated encryption should meet the combined goals of IND–CPA and PA for confidentiality, and INT–RUP for integrity [2, Sect. 8]. Having to satisfy three separate notions may appear needlessly complicated and lacks the elegance a single notion can provide. We propose $RUPAE$ as a natural and neater way of defining Andreeva et al.'s final objective, where we use DI instead of the less direct PA:

$$\mathsf{Adv}_{\Pi}^{\text{RUPAE}} := \Delta_{\$,D_l,\perp}^{E_k,D_k,V_k}$$

This goal may originally have been envisaged by the authors, yet it was not explicitly alluded to (let alone defined). Providing a single succinct security goal is only worthwhile if it properly captures the compound notions, which we show in Theorem 3. The proof is intuitive, based around liberal use of the triangle inequality, see the full version for details.

Theorem 3. *The single term RUPAE notion is equivalent to the triple of goals originally proposed. That is,*

RUPAE $\iff$ CTI–sCPA + DI + IND–CPA $\iff$ INT–RUP + PA + IND–CPA

To relate this to our other notions, we provide the following observation (proven in the full version):

Lemma 3. CTI–sCPA + ERR–CPA $\iff$ CTI–$sCPA$ + DI.

Finally then, we have the reassuring result that security within the RUP framework coincides with our more general definition. To prove it, one simply chains Theorem 1, Lemmas 2 and 3, then Theorem 3 (in that order).

Corollary 2. *RUPAE security is equivalent to SAE security.*

3.3 Robust Authenticated Encryption (RAE, [11])

Robust authenticated encryption, as proposed by Hoang et al. [11, Sect. 3], has robustness against inadvertent leakage of unverified plaintext as one of it goals. A notable difference between traditional notions of AE and RAE is that the latter explicitly targets schemes where the intended level of integrity is specified by the user for each message. To this end, both encryption and decryption algorithms are provided with an additional input, called the stretch parameter τ, leading to the syntax:

$$E : K \times N \times A \times \mathbb{N} \times M \to C$$
$$D : K \times N \times A \times \mathbb{N} \times C \to M.$$

Thus encryption calls are of the form $C = E_k(N, A, \tau, M)$, taking in a nonce N, some associated data A, the stretch parameter τ and a message M, and output some ciphertext C. There is a requirement that τ is indeed the stretch, namely that $|C| = |M| + \tau$. Decryption calls take a similar format, and are allowed to "leak" a string *not* of the correct length when queried with invalid inputs. This length restriction on the leakage implies that valid ciphertexts can easily be determined from their length: if $M = D_k(N, A, \tau, C)$ and $|M| = |C| - \tau$, then it follows that $E_k(N, A, \tau, M) = C$.

The Security Game. The RAE security game aims to describe the *best possible* security for an object with the given syntax. Comparison to ideal objects is not new: it is the standard notion for blockciphers (namely a strong pseudorandom permutation) and has appeared previously as an alternative for deterministic authenticated encryption (namely strong pseudorandom injections).

For given stretch τ, the ideal object is a random element of $\mathrm{Inj}(\tau)$, the set of all injective functions whose outputs are always τ bits longer than their input. The inverse of an element $\pi \in \mathrm{Inj}(\tau)$ is not well defined (for $\tau > 0$) for strings outside of the range π. Since decryption may leak on these incorrect ciphertexts, returning $\perp$ in that case is no longer an option. Hoang et al. solve this problem by introducing a simulator S_π which has very restricted "access" to the ideal encryption π, as explained below.

Security is then defined relative to a simulator S and in terms of distinguishing between two worlds, with

$$\mathrm{Adv}_{\Pi,S}^{\mathrm{RAE}} := \mathbb{P}\left[k \leftarrow_{\$} K : \mathcal{A}^{E_k, D_k} \to 1\right] - \mathbb{P}\left[\pi_{N,A,\tau} \leftarrow_{\$} \mathrm{Inj}(\tau) : \mathcal{A}^{\pi, S_\pi} \to 1\right].$$

Here the injections $\pi_{N,A,\tau}$ are tweaked by the nonce, associated data, and stretch τ. Decryption queries in the ideal world are answered by S_π which exhibits the following behaviour. If a decryption query is valid, then it is of the form (N, A, τ, C) where $C \in \mathsf{Image}(\pi_{N,A,\tau})$ and the simulator S_π returns the preimage M. Otherwise, the ciphertext is invalid, or $C \notin \mathsf{Image}(\pi_{N,A,\tau})$. In this case, the oracle calls a stateful simulator S, which must simulate the decryption oracle and output a bitstring of any length other than $|C| - \tau$, *without* access to the injections $\pi_{.,.,.}$ (and the S_π oracle will simply forward S's output). A code-based description of this can be found in the original paper, where it is referred to as world $\mathbf{RAE}_{\Pi,S}$ [11, Fig. 2].

Fixing the Stretch. The variable, user-defined stretch sets RAE apart from the notions discussed in this paper so far. Although Hoang et al. insist that all values of stretch should be allowed for a scheme to be RAE, including $\tau = 0$, they hasten to add that this does make forging trivial, making it impossible to get a good (generic) upper bound on the CTI-CPA advantage. However, there is no intrinsic reason not to let a scheme restrict which values of τ it supports. Certainly their security definition still makes perfect sense if the stretch is no longer user defined and depends only on the length of the input message.

To ease comparison with previous security notions, we will henceforth restrict attention to fixed stretch schemes. This makes the mapping that takes an RAE scheme to an SAE scheme rather intuitive, and analogous to that used in Sect. 3.2. Explicitly, let (E, D) be an RAE scheme, and (inspired by RUP) let V_k be the associated validity function, where $\mathsf{V}_k^{N,A,\tau}(C) = \top \iff |\mathsf{D}_k^{N,A,\tau}(C)| - |C| = \tau$. Then $(\mathcal{E}, \mathcal{D}, \Lambda)$ is an SAE scheme, where $\mathcal{E}_k^{N,A}(M) := \mathsf{E}_k^{N,A,\tau}(M)$ and

$$\mathcal{D}_k^{N,A}(C) := \begin{cases} \mathsf{D}_k^{N,A}(C) \\ \bot \end{cases} \quad , \quad \Lambda_k^{N,A}(C) := \begin{cases} \top & \text{if } \mathsf{V}_k^{N,A,\tau}(C) = \top \\ \mathsf{D}_k^{N,A}(C) & \text{if } \mathsf{V}_k^{N,A,\tau}(C) \neq \top \end{cases}$$

Clearly this security game is similar to those presented above.

Comments and Comparisons. Following Rogaway's definitional papers [14–16], most recent symmetric results have been given in terms of indistinguishability from the ideal world ($\$, \bot$): an ideal encryption oracle that outputs random bits and an ideal decryption oracle that never accepts. Hoang et al. instead opt for an ideal world that corresponds to the "best achievable", a contrast they emphasize: "Before, AE–quality was always measured with respect to an aspirational goal; now we're suggesting to employ an achievable one." [11, Sect. 1: Discussion].

One feature, possibly by design, of RAE is that it accurately describes the security attainable from a PRP through the encode-then-encipher paradigm. Leakage is envisaged as being an invalid final buffer: one that has been deciphered but did not decode. This leads to the slightly artificial restriction that leakage cannot be a string of valid length.

Fixed–stretch RAE as an SAE Goal. Having applied the transform (which has no bearing on security), it is not surprising to find RAE and SAE security essentially coincides, with the only complication a generic attacks term, reflecting the difference between ideal and best possible security. After providing the RAE[τ] analogue of Lemma 1, we provide an explicit relationship between the games.

Lemma 4. *For any simulator* S, $\mathsf{Adv}_{\Pi,\Lambda}^{\mathrm{RAE}[\tau]}(\mathcal{A}) \leq 2 \cdot \mathsf{Adv}_{\Pi,\mathsf{S}}^{\mathrm{RAE}[\tau]}(\mathcal{A})$, *where* Λ *is to the simulator that first samples* $l \leftarrow_{\$} \mathsf{K}$, *then for all queries evaluates* Λ_l.

Theorem 4. *RAE[τ] and SAE security are equivalent. Explicitly, for an adversary $\mathcal{A}$ making at most q queries, and using a repeated nonces r times,*

$$\left| \mathsf{Adv}_{\Pi,\Lambda}^{\mathrm{RAE}[\tau]}(\mathcal{A}) - \mathsf{Adv}_{\Pi}^{\mathrm{SAE}}(\mathcal{A}) \right| \leq \frac{q}{2^{\tau-1}} + \frac{r^2+r}{2^{\tau+m+1}}.$$

4   Conclusions

By defining SAE we provided a framework useful to compare prior notions all addressing the same problems, but from slightly differing perspectives. BDPS provides the most generalised syntax, although a (seemingly unnecessary) condition that the error space be finite limits the applicability of their results. RUP presents the material in a very practical way, with definitions and models that clearly describe how decrypt-then-verify schemes behave, but in doing so yield a scheme that does not readily generalise to handling alternative leakage sources. RAE on the other hand defines a goal that, at first glance, appears to be the strongest of them all, but upon further inspection is rather more nuanced. Overall, the three recent works have more in common than the original authors (esp. of RUP and RAE) might have indicated.

Acknowledgements. We thank Dan Martin and Elisabeth Oswald for fruitful discussions regarding leakage-resilience and the anonymous referees of the IMA International Conference on Cryptography and Coding 2015 for their constructive feedback.

This work was conducted whilst Guy Barwell was a Ph.D. student at the University of Bristol, supported by an EPSRC grant.

References

1. Abed, F., Forler, C., List, E., Lucks, S., Wenzel, J.: Don't panic! the cryptographers' guide to robust authenticated (On-line) encryption. Comments to CAESAR mailing list (2015)
2. Andreeva, E., Bogdanov, A., Luykx, A., Mennink, B., Mouha, N., Yasuda, K.: How to securely release unverified plaintext in authenticated encryption. In: Sarkar, P., Iwata, T. (eds.) ASIACRYPT 2014. LNCS, vol. 8873, pp. 105–125. Springer, Heidelberg (2014)
3. Barwell, G., Page, D., Stam, M.: Rogue decryption failures: reconciling AE robustness notions (2015). http://eprint.iacr.org/2015/895

4. Bernstein, D.J.: CAESAR competition call (2013). http://competitions.cr.yp.to/caesar-call-3.html
5. Bleichenbacher, D.: Chosen ciphertext attacks against protocols based on the RSA encryption standard PKCS #1. In: Krawczyk, H. (ed.) CRYPTO 1998. LNCS, vol. 1462, pp. 1–12. Springer, Heidelberg (1998)
6. Boldyreva, A., Degabriele, J.P., Paterson, K.G., Stam, M.: On symmetric encryption with distinguishable decryption failures. In: Moriai, S. (ed.) FSE 2013. LNCS, vol. 8424, pp. 367–390. Springer, Heidelberg (2014)
7. Davies, G.T., Stam, M.: KDM security in the hybrid framework. In: Benaloh, J. (ed.) CT-RSA 2014. LNCS, vol. 8366, pp. 461–480. Springer, Heidelberg (2014)
8. Dent, A.W.: A designer's guide to KEMs. In: Paterson, K.G. (ed.) Cryptography and Coding 2003. LNCS, vol. 2898, pp. 133–151. Springer, Heidelberg (2003)
9. Dodis, Y., Pietrzak, K.: Leakage-resilient pseudorandom functions and side-channel attacks on Feistel networks. In: Rabin, T. (ed.) CRYPTO 2010. LNCS, vol. 6223, pp. 21–40. Springer, Heidelberg (2010)
10. Dziembowski, S., Pietrzak, K.: Leakage-resilient cryptography. In: 49th FOCS, pp. 293–302. IEEE Computer Society Press (2008)
11. Hoang, V.T., Krovetz, T., Rogaway, P.: Robust authenticated-encryption AEZ and the problem that it solves. In: Oswald, E., Fischlin, M. (eds.) EUROCRYPT 2015. LNCS, vol. 9056, pp. 15–44. Springer, Heidelberg (2015)
12. Hoang, V.T., Reyhanitabar, R., Rogaway, P., Vizár, D.: Online authenticated-encryption and its nonce-reuse misuse-resistance. In: Proceedings of CRYPTO (2015). http://eprint.iacr.org/2015/189
13. Namprempre, C., Rogaway, P., Shrimpton, T.: Reconsidering generic composition. In: Nguyen, P.Q., Oswald, E. (eds.) EUROCRYPT 2014. LNCS, vol. 8441, pp. 257–274. Springer, Heidelberg (2014)
14. Rogaway, P.: Authenticated-encryption with associated-data. In: Atluri, V. (ed.) ACM CCS 2002, pp. 98–107. ACM Press (2002)
15. Rogaway, P.: Nonce-based symmetric encryption. In: Roy, B., Meier, W. (eds.) FSE 2004. LNCS, vol. 3017, pp. 348–359. Springer, Heidelberg (2004)
16. Rogaway, P., Shrimpton, T.: A provable-security treatment of the key-wrap problem. In: Vaudenay, S. (ed.) EUROCRYPT 2006. LNCS, vol. 4004, pp. 373–390. Springer, Heidelberg (2006)
17. Sleevi, R., Watson, M.: Web cryptography API. W3C candidate recommendation (2014). http://www.w3.org/TR/WebCryptoAPI/
18. Vaudenay, S.: Security flaws induced by CBC padding - applications to SSL, IPSEC, WTLS. In: Knudsen, L.R. (ed.) EUROCRYPT 2002. LNCS, vol. 2332, pp. 534–546. Springer, Heidelberg (2002)

Robust Authenticated Encryption
and the Limits of Symmetric Cryptography

Christian Badertscher[1]($\boxtimes$), Christian Matt[1], Ueli Maurer[1], Phillip Rogaway[2],
and Björn Tackmann[3]

[1] Department of Computer Science, ETH Zurich, Zurich, Switzerland
{badi,mattc,maurer}@inf.ethz.ch
[2] Department of Computer Science, University of California, Davis, USA
rogaway@cs.ucdavis.edu
[3] Department of Computer Science and Engineering, University of California,
San Diego, USA
btackmann@eng.ucsd.edu

Abstract. Robust authenticated encryption (RAE) is a primitive for symmetric encryption that allows to flexibly specify the ciphertext expansion, i.e., how much longer the ciphertext is compared to the plaintext. For every ciphertext expansion, RAE aims at providing the best-possible authenticity and confidentiality. To investigate whether this is actually achieved, we characterize exactly the guarantees symmetric cryptography can provide for any given ciphertext expansion. Our characterization reveals not only that RAE reaches the claimed goal, but also, contrary to prior belief, that one cannot achieve full confidentiality without ciphertext expansion. This provides new insights into the limits of symmetric cryptography.

Moreover, we provide a rigorous treatment of two previously only informally stated additional features of RAE; namely, we show how redundancy in the message space can be exploited to improve the security and we analyze the exact security loss if multiple messages are encrypted with the same nonce.

1 Introduction

Authenticity and confidentiality are arguably among the most important cryptographic objectives. Authenticated encryption is a symmetric primitive that aims to achieve both at the same time, allowing efficiency gains and reducing the risk of misuse compared to combined schemes. Several notions of authenticated encryption have emerged over a series of works [2,3,5,7,8,13], including authenticated encryption with associated data [4,12] and misuse-resistant authenticated encryption [14]. In this development, *robust authenticated encryption (RAE)*, introduced by Hoang, Krovetz, and Rogaway [6], is the latest and most ambitious notion. Robust authenticated encryption allows to specify the ciphertext expansion λ that determines how much longer ciphertexts are compared to the corresponding plaintexts. Its self-declared goal in [6] is to provide the best-possible authenticity and confidentiality for every choice of λ. This raises the

© Springer International Publishing Switzerland 2015
J. Groth (Ed.): IMACC 2015, LNCS 9496, pp. 112–129, 2015.
DOI: 10.1007/978-3-319-27239-9_7

question of what best-possible authenticity and confidentiality is, and whether RAE actually achieves it. We provide a formal model that allows us to investigate this question and answer it in the affirmative. We further show how to use verifiable redundancy to improve security, and we show what security guarantees remain if values intended as nonces are reused. Both questions were addressed in [6] but not proven formally.

1.1 Robust Authenticated Encryption

An RAE scheme consists of a key distribution $\mathcal{K}$, a deterministic encryption algorithm $\mathcal{E}$, and a deterministic decryption algorithm $\mathcal{D}$. The encryption algorithm takes as input a key K, a nonce N, associated data A, the ciphertext expansion λ, and a message M. It outputs a ciphertext C. The decryption algorithm takes as input K, N, A, λ, and C, and returns the corresponding message M (or $\perp$ if C is an invalid ciphertext). In [6], the security of an RAE scheme is defined via a game in which an adversary has access to two oracles and has to distinguish between two possible settings. In the first setting, the oracles correspond to the encryption and decryption algorithm of the RAE scheme, where the key is fixed in the beginning and chosen according to $\mathcal{K}$. In the second setting, the first oracle chooses for each N, A, λ, and message length ℓ an injective function that maps strings of length ℓ to strings of length $\ell + \lambda$. On input (N, A, λ, M), the oracle answers by evaluating the corresponding function. The second oracle corresponds to the partially defined inverse of that function that answers $\perp$ if the given value has no preimage. An RAE scheme is secure if these two settings are indistinguishable for efficient adversaries. While this seems to be a strong guarantee, it is not clear which security such a scheme actually provides in a specific application and whether it is best-possible.

1.2 Security Definitions and Constructive Cryptography

Since game-based security definitions only capture what an adversary can do in a specific attack-scenario, they inherently fall short of providing guarantees that hold in any possible application of the scheme. To capture what RAE schemes achieve, we formulate our results using the constructive cryptography framework by Maurer and Renner [9,10]. The central idea of this framework is that the resources available to the parties, such as communication channels or shared randomness like cryptographic keys, are made explicit. The goal of a cryptographic protocol is then to construct, from certain existing resources, another resource that can again be used by higher-level protocols or applications. For example, the goal of an authentication scheme can be formalized as constructing an authenticated channel from a shared secret key and an insecure channel. The insecure channel allows a sender, say Alice, to send messages to a receiver, say Bob, but entirely leaks the transmitted messages to the adversary and additionally allows the adversary to delete messages and inject arbitrary messages; an authenticated channel still leaks the messages but only allows the adversary to delete messages and to deliver the messages originally sent. A conventional encryption scheme is supposed to construct a secure channel from a shared secret key and an authenticated channel, where the secure channel restricts the leakage to the length of the transmitted messages. The composition theorem of constructive cryptography guarantees that if two protocols achieve these constructions, the composed protocol constructs a secure channel from two shared secret keys and an insecure channel, i.e., the security of the overall construction follows from the security of the individual construction steps. On the other hand, authenticated encryption directly achieves the overall construction.

1.3   Our Contributions

In the vein of [1] and accounting for the associated data RAE schemes support, we formalize the goal of RAE as constructing an *augmented secure channel (ASC)* from a shared secret key and an insecure channel. An ASC takes as input from the sender a tuple (A, M), leaks A and the length of M to the adversary, and allows the adversary to either deliver the pair (A, M) or to terminate the channel. This channel provides authenticity for both A and M, but confidentiality is only guaranteed for the message M. The value A can for example be used to authenticate non-private header information; see [1] for an application of ASC in the context of TLS 1.3.

Uniform Random Injection Resource. Instead of directly constructing channels from a shared secret key and an insecure channel, we introduce an intermediate system *URI (uniform random injection)* that provides the sender and receiver access to the same uniform random injections and their inverses chosen as follows: For each combination of N, A, λ, and message length ℓ, an injective function that maps strings of length ℓ to strings of length $\ell + \lambda$ is chosen uniformly at random.

As we shall see, this resource can be constructed from a shared secret key using an RAE scheme in a straightforward manner. We then construct several channels from URI and an insecure channel. The advantage of this approach is that all further constructions in this paper are information-theoretic, i.e., we do not have to relate the security of each construction step to the RAE security game. Instead, we can rely on the composition theorem to guarantee the security of the overall construction.

Random Injection Channel. We show that one can construct a channel we call *RIC (random injection channel)* from URI and an insecure channel by fixing λ and using a counter as the nonce. RIC can be seen as a further intermediate step towards constructing ASC, that in addition allows us to analyze best-possible security.

The channel RIC takes as input a pair (A, M) from the sender and leaks A and the length of M to the adversary. The adversary can deliver the pair (A, M), and further at any point in time try to inject a new message of length ℓ and some value A. The probability with which such an injection is successful depends on

λ and ℓ. In case of a success, an almost uniform message of length ℓ from the message space together with A is delivered to the receiver. If an injection was successful and the tuple (A, M) was received, and if the sender subsequently sends exactly the pair (A, M), then the adversary is notified about this repetition.

Best Possible Authenticity and Confidentiality. If ASC is considered as the ultimate goal of RAE and authenticated encryption in general, the only shortcomings of RIC are that it is possible to inject messages with positive probability and that, if an attempted message injection was successful, the channel leaks a certain repetition to the adversary. While the first shortcoming is a lack of authenticity, the second one is a lack of confidentiality. While the type of leakage violating confidentiality might seem artificial, we describe an application in which such leakage might be problematic. Briefly, the leakage can reveal hidden information flow from the receiver to the sender.

We then analyze whether RAE really achieves the best-possible authenticity and confidentiality by bounding the probabilities of successful message injections and of leaking this particular repetition pattern for arbitrary schemes for achieving authenticity and confidentiality. While it is straightforward to see that authenticity requires redundancy and therefore a large ciphertext expansion, one might hope that the repetition leakage can be avoided. We prove that this is not the case, i.e., we show that the probability of an adversary being able to observe such a repetition is at least as high as in RIC, no matter what scheme is used or which setup assumptions are made.

To illustrate this lack of confidentiality for a concrete scheme, consider the following scenario in which the one-time pad is used over an insecure channel: Assume the attacker injects a ciphertext to Bob who decrypts it using the shared secret key and outputs the resulting message. Further assume Alice afterwards sends a message to Bob which results in the attacker seeing the same ciphertext. In that case, the attacker learns the fact that the message sent by Alice equals the message output by Bob. This contradicts the understanding of confidentiality as revealing nothing except the length of the transmitted message.[1] Our results generalize this observation to arbitrary schemes. We thereby refine the understanding of what symmetric cryptography can and cannot achieve by showing that confidentiality, quite surprisingly, also requires redundancy in the ciphertexts when only insecure channels and an arbitrary setup are assumed, even if the protocol can keep state.

Augmented Secure Channels and Message Redundancy. Since the probability of successful message injections decreases exponentially with λ, RIC is indistinguishable from ASC for large λ. We further provide a construction that

---

[1] This also contradicts a prior result in [11] that claims that the one-time pad constructs a certain (fully) confidential channel, a so-called XOR-malleable channel, from an insecure channel and a shared key. The proof in that paper is flawed in that the simulation fails if more ciphertexts are injected than messages sent.

incorporates an idea from [6] to exploit the redundancy in messages to achieve a better bound. Our construction reveals the exact trade-off between ciphertext expansion and redundancy to achieve a required security level.

Nonce-Reuse Resistance. It was claimed in [6] that reusing nonces only results in leaking the repetition pattern of messages, but does not compromise security beyond that. However, the claim was neither formalized nor proven. We fill this gap by introducing the channel resource *RASC (Repetition ASC)* that, aside of the length of each message, leaks the repetition pattern of the transmitted messages to the adversary. Furthermore, the adversary can deliver messages out-of-order and arbitrarily replay messages. We show that RASC can be constructed from URI and an insecure channel if the used nonce is always the same. This confirms the informal claim from [6] and makes explicit that some authenticity is lost by allowing the adversary to reorder messages.

2 Preliminaries

2.1 Notation for Systems and Algorithms

We describe our systems with pseudocode using the following conventions: We write $x \leftarrow y$ for assigning the value y to the variable x. For a distribution $\mathcal{X}$ over some set, $x \leftarrow \mathcal{X}$ denotes sampling x according to $\mathcal{X}$. For a finite set X, $x \leftarrow X$ denotes assigning to x a uniformly random value in X.

We denote the empty list by $[\,]$ and for a list L, $L \parallel x$ denotes the list L with x appended. Furthermore, $|L|$ denotes the number of elements in L and the ith element in L is denoted by $L[i]$ for $i \in \{1, \ldots, |L|\}$. For a FIFO queue $\mathcal{Q}$, we write $\mathcal{Q}.\text{enqueue}(x)$ to insert x into the queue and $\mathcal{Q}.\text{dequeue}()$ to retrieve (and remove) the element of the queue that was inserted first among all remaining elements.

For $n, m \in \mathbb{N}$, $\text{Inj}(\Sigma^n, \Sigma^m)$ denotes the set of injective functions $\Sigma^n \to \Sigma^m$. For an injective function $f \colon X \to Y$, we denote by f^{-1} the function $Y \to X \cup \{\bot\}$ that maps y to the preimage of y under f if existing, and to the distinct element $\bot$ otherwise.

Typically queries to systems consist of a suggestive keyword and a list of arguments (e.g., (send, M) to send the message M). We ignore keywords in writing the domains of arguments, e.g., $(\text{send}, M) \in \mathcal{M}$ indicates that $M \in \mathcal{M}$.

2.2 Constructive Cryptography

Constructive cryptography makes statements about *constructions* of *resources* from other resources. A resource is a system with interfaces via which the resource interacts with its environment and which can be thought of as being assigned to parties. All resources in this paper have an interface A for the sender (Alice), an interface B for the receiver (Bob), and an interface E for the adversary (Eve). In our security statements, we are interested in the advantage of a distinguisher **D** in distinguishing two resources, say **R** and **S** which is defined as

$$\Delta^{\mathbf{D}}(\mathbf{R}, \mathbf{S}) = \Pr[\mathbf{DR} = 1] - \Pr[\mathbf{DS} = 1],$$

where $\Pr[\mathbf{DR} = 1]$ denotes the probability that $\mathbf{D}$ outputs 1 when connected to resource $\mathbf{R}$. More concretely, $\mathbf{DR}$ is a random experiment, where the distinguisher repeatedly provides an input to one of the interfaces A, B, or E and observes the output generated in reaction to that input before it decides on its output bit.

Converters are systems that can be attached to an interface of a resource to change the inputs and outputs at that interface, which yields another resource. A converter is a system with two interfaces: the *inner interface* in is connected to an interface of a resource and the *outer interface* out, becomes the new connection point of that resource towards the environment. The protocols of the honest parties and simulators correspond to converters.

We directly state the central definition of a construction of [9] and briefly explain the relevant conditions.

Definition 1. *Let* $\mathbf{R}$ *and* $\mathbf{S}$ *be resources and let* $\mathsf{noAtck_R}$ *and* $\mathsf{noAtck_S}$ *be converters that describe the default behavior at interface* E *when no attacker is present. Let* ε *be a function that maps distinguishers to a value in* $[-1, 1]$ *and let* sim *be a converter (the simulator). A protocol, i.e., a pair* $(\mathsf{conv_1}, \mathsf{conv_2})$ *of converters, constructs resource* $\mathbf{S}$ *from resource* $\mathbf{R}$ *within* ε *and with respect to the pair* $(\mathsf{noAtck_R}, \mathsf{noAtck_S})$ *and the simulator* sim, *if for all distinguishers* $\mathbf{D}$,

$$\Delta^{\mathbf{D}} \left(\mathsf{conv_1}^{\mathsf{A}} \mathsf{conv_2}^{\mathsf{B}} \mathsf{noAtck_R}^{\mathsf{E}} \mathbf{R}, \mathsf{noAtck_S}^{\mathsf{E}} \mathbf{S} \right) \leq \varepsilon(\mathbf{D}) \qquad \text{(Availability)}$$

$$\Delta^{\mathbf{D}} \left(\mathsf{conv_1}^{\mathsf{A}} \mathsf{conv_2}^{\mathsf{B}} \mathbf{R}, \mathsf{sim}^{\mathsf{E}} \mathbf{S} \right) \leq \varepsilon(\mathbf{D}). \qquad \text{(Security)}$$

The first condition ensures that the protocol implements the required functionality if there is no attacker. For example, for communication channels, all sent messages have to be delivered when no attacker interferes with the protocol.

The second condition ensures that whatever Eve can do with the assumed resource, she could do as well with the constructed resource by using the simulator sim. Turned around, if the constructed resource is secure by definition, there is no successful attack on the protocol.

The notion of construction is composable, which intuitively means that the constructed resource can be replaced in any context by the assumed resource with the protocol attached without affecting the security. This is proven in [9].

2.3   Robust Authenticated Encryption

Let Σ be an alphabet (a finite nonempty set). Typically an element of Σ is a bit ($\Sigma = \{0, 1\}$) or a byte ($\Sigma = \{0, 1\}^8$). For a string $x \in \Sigma^*$, $|x|$ denotes its length. We define the syntax of a robust authenticated encryption scheme following [6].

Definition 2. *A robust authenticated encryption (RAE) scheme* $\Pi = (\mathcal{K}, \mathcal{E}, \mathcal{D})$ *consists of a key distribution* $\mathcal{K}$, *a deterministic encryption algorithm* $\mathcal{E}$ *that maps a key* $K \in \mathcal{K}$, *a nonce* $N \in \mathcal{N}$, *associated data* $A \in \mathcal{A}$, *ciphertext expansion* $\lambda \in \mathbb{N}$, *and a message* $M \in \mathcal{M}$ *to a ciphertext* $C \in \mathcal{C}$, *and a deterministic* decryption algorithm $\mathcal{D}$ that maps (K, N, A, λ, C) to an element in $\mathcal{M} \cup \{\bot\}$. We assume the domains $\mathcal{N}$, $\mathcal{A}$, $\mathcal{M}$, and $\mathcal{C}$ are equal to Σ^*. We write $\mathcal{E}_K^{N,A,\lambda}$ and $\mathcal{D}_K^{N,A,\lambda}$ for the functions $\mathcal{E}(K, N, A, \lambda, \cdot)$ and $\mathcal{D}(K, N, A, \lambda, \cdot)$, respectively. We require that $\mathcal{D}_K^{N,A,\lambda}\left(\mathcal{E}_K^{N,A,\lambda}(M)\right) = M$ for all K, N, A, λ, M.

3 Shared Uniform Random Injections and RAE Security

In this section, we describe the resource **URI** that grants access to shared uniform random injections and their inverses at interfaces A and B, and no access at interface E. We then use **URI** to define the security of RAE schemes and show that any RAE scheme that satisfies this definition can be used to construct **URI** from a shared secret key. Though syntactically different, it is easy to see that our definition is equivalent to the security definition from [6]. We recall that definition and prove the equivalence in the full version of this paper.

We first give a definition for the uniform random injection system **URI**.

Definition 3. *The resource **URI** has interfaces* A, B, *and* E *and takes inputs of the form* $(\mathsf{fun}, N, A, \lambda, x)$ *and* $(\mathsf{inv}, N, A, \lambda, y)$ *at interfaces* A *and* B *for* $N \in \mathcal{N}$, $A \in \mathcal{A}$, $\lambda \in \mathbb{N}$, $x \in \mathcal{M}$, *and* $y \in \mathcal{C}$. *Any input at interface* E *is ignored. We assume the domains* $\mathcal{N}$, $\mathcal{A}$, $\mathcal{M}$, *and* $\mathcal{C}$ *are equal to* Σ^*. *On input* $(\mathsf{fun}, N, A, \lambda, x)$ *at interface* A *or* B, *it returns* $f_{N,A,\lambda,|x|}(x)$ *at the same interface. On input* $(\mathsf{inv}, N, A, \lambda, y)$, *it returns* $f_{N,A,\lambda,|y|-\lambda}^{-1}(y)$ *if* $|y| > \lambda$, *and* $\bot$ *otherwise. The function* $f_{N,A,\lambda,\ell}$ *is chosen uniformly at random from the set* $\mathrm{Inj}\left(\Sigma^\ell, \Sigma^{\ell+\lambda}\right)$ *when needed for the first time and reused for later inputs.*

3.1 Definition of RAE Security and Construction of URI

We define a shared key resource $\mathbf{SK}_\mathcal{K}$ for some key distribution $\mathcal{K}$. The resource initially chooses a key according to $\mathcal{K}$ and outputs this key to interfaces A and B while interface E remains inactive, see Fig. 1. Slightly abusing notation, we will also refer to the key space by $\mathcal{K}$ whenever no confusion can arise. We further define the converter rae_Π that is based on an RAE scheme $\Pi = (\mathcal{K}, \mathcal{E}, \mathcal{D})$. First, rae_Π requests the key from $\mathbf{SK}_\mathcal{K}$. For any input at the outer interface, it evaluates $\mathcal{E}$ or $\mathcal{D}$ using that key (and the arguments provided in the input) and returns the result. The code is given in Fig. 1.

We consider an RAE scheme secure if all efficient distinguishers have poor advantage with respect to the following definition.

Definition 4. *The advantage of a distinguisher* **D** *for an RAE scheme* Π *is quantified as*

$$\mathbf{Adv}_\Pi^{\mathrm{rae}}(\mathbf{D}) := \Delta^\mathbf{D}\left(\mathsf{rae}_\Pi{}^\mathsf{A}\,\mathsf{rae}_\Pi{}^\mathsf{B}\,\mathbf{SK}_\mathcal{K}, \mathbf{URI}\right).$$

It is straightforward to see that the definition implies the following construction statement, where the converters sim and noAtck are defined as the converter that blocks any interaction at the interface it is connected to.

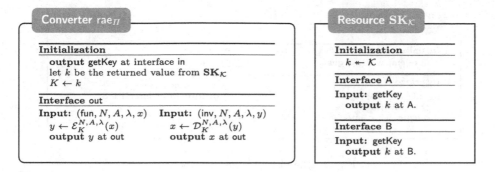

Fig. 1. Protocol that constructs **URI** from a shared secret key (left) and the shared secret key resource (right). For the shared key resource, interface E remains inactive.

Lemma 1. *The protocol* $(\mathrm{rae}_\Pi, \mathrm{rae}_\Pi)$ *constructs* **URI** *from* $\mathbf{SK}_\mathcal{K}$ *within* $\mathbf{Adv}_\Pi^{\mathrm{rae}}$ *with respect to* $(\mathsf{noAtck}, \mathsf{noAtck})$ *and simulator* sim *defined above.*

Proof. Since interface E of $\mathbf{SK}_\mathcal{K}$ and **URI** are inactive, the converters sim and noAtck have no effect when connected to that interface, i.e., $\mathsf{noAtck}^\mathsf{E} \mathbf{SK}_\mathcal{K} = \mathbf{SK}_\mathcal{K}$ and $\mathsf{noAtck}^\mathsf{E} \mathbf{URI} = \mathsf{sim}^\mathsf{E} \mathbf{URI} = \mathbf{URI}$. Thus, both the availability and the security condition of the construction are equivalent to

$$\Delta^\mathbf{D}\left(\mathrm{rae}_\Pi{}^\mathsf{A} \, \mathrm{rae}_\Pi{}^\mathsf{B} \, \mathbf{SK}_\mathcal{K}, \mathbf{URI}\right) \leq \mathbf{Adv}_\Pi^{\mathrm{rae}}(\mathbf{D})$$

for all distinguishers $\mathbf{D}$, which trivially holds by definition of $\mathbf{Adv}_\Pi^{\mathrm{rae}}$. $\qquad\square$

4 Random Injection Channels: Security for any Expansion

The goal of the current section is to examine the exact security achieved by RAE schemes when used to protect communication. We present constructions of specific secure channels from insecure channels and resource **URI** where each type of secure channel precisely captures the amount of leakage to an eavesdropper and the possible influence of an adversary interfering with the protocol execution. As an additional result, we are able to answer what best-possible communication security is and observe that RAE schemes achieve this level of security.

The insecure channel **IC** allows messages $m \in \mathcal{M}$ to be input repeatedly at interface A. Each message is subsequently leaked at the E-interface. At interface E, arbitrary messages (including those that were previously input at interface A) can be injected such that they are delivered to B. This channel does not give any security guarantees to Alice and Bob. A formal description is provided in Fig. 2. For the rest of this paper, the message space of the insecure channel is Σ^*.

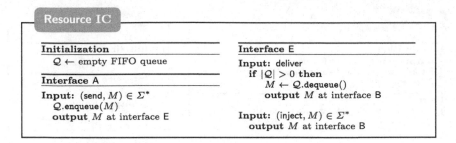

Fig. 2. The insecure channel resource.

4.1   Constructing Random Injection Channels

The Constructed Channel. The channel we construct in this section is defined in Fig. 3 and can be roughly described as follows: It allows to repeatedly send pairs (A_i, M_i) in an ordered fashion from a sender to a receiver. Each pair consists of the associated data A_i and the message M_i. The attacker is limited to seeing the associated data A_i and the length of the message $|M_i|$ of each transmitted pair. Additionally, the attacker learns whether the ith injected pair equals the one that is currently sent.

The attacker can either deliver the next legitimate pair (A_i, M_i) or try to inject a pair (A, M) that is different from (A_i, M_i). Such an injection is only successful with a certain probability. The associated data A and the length ℓ of the message are chosen by the attacker and M is a uniformly random message of length ℓ if $A \neq A_i$. Otherwise, M is a uniformly random message $M \neq M_i$ of length ℓ. If an injection attempt is not successful, the resource does not deliver messages at interface B any more and signals an error by outputting $\perp$. If the adversary injects the ith message, the legitimate ith message cannot be delivered anymore.[2]

The success probability of an injection attempt depends on the expansion λ and the specified message length ℓ and whether the sender's queue $\mathcal{S}$ is empty or not. The exact probabilities are quantified by the two sampling functions SAMPLE and SAMPLEEXCL. The function SAMPLE first samples a bit according to the probability that a fixed element from $\Sigma^{\ell+\lambda}$ has a preimage under a uniform random injection $\Sigma^\ell \to \Sigma^{\ell+\lambda}$. If the bit is 1, a uniform random preimage is returned. The function SAMPLEEXCL essentially does the same, but the domain and codomain are both reduced by one element.[3]

Protocol. We construct resource $\mathbf{RIC}_\lambda$ from $[\mathbf{URI}, \mathbf{IC}]$ which denotes the resource that provides at each interface access to the corresponding interface

---

[2] This relates to the security of RAE schemes which ensures that the message cannot be decrypted using a wrong nonce. In our construction, the nonce is implemented as the sequence number.

[3] This ensures that the injected message is different from the one that the sender provided.

Resource RIC$_\lambda$

Initialization

$\quad S \leftarrow$ empty FIFO queue
$\quad i \leftarrow 0$
$\quad \mathcal{R} \leftarrow []$

Interface A

Input: (send, $A, M) \in \mathcal{A} \times \mathcal{M}$
$\quad i \leftarrow i + 1$
$\quad$ if $i > |\mathcal{R}|$ then
$\quad\quad S.$enqueue$((A, M))$
$\quad$ if $i \leq |\mathcal{R}|$ and $\mathcal{R}[i] = (A, M)$ then
$\quad\quad$ **output** repeat at interface E
$\quad$ else
$\quad\quad$ **output** $(A, |M|)$ at interface E

Interface E

Input: deliver
$\quad$ if $|S| > 0$ and halt $= 0$ then
$\quad\quad (A, M) \leftarrow S.$dequeue$()$
$\quad\quad \mathcal{R} \leftarrow \mathcal{R} \parallel (A, M)$
$\quad\quad$ **output** (A, M) at interface B

Input: (inject, $A, \ell) \in \mathcal{A} \times \mathbb{N}_{>0}$
$\quad$ if halt $= 0$ then
$\quad\quad$ if $|S| > 0$ then
$\quad\quad\quad (A', M') \leftarrow S.$dequeue$()$
$\quad\quad\quad$ if $A = A'$ and $\ell = |M'|$ then
$\quad\quad\quad\quad M \leftarrow \textsc{SampleExcl}(\ell, \lambda, M')$
$\quad\quad\quad$ else
$\quad\quad\quad\quad M \leftarrow \textsc{Sample}(\ell, \lambda)$
$\quad\quad$ else
$\quad\quad\quad M \leftarrow \textsc{Sample}(\ell, \lambda)$
$\quad\quad$ if $M \neq \bot$ then
$\quad\quad\quad \mathcal{R} \leftarrow \mathcal{R} \parallel (A, M)$
$\quad\quad\quad$ **output** (A, M) at B
$\quad\quad$ else
$\quad\quad\quad$ **output** $\bot$ at B
$\quad\quad\quad$ halt $\leftarrow 1$

---

function $\textsc{Sample}(\ell, \lambda)$

$\quad B \twoheadleftarrow$ Bernoulli $\left(|\Sigma|^{-\lambda}\right)$

$\quad$ if $B = 1$ then
$\quad\quad M \twoheadleftarrow \Sigma^\ell$
$\quad\quad$ **return** M
$\quad$ else
$\quad\quad$ **return** $\bot$

function $\textsc{SampleExcl}(\ell, \lambda, m)$

$\quad B \twoheadleftarrow$ Bernoulli $\left(\frac{|\Sigma|^\ell - 1}{|\Sigma|^{\ell+\lambda} - 1}\right)$

$\quad$ if $B = 1$ then
$\quad\quad M \twoheadleftarrow \Sigma^\ell \setminus \{m\}$
$\quad\quad$ **return** M
$\quad$ else
$\quad\quad$ **return** $\bot$

Fig. 3. Description of **RIC$_\lambda$**. In the description, Bernoulli(p) denotes the distribution over $\{0, 1\}$, where 1 has probability p and 0 has probability $1 - p$.

of both resources. Our protocol specifies a particular but very natural usage of **URI** where the nonce is implemented as a counter value.[4] We present the protocol as pseudocode in Fig. 4. The converter for the sender, snd$_\lambda$, accepts inputs of the form (send, A, M) at its outer interface. It outputs (fun, i, A, λ, M) at the inner interface to **URI**. The nonce is implemented as a counter and λ is the parameter of the protocol. Once a ciphertext is received as a return value from **URI**, it is output together with its associated data at the inner interface for the insecure channel **IC**. The receiver converter rcv$_\lambda$ receives ciphertexts together with the associated data at its inner interface from **IC** and decrypts C using parameters A, i and λ. On success, the corresponding plaintext is output at the outer interface. If decryption fails, the converter stops and signals an error by outputting $\bot$.

Construction Statement. In order to show that the protocol (snd$_\lambda$, rcv$_\lambda$) constructs **RIC$_\lambda$** from [**URI, IC**], we prove both conditions of Definition 1.

---

[4] Implementing the nonce as a counter allows to maintain the order of messages.

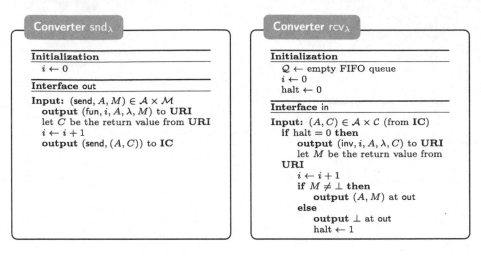

Fig. 4. The converters for the sender (left) and the receiver (right) to construct $\mathbf{RIC}_\lambda$.

For all channels, the converter noAtck corresponds to the converter dlv that on any input at its inner interface, outputs deliver to the channel connected to its inner interface and blocks any interaction at its outer interface.

Theorem 1. *Let* $\lambda \in \mathbb{N}$. *The protocol* $(\mathsf{snd}_\lambda, \mathsf{rcv}_\lambda)$ *constructs resource* $\mathbf{RIC}_\lambda$ *from* $[\mathbf{URI}, \mathbf{IC}]$ *with respect to* $(\mathsf{dlv}, \mathsf{dlv})$ *and simulator* $\mathsf{sim}_{\mathrm{RIC}}$ *as defined in Fig. 5. More specifically, for all distinguishers* $\mathbf{D}$

$$\Delta^{\mathbf{D}} \left(\mathsf{snd}_\lambda^{\mathsf{A}} \mathsf{rcv}_\lambda^{\mathsf{B}} \mathsf{dlv}^{\mathsf{E}} [\mathbf{URI}, \mathbf{IC}], \mathsf{dlv}^{\mathsf{E}} \mathbf{RIC}_\lambda \right) = 0 \tag{1}$$

$$and \quad \Delta^{\mathbf{D}} \left(\mathsf{snd}_\lambda^{\mathsf{A}} \mathsf{rcv}_\lambda^{\mathsf{B}} [\mathbf{URI}, \mathbf{IC}], \mathsf{sim}_{\mathrm{RIC}}^{\mathsf{E}} \mathbf{RIC}_\lambda \right) = 0. \tag{2}$$

We prove Theorem 1 in the full version of this paper.

4.2 What is Best-Possible Security?

We observe that $\mathbf{RIC}_\lambda$ has two undesirable properties: messages can be injected and the output at interface E leaks more than only the length of the payload in that it reveals whether Alice sends the pair (A, M) that has been output by Bob upon an adversarial injection. In contrast, a channel that only leaks the message length is considered fully confidential.

We first illustrate an application in which this lack of full confidentiality is problematic. The main purpose of our example is to show that one cannot exclude the existence of an application where exactly this (intuitively small) difference to full confidentiality yields a security problem.

Second, we show that a successful injection followed by the undesired information leakage about the repetition is possible for any scheme, even if it is stateful and uses an arbitrary setup before starting communication, and that the probability of this is minimized if $\mathbf{RIC}_\lambda$ is used.

Converter sim$_{\mathrm{RIC}}$

Initialization
$\mathcal{Q}_1, \mathcal{Q}_2 \leftarrow$ empty FIFO queues
$\mathcal{R} \leftarrow []$
$i \leftarrow 0$
diff $\leftarrow 0$ ▷ **invariant:** diff $= \max\{|\mathcal{R}| - i, 0\}$

Interface out

Input: (inject, $A, C) \in \mathcal{A} \times \mathcal{C}$
$\mathcal{R} \leftarrow \mathcal{R} \parallel (A, C)$
if $|\mathcal{Q}_2| = 0$ **then**
    diff $\leftarrow$ diff $+ 1$
    output (inject, $A, |C| - \lambda$) at in
else
    $(A', C') \leftarrow \mathcal{Q}_2$.dequeue()
    if $A = A$ and $C = C'$ **then**
        output deliver at in
    else
        output (inject, $A, |C| - \lambda$) at in

Input: deliver
if $|\mathcal{Q}_1| > 0$ **then**
    $(E, C) \leftarrow \mathcal{Q}_1$.dequeue()
    execute commands for (inject, A, C)

Interface in

Input: $(A, \ell) \in \mathcal{A} \times \mathbb{N}_{>0}$
$i \leftarrow i + 1$
if diff > 0 **then**
    diff $\leftarrow$ diff $- 1$
    $(A_i, C_i) \leftarrow \mathcal{R}[i]$
    if $A = A_i$ **then**
        $C \twoheadleftarrow \Sigma^{\ell+\lambda} \setminus \{C_i\}$
    else
        $C \twoheadleftarrow \Sigma^{\ell+\lambda}$
else
    $C \twoheadleftarrow \Sigma^{\ell+\lambda}$
    $\mathcal{Q}_2$.enqueue$((A, C))$
    $\mathcal{Q}_1$.enqueue$((A, C))$
    output (A, C) at out

Input: Repeat
$i \leftarrow i + 1$
$(A_i, C_i) \leftarrow \mathcal{R}[i]$
if diff > 0 **then**
    diff $\leftarrow$ diff $- 1$
else
    $\mathcal{Q}_2$.enqueue$((A_i, C_i))$
    $\mathcal{Q}_1$.enqueue$((A_i, C_i))$
    output (A_i, C_i) at out

Fig. 5. Simulator for the security condition of the construction of $\mathbf{RIC}_\lambda$.

Sample Scenario: On the Difference to Full Confidentiality. Assume a setting in which party A is allowed to send information to party B via a fully confidential channel but not vice versa. Suppose now that B finds a possibility to send information to A via a covert channel and the two parties use the confidential channel for messages from A to B and the covert channel for messages from B to A. Suppose now that the confidential channel is in fact a channel that leaks the above repetition event instead of only the message length. This gives an investigator E a means to test for the existence of a covert channel from B to A as follows: At some point, E injects a random message M to B. Assuming information flow from B to A, party B might start a discussion about M with party A. As part of this conversation, A might send M to B, which would signal a repetition-event to E. For large message spaces, it is very unlikely that A comes up with the exact same message that was randomly injected to B before, unless there is a (hidden) flow of information. The occurrence of the event is therefore a witness for the existence of a channel from B to A. In contrast, a fully confidential channel would not reveal the existence of the covert channel.

RIC Provides Best-Possible Security. In $\mathbf{RIC}_\lambda$, an injection attempt is successful with probability at most $\frac{|\mathcal{M}|}{|\mathcal{C}|} = |\Sigma|^{-\lambda}$ and given a successful injection and that Alice subsequently sends the corresponding output of Bob, the above described leakage occurs with probability 1. Overall, the total probability that an undesired property can be observed is bounded by $|\Sigma|^{-\lambda}$.

We show that this probability is optimal and that no protocol can achieve a better bound. Hence, $\mathbf{RIC}_\lambda$ maximizes authenticity and confidentiality. We first prove the following general lemma.

Lemma 2. *Let $\mathcal{M}$ and $\mathcal{C}$ be finite nonempty sets and let E and D be random variables over functions $\mathcal{A} \times \mathcal{M} \to \mathcal{C}$ and $\mathcal{A} \times \mathcal{C} \to \mathcal{M} \cup \{\bot\}$, respectively, such that*

$$\forall m \in \mathcal{M}, a \in \mathcal{A}: \quad \Pr[D(a, E(a, m)) = m] \geq p$$

for some $p \in [0, 1]$. We then have for all $a \in \mathcal{A}$ and any random variable C that is distributed uniformly over $\mathcal{C}$ and independent from E and D,

$$\Pr[D(a, C) \neq \bot \wedge E(a, D(a, C)) = C] \geq p \cdot \frac{|\mathcal{M}|}{|\mathcal{C}|}.$$

Proof. We have for all $a \in \mathcal{A}$

$$\Pr[D(a, C) \neq \bot \wedge E(a, D(a, C)) = C]$$
$$= \sum_{m \in \mathcal{M}} \sum_{c \in \mathcal{C}} \Pr[D(a, c) = m \wedge E(a, m) = c \wedge C = c]$$
$$= \frac{1}{|\mathcal{C}|} \sum_{m \in \mathcal{M}} \underbrace{\sum_{c \in \mathcal{C}} \Pr[D(a, c) = m \wedge E(a, m) = c]}_{= \Pr[D(a, E(a,m))=m] \, \geq \, p}$$
$$\geq p \cdot \frac{|\mathcal{M}|}{|\mathcal{C}|},$$

where we used in the second step that C is distributed uniformly over $\mathcal{C}$ and independent from E and D. $\qquad\square$

Lemma 2 can be applied to our usual setting $\mathrm{enc}_\lambda^A \mathrm{dec}_\lambda^B[\mathbf{SK}_\mathcal{K}, \mathbf{IC}]$ for a generic protocol $(\mathrm{enc}_\lambda, \mathrm{dec}_\lambda)$ in a straightforward manner: we only have to observe that for the ith input $(\mathsf{send}, A_i, M_i)$, for all $i \in \mathbb{N}$, converter enc_λ is characterized by a probabilistic map $\mathcal{A} \times \mathcal{M} \to \mathcal{C}$, that may depend on previous inputs and outputs and on the key k. Similarly, the converter dec_λ is characterized by a probabilistic map $\mathcal{A} \times \mathcal{C} \to \mathcal{M} \cup \{\bot\}$ for any $i \in \mathbb{N}$.

Correctness of the protocol implies that if $(\mathsf{send}, A_i, M_i)$ is input to enc_λ as the ith query and yields ciphertext C_i, then the probability that on the ith input (A_i, C_i) to dec_λ, and if dec_λ has not halted yet, the converter decrypts the ciphertext to M_i with probability p; note that $p = 1$ for RAE schemes. Hence, Lemma 2 implies that the probability that any of the two undesirable properties can be observed during protocol execution is at least $|\Sigma|^{-\lambda}$.

5   Augmented Secure Channels and Verifiable Redundancy

Looking at the specification of $\mathbf{RIC}_\lambda$, we observe that for large values λ, the probability of successful injections becomes exponentially small, and so are the repetition events at interface E. We are particularly interested in the resource that specifies this abstraction: a channel that allows to securely transmit messages consisting of an associated data and a payload part such that an attacker is limited to seeing the associated data and the length of the payload, to deliver the next message, or to abort the whole communication. This channel abstraction corresponds to an augmented secure channel. Such channels were introduced in [1] and shown to be achievable by the AEAD notion of [12]. Not surprisingly, this confirms that RAE and AEAD achieve the same security goals for large ciphertext expansion.

Additionally, we formally show how redundancy in messages can be exploited to improve authenticity, where redundancy restricts the set of valid messages to a subset of $\mathcal{M} = \Sigma^*$.

The following theorem provides the exact security bound in terms of redundancy in the message space and ciphertext expansion λ. We thereby confirm a conjecture of [6]. Let $v : \mathcal{M} \mapsto \{\mathsf{true}, \mathsf{false}\}$ be a predicate on the message space. We define the subset $\mathcal{M}_v := \{M \mid M \in \mathcal{M} \wedge v(M)\}$ which we call the set of valid messages. Following [6], the density of $\mathcal{M}_v$ is defined as

$$d_v := \max_{\ell \in \mathbb{N}} \frac{|\Sigma^\ell \cap \mathcal{M}_v|}{|\Sigma^\ell|}.$$

The Constructed Channel. The augmented secure channel $\mathbf{ASC}_{\mathcal{M}_v}$ is described in Fig. 6. The channel is derived from $\mathbf{RIC}_\lambda$ by requiring that $M \in \mathcal{M}_v$ and by removing undesired capabilities that vanish due to the exponentially small success probability for large λ.

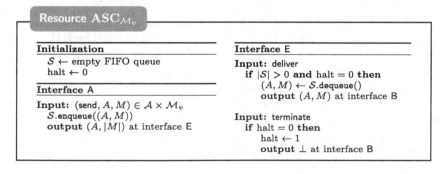

Fig. 6. Description of **ASC**, an augmented secure channel.

Protocol. The protocol for the sender, sndChk_v, accepts inputs of the form (send, A, M) at its outer interface and forwards the pair to the channel $\mathbf{RIC}_\lambda$ if and only if $v(M)$ (and otherwise ignores the request). The receiver converter rcvChk_v, on receiving the pair (A, M) from $\mathbf{RIC}_\lambda$, outputs (A, M) at its outer interface if and only if $v(M)$. If rcvChk_v receives $\perp$ from $\mathbf{RIC}_\lambda$ or if $\neg v(M)$, it outputs $\perp$ at its outer interface and halts.

Theorem 2. *Let $\lambda \in \mathbb{N}$. The protocol $(\mathsf{sndChk}_v, \mathsf{rcvChk}_v)$ constructs $\mathbf{ASC}_{\mathcal{M}_v}$ from $\mathbf{RIC}_\lambda$ with respect to $(\mathsf{dlv}, \mathsf{dlv})$ and simulator $\mathsf{sim}_{\mathrm{ASC}}$ defined in Fig. 7. More specifically, for all distinguishers $\mathbf{D}$*

$$\Delta^{\mathbf{D}}\left(\mathsf{sndChk}_v^{\mathsf{A}}\mathsf{rcvChk}_v^{\mathsf{B}}\mathsf{dlv}^{\mathsf{E}}\mathbf{RIC}_\lambda, \mathsf{dlv}^{\mathsf{E}}\mathbf{ASC}_{\mathcal{M}_v}\right) = 0 \tag{3}$$

$$and \quad \Delta^{\mathbf{D}}\left(\mathsf{sndChk}_v^{\mathsf{A}}\mathsf{rcvChk}_v^{\mathsf{B}}\mathbf{RIC}_\lambda, \mathsf{sim}_{\mathrm{ASC}}^{\mathsf{E}}\mathbf{ASC}_{\mathcal{M}_v}\right) \leq d_v \cdot |\Sigma|^{-\lambda}. \tag{4}$$

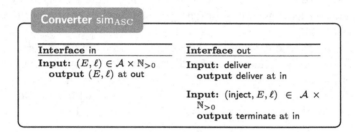

Fig. 7. Simulator for the security condition of the construction of $\mathbf{ASC}_{\mathcal{M}_v}$.

Proof. The availability condition (3) is straightforward to verify. We only prove the security condition (4). It is easy to see that the two systems behave identically as long as no injection attempt is successful. This is because successful injections are necessary for observing **repeat**: as long as no injection is successful, for any **send**-query to $\mathbf{RIC}_\lambda$, the condition $i \leq |\mathcal{R}|$ is not satisfied after incrementing i. We thus only have to bound this probability. We hence consider the event that in an interaction of a distinguisher with the real system $\mathsf{sndChk}_v^{\mathsf{A}}\mathsf{rcvChk}_v^{\mathsf{B}}\mathbf{RIC}_\lambda$ the first attempt to inject a random message is successful (since in case of an unsuccessful attempt, both channels stop delivering messages). In any interaction of $\mathbf{D}$ with the resource, the probability of the event is determined by $\mathbf{RIC}_\lambda$ as one out of two possibilities, see Fig. 3. For any $i \in \mathbb{N}$, if the ith query at interface E is the first attempt to inject a message, then the probability depends on whether the specified associated data and the length coincides with the length of the message and the associated data of the ith input at interface A. Both probabilities are upper bounded by

$$\max\left\{\frac{|\Sigma^{|C_i|-\lambda} \cap \mathcal{M}_v| - 1}{|\Sigma^{|C_i|}| - 1}, \frac{|\Sigma^{|C_i|-\lambda} \cap \mathcal{M}_v|}{|\Sigma^{|C_i|}|}\right\} \leq \frac{|\Sigma^{|C_i|-\lambda} \cap \mathcal{M}_v|}{|\Sigma^{|C_i|}|}$$

$$= \frac{|\Sigma^{|C_i|-\lambda} \cap \mathcal{M}_v|}{|\Sigma^{|C_i|-\lambda}|} \cdot |\Sigma|^{-\lambda} \leq d_v \cdot |\Sigma|^{-\lambda},$$

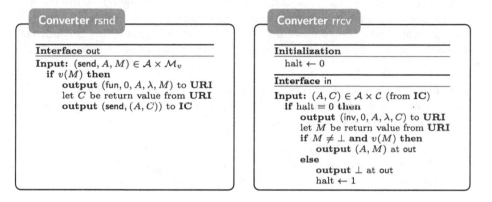

Resource RASC$_{\mathcal{M}_v}$

Initialization
$\mathcal{R} \leftarrow$ empty FIFO queue
$\mathcal{S} \leftarrow [\ ]$; halt $\leftarrow 0$

Interface A
Input: (send, $A, M) \in \mathcal{A} \times \mathcal{M}_v$
$\quad \mathcal{S} \leftarrow \mathcal{S} \parallel (A, M)$
$\quad i \leftarrow \min\{n \in \mathbb{N} \mid \mathcal{S}[n] = (A, M)\}$
$\quad$**output** $(A, |M|, i)$ at interface E

Interface E
Input: (deliver, $i) \in \mathbb{N}_{>0}$
$\quad$if $|\mathcal{S}| \geq i$ **and** halt $= 0$ **then**
$\quad\quad (E, M) \leftarrow \mathcal{S}[i]$
$\quad\quad$**output** (A, M) at interface B

Input: terminate
$\quad$if halt $= 0$ **then**
$\quad\quad$halt $\leftarrow 1$
$\quad\quad$**output** $\perp$ at interface B

Fig. 8. Description of **RASC**.

Converter rsnd

Interface out
Input: (send, $A, M) \in \mathcal{A} \times \mathcal{M}_v$
$\quad$if $v(M)$ **then**
$\quad\quad$**output** (fun, $0, A, \lambda, M$) to **URI**
$\quad\quad$let C be return value from **URI**
$\quad\quad$**output** (send, (A, C)) to **IC**

Converter rrcv

Initialization
$\quad$halt $\leftarrow 0$

Interface in
Input: $(A, C) \in \mathcal{A} \times \mathcal{C}$ (from **IC**)
$\quad$if halt $= 0$ **then**
$\quad\quad$**output** (inv, $0, A, \lambda, C$) to **URI**
$\quad\quad$let M be return value from **URI**
$\quad\quad$if $M \neq \perp$ **and** $v(M)$ **then**
$\quad\quad\quad$**output** (A, M) at out
$\quad\quad$**else**
$\quad\quad\quad$**output** $\perp$ at out
$\quad\quad\quad$halt $\leftarrow 1$

Fig. 9. The converters for the sender (left) and the receiver (right).

Converter sim$_{\text{RASC}}$

Initialization
$\mathcal{Q} \leftarrow$ empty FIFO queue
$\mathcal{S} \leftarrow [\ ]$

Interface in
Input: $(A, \ell, i) \in \mathcal{A} \times \mathbb{N}_{>0} \times \mathbb{N}_{>0}$
$\quad$if $i = |\mathcal{S}| + 1$ **then**
$\quad\quad C \leftarrow \Sigma^{\ell+\lambda}$
$\quad$**else**
$\quad\quad C \leftarrow \mathcal{S}[i]$
$\quad \mathcal{S} \leftarrow \mathcal{S} \parallel (A, C)$
$\quad \mathcal{Q}$.enqueue$((A, C))$
$\quad$**output** (A, C) at out

Interface out
Input: (inject, $A, C) \in \mathcal{A} \times \mathcal{C}$
$\quad$if $\exists n \in \mathbb{N} : \mathcal{S}[n] = (A, C)$ **then**
$\quad\quad i \leftarrow \min\{n \in \mathbb{N} \mid \mathcal{S}[n] = (A, C)\}$
$\quad\quad$**output** (deliver, i) at in
$\quad$**else**
$\quad\quad$**output** terminate at in

Input: deliver
$\quad$if $|\mathcal{Q}| > 0$ **then**
$\quad\quad (A, C) \leftarrow \mathcal{Q}$.dequeue()
$\quad\quad$execute instructions for (Inject, A, C)

Fig. 10. Simulator for the security condition of the construction of **RASC$_{\mathcal{M}_v}$**.

where we used $\frac{x-1}{y-1} \leq \frac{x}{y}$ for $x \leq y$ in the first step, and the definition of d_v in the last step. $\square$

6   Guarantees for Nonce-Reuse

One goal of robust authenticated encryption is to provide resilience to the misuse when nonces are repeated. While the expected security loss was only informally stated in [6], we rigorously derive the exact guarantees that can still be expected in such a scenario. To this end, we consider the extreme case where the nonce is a constant value.

Repetition ASC. The channel that is achieved if the nonce is repeating is denoted $\mathbf{RASC}_{\mathcal{M}_v}$ and its description is given in Fig. 8. There are two differences to $\mathbf{ASC}_{\mathcal{M}_v}$: First, not only the length of the message is leaked at interface E but also the number i of the first transmitted message that equals the current message. This leaks the repetition pattern of transmitted values. Second, the adversary can replay messages and induce arbitrary out-of-order delivery.

Protocol. The protocol, which we denote by $(\mathsf{rsnd}, \mathsf{rrcv})$, invokes **URI** using the constant nonce 0. Furthermore, the protocol verifies that all messages are from the set $\mathcal{M}_v$. The protocol is specified in Fig. 9.

Theorem 3. *Let* $\lambda \in \mathbb{N}$. *The protocol* $(\mathsf{rsnd}_\lambda, \mathsf{rrcv}_\lambda)$ *constructs* $\mathbf{RASC}_{\mathcal{M}_v}$ *from* $[\mathbf{URI}, \mathbf{IC}]$ *with respect to* $(\mathsf{dlv}, \mathsf{dlv})$ *and simulator* $\mathsf{sim}_{\mathrm{RASC}}$ *defined in Fig. 10. More specifically, we have for all distinguishers* $\mathbf{D}$

$$\Delta^{\mathbf{D}}\left(\mathsf{rsnd}_\lambda^{\mathsf{A}}\mathsf{rrcv}_\lambda^{\mathsf{B}}\mathsf{dlv}^{\mathsf{E}}[\mathbf{URI}, \mathbf{IC}], \mathsf{dlv}^{\mathsf{E}}\mathbf{RASC}_{\mathcal{M}_v}\right) \leq \frac{2d_v + q \cdot (q-1)}{2} \cdot |\Sigma|^{-\lambda} \quad (5)$$

and

$$\Delta^{\mathbf{D}}\left(\mathsf{rsnd}_\lambda^{\mathsf{A}}\mathsf{rrcv}_\lambda^{\mathsf{B}}[\mathbf{URI}, \mathbf{IC}], \mathsf{sim}_{\mathrm{RASC}}^{\mathsf{E}}\mathbf{RASC}_{\mathcal{M}_v}\right) \leq \frac{2d_v + q \cdot (q-1)}{2} \cdot |\Sigma|^{-\lambda}, \quad (6)$$

where q *is the total number of inputs made by* $\mathbf{D}$.

We prove Theorem 3 in the full version of this paper.

Acknowledgments. Ueli Maurer was supported by the Swiss National Science Foundation (SNF), project no. 200020-132794. Björn Tackmann was supported by the Swiss National Science Foundation (SNF) via Fellowship no. P2EZP2_155566 and the NSF grants CNS-1228890 and CNS-1116800. Much of the work on this paper was done while Phil Rogaway was visiting Ueli Maurer's group at ETH Zurich. Many thanks to Ueli for hosting that sabbatical. Rogaway was also supported by NSF grants CNS-1228828 and CNS-1314885

References

1. Badertscher, C., Matt, C., Maurer, U., Rogaway, P., Tackmann, B.: Augmented secure channels as the goal of the TLS 1.3 record layer. Cryptology ePrint Archive, report 2015/394 (2015)
2. Bellare, M., Namprempre, C.: Authenticated encryption: relations among notions and analysis of the generic composition paradigm. In: Okamoto, T. (ed.) ASIACRYPT 2000. LNCS, vol. 1976, pp. 531–545. Springer, Heidelberg (2000)
3. Bellare, M., Rogaway, P.: Encode-then-encipher encryption: how to exploit nonces or redundancy in plaintexts for efficient cryptography. In: Okamoto, T. (ed.) ASIACRYPT 2000. LNCS, vol. 1976, pp. 317–330. Springer, Heidelberg (2000)
4. Bellare, M., Rogaway, P., Wagner, D.: The EAX mode of operation. In: Roy, B., Meier, W. (eds.) FSE 2004. LNCS, vol. 3017, pp. 389–407. Springer, Heidelberg (2004)
5. Gligor, V.D., Donescu, P.: Fast encryption and authentication: XCBC encryption and XECB authentication modes. In: Matsui, M. (ed.) FSE 2001. LNCS, vol. 2355, pp. 92–108. Springer, Heidelberg (2002)
6. Hoang, V.T., Krovetz, T., Rogaway, P.: Robust authenticated-encryption AEZ and the problem that it solves. In: Oswald, E., Fischlin, M. (eds.) EUROCRYPT 2015. LNCS, vol. 9056, pp. 15–44. Springer, Heidelberg (2015)
7. Jutla, C.S.: Encryption modes with almost free message integrity. In: Pfitzmann, B. (ed.) EUROCRYPT 2001. LNCS, vol. 2045, pp. 529–544. Springer, Heidelberg (2001)
8. Katz, J., Yung, M.: Unforgeable encryption and chosen ciphertext secure modes of operation. In: Schneier, B. (ed.) FSE 2000. LNCS, vol. 1978, pp. 284–299. Springer, Heidelberg (2001)
9. Maurer, U.: Constructive cryptography – a new paradigm for security definitions and proofs. In: Mödersheim, S., Palamidessi, C. (eds.) TOSCA 2011. LNCS, vol. 6993, pp. 33–56. Springer, Heidelberg (2012)
10. Maurer, U., Renner, R.: Abstract cryptography. In: Chazelle, B. (ed.) The Second Symposium on Innovations in Computer Science, ICS 2011, pp. 1–21. Tsinghua University Press, Beijing (2011)
11. Maurer, U., Rüedlinger, A., Tackmann, B.: Confidentiality and integrity: a constructive perspective. In: Cramer, R. (ed.) TCC 2012. LNCS, vol. 7194, pp. 209–229. Springer, Heidelberg (2012)
12. Rogaway, P.: Authenticated-encryption with associated-data. In: Proceedings of the 9th ACM Conference on Computer and Communications Security, pp. 98–107. ACM (2002)
13. Rogaway, P., Bellare, M., Black, J.: OCB: a block-cipher mode of operation for efficient authenticated encryption. ACM Trans. Inf. Syst. Secur. (TISSEC) 6(3), 365–403 (2003)
14. Rogaway, P., Shrimpton, T.: A provable-security treatment of the key-wrap problem. In: Vaudenay, S. (ed.) EUROCRYPT 2006. LNCS, vol. 4004, pp. 373–390. Springer, Heidelberg (2006)

2-Party Computation

A Compiler of Two-Party Protocols
for Composable and Game-Theoretic Security,
and Its Application to Oblivious Transfer

Shota Goto[1]([✉]) and Junji Shikata[1,2]

[1] Graduate School of Environment and Information Sciences,
Yokohama National University, Yokohama, Japan
`goto-shota-zx@ynu.jp, shikata@ynu.ac.jp`
[2] Institute of Advanced Sciences, Yokohama National University, Yokohama, Japan

Abstract. In this paper, we consider the following question: Does composing protocols having game-theoretic security result in a secure protocol in the sense of game-theoretic security? In order to discuss the composability of game-theoretic properties, we study security of cryptographic protocols in terms of the universal composability (UC) and game theory simultaneously. The contribution of this paper is the following: (i) We propose a compiler of two-party protocols in the local universal composability (LUC) framework such that it transforms any two-party protocol secure against semi-honest adversaries into a protocol secure against malicious adversaries in the LUC framework; (ii) We consider the application of our compiler to oblivious transfer (OT) protocols, by which we obtain a construction of OT meeting both UC security and game-theoretic security.

1 Introduction

1.1 Background

In recent years, game-theoretic security of cryptographic protocols has been studied. Generally, cryptographic security is defined so as to guarantee some basic concrete properties when participants follow the designed algorithms, even if facing an adversarial behavior. In contrast, game-theoretic security is defined such that, by considering behaviors of rational participants in a protocol whose goal is to achieve their best satisfactions, following the specifications of the protocol honestly is the most reasonable for the rational participants. This security notion enables us to design protocols more realistically. In this way, these concepts capture situations from different perspectives and it seems that there is great difference between the cryptographic security and game-theoretic security. Up to date, there are several works aiming at bridging the gaps between the two kinds of security [9,13,17–19]. Recently, Asharov et al. [4] studied two-party protocols in the fail-stop model in terms of game-theoretic security and showed how the notion of game theory can capture cryptographic properties.

© Springer International Publishing Switzerland 2015
J. Groth (Ed.): IMACC 2015, LNCS 9496, pp. 133–151, 2015.
DOI: 10.1007/978-3-319-27239-9_8

Furthermore, the game-theoretic security for oblivious transfer [14] and bit commitment [15] has been studied in the malicious model.

In addition to cryptographic security and game-theoretic security, composable security has also been studied in order to guarantee security of protocols even if they are composed with other ones. The previous frameworks of this line of research are based on the ideal-world/real-world paradigm, and the paradigm underlies universal composability (UC) by Canetti [6] and reactive simulatability by Backes, Pfitzmann and Waidner [5]. In addition, a simple paradigm for composable security was given by Maurer [21], and this approach is called constructive cryptography. In this paper, we utilize the UC framework [6] to consider composable security of cryptographic protocols, since this approach has been utilized in discussing composability of protocols in many papers.

In this paper, we consider the following question: Does composing protocols having game-theoretic security result in a *secure* protocol in the sense of game-theoretic security? In order to discuss the composability of game-theoretic properties, we need to consider protocols having both universally composable (UC) and game-theoretic security. Although the UC framework achieves guarantee of composability of protocols, the framework models the attacker as a centralized entity so that it can capture only the situation that the attacker is like a dictator and corrupted parties are all cooperative. For this reason, some other formalisms have been proposed in [1–3,8,16,20,22]. In these formalizations, the centralized adversary is shattered to plural adversaries and each of them is limited to obtain only local information. This modeling seems to be more realistic than existing ones and can capture many settings that are not captured by centralized adversary approach. In particular, we focus on the local universal composability (LUC) framework in [8] in this paper, and we try to answer the question mentioned above.

1.2   Our Approach

In this paper, we study security of cryptographic protocols in terms of composability and game theory simultaneously. In particular, we consider realizing a compiling mechanism which transforms a protocol that is not game-theoretically secure into a protocol that achieves the composable and game-theoretic security. Although the UC framework is a powerful theory to consider composability of protocols, it cannot cover game-theoretic security since the UC framework considers a centralized adversary and cannot deal with protocols as games among plural rational participants. However, if we switch the framework to the local universal composability (LUC) framework [8], we can analyze protocols in terms of game-theoretic security by clarifying which strategy is in Nash equilibrium.

Besides the LUC framework, there is also a well-established framework with a composition theorem and an application to game theory, called collusion-preserving (CP) framework [2]. The reason for our choice of the LUC framework over the CP framework is that, the compiler of two-party protocols which we focus on in this paper was originally proposed on the basis of the UC framework [7]. On that point, choosing the LUC framework whose modeling is a direct extension of the UC framework enables us to discuss the whole aspect of the compiler simply and similarly to the case of UC.

Furthermore, we refer to a connection between the LUC framework and game-theoretic security. At first sight, one may think that these two notions are not well connected, since there is a difference in the requirement for security definitions: game-theoretic security requires that all participants can get the highest utility when each of them acts honestly, while LUC security requires the indistinguishability between the real-world and the ideal-world. However, there is an important point common to these two notions, namely, all participants are allowed to behave in a malicious (or rational) way. Considering this point, if we define an ideal functionality in the LUC framework accurately so that it captures the correct actions which each participant should essentially take, LUC security will satisfy the desirable property that following the protocol specifications honestly is the most reasonable for rational participants. However, in general, defining an ideal functionality in such a reasonable way may be a hard work, if we target complicated protocols where participants communicate intricately. If we can do it in such a way, we can say that LUC security implies game-theoretic security. As an illustration, in this paper we target oblivious transfer (OT), since its functionality is traditional and relatively simple. Specifically, we explicitly formalize the functionality of OT in the LUC framework in a way mentioned above, and our resulting OT protocol will be proven to be secure even in terms of game-theoretic security.

1.3   Our Results

(i) A Compiler for Both UC Security and Game-Theoretic Security.
First, we propose a compiler of two-party protocols in the LUC framework such that it transforms any two-party protocol secure against semi-honest adversaries into a protocol secure against malicious adversaries in the LUC framework. Our compiler is constructed based on the compiler of [7] in the UC framework. In other words, we try to adapt the compiler of [7] to the LUC framework. For doing it, we define a commit-and-prove functionality, denoted by $\hat{\mathcal{F}}_{CP}$, which is a slight modification of the commit-and-prove functionality $\mathcal{F}_{CP}$ in the UC framework. And, we show that the compiled protocol is secure against malicious adversaries in the $\hat{\mathcal{F}}_{CP}$-hybrid model in the LUC framework (in Theorem 1 in Sect. 3.2).

(ii) Application of the Compiler to Oblivious Transfer.
Second, we consider the application of our compiler to oblivious transfer (OT) protocols. Since, OT is an important primitive for secure multi-party computation, it is worth exploring a practical construction. In particular, we consider the construction of the OT protocol, denoted by SOT, in [7,11,12] which UC-realizes the OT functionality in static and semi-honest adversarial model. For the protocol SOT, we show that:

(1) SOT LUC-realizes $\hat{\mathcal{F}}_{OT}$ in the presence of semi-honest and static adversaries, where $\hat{\mathcal{F}}_{OT}$ is the OT functionality in the LUC framework (in Theorem 2 in Sect. 4.2);

(2) SOT is not game-theoretically secure in the presence of rational parties (in Theorem 3 in Sect. 4.3); and

(3) The compiled protocol of SOT by our compiler is game-theoretically secure in the presence of rational parties (in Theorem 3 in Sect. 4.3).

Since the functionality of OT is relatively simple, we will be able to define it in the LUC framework so that (3) follows from (i) and (1). However, we directly prove (3) in terms of the game theory in order to confirm that the compiled protocol of SOT actually meets game-theoretic security, and the analysis of the compiled protocol from the viewpoint of the game theory enables us to see how Nash equilibrium is achieved in it.

2   Preliminaries

2.1   Framework of Universally Composable Security

In this section, we provide an overview of the universal composability framework (UC framework for short) in [6]. This framework allows us to define the security properties of given tasks, as follows. First, the process of executing a protocol π with a realistic adversary $\mathcal{A}$ is formalized. Next, an ideal process with a simulator $\mathcal{S}$ is formalized. In this process, the parties hand their inputs to a trusted party that is programmed to capture the appropriate functionality and obtain their outputs from it with no interaction. A protocol is said to securely realize an ideal functionality $\mathcal{F}$ if the process of executing the protocol amounts to emulating the ideal process. Formally, there is an environment $\mathcal{Z}$ whose task is to distinguish these two worlds. We refer to [6] for a complete overview of this framework, as well as the definition of the real-world ensemble $\text{REAL}_{\pi,\mathcal{A},\mathcal{Z}}$ and the ideal-world ensemble $\text{IDEAL}_{\mathcal{F},\mathcal{S},\mathcal{Z}}$, and also the composition theorem. In this paper, we only write down the basic definitions due to lack of spaces.

Definition 1. *Two binary distribution ensembles X and Y are computationally indistinguishable (written as $X \overset{c}{\approx} Y$), if for any $c \in \mathbf{N}$ there exists $k_0 \in \mathbf{N}$ such that for all $k > k_0$ and for all a we have $|\Pr(X(k,a) = 1) - \Pr(Y(k,a) = 1)| < k^{-c}$.*

Definition 2. *Let $n \in \mathbf{N}$. Let $\mathcal{F}$ be an ideal functionality and let π be an n-party protocol. We say that π UC-realizes $\mathcal{F}$, if for any adversary $\mathcal{A}$ there exists an ideal-process simulator $\mathcal{S}$ such that for any environment $\mathcal{Z}$, $\text{IDEAL}_{\mathcal{F},\mathcal{S},\mathcal{Z}} \overset{c}{\approx} \text{REAL}_{\pi,\mathcal{A},\mathcal{Z}}$.*

In general, a protocol is designed by considering a model of adversaries, which depends on to what extent the designer wants to achieve security against the adversaries. We outline the model of corruptions and adversarial behaviors as follows.

1. The model of corruptions.
   (a) **Static corruption model.** The set of parties who are to be corrupted by an adversary is fixed at the beginning of the computation and no more corruptions will be happen after that.
   (b) **Adaptive corruption model.** In contrast to the static corruption model, an adversary is allowed to corrupt parties at any time throughout the computation.
2. The model of adversarial behaviors.
   (a) **Semi-honest adversarial model.** Even if parties are corrupted, they follow the specification of the protocol. Therefore, the adversary is restricted only to get read access to the states of corrupted parties.
   (b) **Malicious adversarial model.** Once the adversary corrupts parties, they follow all the instruction of the adversary. In particular, the adversary can make the corrupted parties deviate from the specification of the protocol.

2.2 Framework of Local Universally Composable Security

The notion of the local universal composability (LUC for short) was proposed by Canetti and Vald [8]. Roughly speaking, instead of setting a single adversary as in the UC framework, there can be plural local adversaries who can corrupt only a single party according to their party IDs. In the ideal process, the simulator is also shattered to plural local simulators, therefore, the simulation is done by relying only on each entity's local information. We describe the LUC model of protocol execution as follows, and aside from some modifications, the underlying computational model is identical to the UC model.

Protocol Execution in the LUC Framework. At first, a set $\mathcal{P}$ of party IDs and session ID, denoted by pid and sid respectively, are chosen by the environment. This is different from the UC model where the party IDs can be chosen arbitrarily during protocol execution. Next, the adversaries are invoked with identity $id = ((i, j), \bot)$ and denoted by $\mathcal{A}_{(i,j)}$ for ordered pairs $(i, j) \in \mathcal{P}^2$. The purpose of this modeling is to capture locality properly. Each local adversary comes to take charge of a different side of the communication line, and can interfere with the parties' communication only via this line. This means that the centralized adversary no longer exists and many situations, in real-life, where an adversary can only rely on restricted information are capturable.

Once an adversary $\mathcal{A}_{(i,j)}$ is activated, it can send information to $\mathcal{Z}$ or deliver a message to a party with $pid = i$ where the sender's pid must be j. The adversary is also allowed to corrupt parties with $pid = i$ throughout the computation. An important point is that adversaries cannot communicate each other directly and their communications must be done through the environment $\mathcal{Z}$ (or an ideal functionality if any). This formalization enables us to represent different subsets of adversaries, if there exists a trusted party (an ideal functionality in the hybrid model) and it provides a specific communication interface.

Once a party is activated, it basically follows its code and may write an output on its output tape or send a message to the adversary where the pid of the adversary must be sender's $pid = i$ and receiver's $pid = j$. As in the UC model, the protocol execution ends when the environment halts. Let $\mathrm{LREAL}_{\pi,\mathcal{A},\mathcal{Z}}(k,z)$ be a random variable taking the output of $\mathcal{Z}$ and $\mathrm{LREAL}_{\pi,\mathcal{A},\mathcal{Z}}$ be the ensemble as in the UC model. The random variable $\mathrm{LIDEAL}_{\mathcal{F},\mathcal{S},\mathcal{Z}}(k,z)$ and the ensemble $\mathrm{LIDEAL}_{\mathcal{F},\mathcal{S},\mathcal{Z}}$ in the ideal process are defined as well. Then, we have the following definition in the LUC model as well as that of the UC model.

Definition 3 ([8]). *Let $n \in \mathbf{N}$. Let $\mathcal{F}$ be an ideal functionality and let π be an n-party protocol. We say that π LUC-realizes $\mathcal{F}$, if for every adversary $\mathcal{A}$ there exists an ideal-process simulator $\mathcal{S}$ such that for any environment $\mathcal{Z}$,*
$$\mathrm{LIDEAL}_{\mathcal{F},\mathcal{S},\mathcal{Z}} \overset{c}{\approx} \mathrm{LREAL}_{\pi,\mathcal{A},\mathcal{Z}}$$

3   A Compiler in the LUC Framework

In this section, we analyze the protocol-compiler of [7] (i.e., the compiler in the UC model) in the LUC framework. At the beginning, we describe it, and then we point out that it does not work well in the LUC framework in general, and show a condition that it works well even in LUC framework.

3.1   Previous Compiler in the UC Framework

In order to transform a protocol into one that is secure against malicious adversaries, it is necessary to enforce malicious corrupted parties to follow the prescribed protocol in a semi-honest way. Canetti et al. [7] proposed a universally composable compiler based on the work of [12]. The compiler uses the commit-and-prove functionality $\mathcal{F}_{CP}$ which is defined so that only correct statements are received by a receiver and incorrect statements are rejected. In a nutshell, the committer commits its input value w as a witness and forwards a statement x to the verifier by using $\mathcal{F}_{CP}$. The statement x is received by the verifier only when $R(x,w)$ holds, where R is a predetermined relation. In the compiled protocol, there are two copies of the functionality, one for the case where P_1 is the committer and the other one for the case where P_2 is the committer, denoted by $\mathcal{F}_{CP}^1$ and $\mathcal{F}_{CP}^2$ respectively, and these are identified by session-identifiers sid_1 and sid_2. The definition of the ideal functionality $\mathcal{F}_{CP}$ and the protocol-compiler Comp() are given as follows (Fig. 1).

Description of Comp($\cdot$): A party P_1 proceeds as follows (the code for a party P_2 is analogous).

1. **Random tape generation.** When activating $\mathrm{Comp}(\pi)$ for a protocol π for the first time with a session-identifier sid, the party P_1 (and P_2) proceeds as follows.
   (a) *Choosing a random tape for P_1.*
      i. P_1 chooses $r_1^1 \in_R \{0,1\}^k$ and sends (**commit**,sid_1,r_1^1) to $\mathcal{F}_{CP}^1$. Then, P_2 receives a (**receipt**, sid_1), and P_2 chooses $r_1^2 \in_R \{0,1\}^k$ and sends (sid, r_1^2) to P_1.

Functionality $\mathcal{F}_{CP}$

$\mathcal{F}_{CP}$, which is running with a committer C, a receiver V and an adversary $\mathcal{S}$, and is parameterized by a value k and a relation R, proceeds as follows:

• **Commit phase.** Upon receiving a message (**commit**, sid, w) from C where $w \in \{0,1\}^k$, append the value w to the list $\overline{w}$, and send the message (**receipt**, sid) to V and $\mathcal{S}$. (Initially, the list $\overline{w}$ is empty.)

• **Prove phase.** Upon receiving a message (**CP-prover**, sid, x) from C where $x \in \{0,1\}^{poly(k)}$, compute $R(x, \overline{w})$: If $R(x, \overline{w}) = 1$, then send V and S the message (**CP-proof**, sid, x). Otherwise, ignore the message.

Fig. 1. The commit-and-prove functionality in the UC model.

ii. When P_1 receives a message (sid, r_1^2) from P_2, it sets $r_1 := r_1^1 \oplus r_1^2$ (r_1 serves as P_1's random tape for execution of π).

(b) *Choosing a random tape for P_2.*

i. P_1 waits to receive a message (**receipt**,sid_2) from $\mathcal{F}_{CP}^2$ (this occurs after P_2 sends a commit message (**commit**,sid_2,r_2^2) to $\mathcal{F}_{CP}^2$). It then chooses $r_2^1 \in_R \{0,1\}^k$ and sends (sid,r_2^1) to P_2.

2. **Activation due to a new input.** When activated with an input (sid,x), the party P_1 proceeds as follows.

(a) *Input commitment.* P_1 sends (**commit**,sid_1,x) to $\mathcal{F}_{CP}^1$ and adds x to the list of inputs $\overline{x}$ (this list is initially empty and contains P_1's inputs from all the previous activations of π).

(b) *Protocol computation.* Let $\overline{m}_1$ be the series of π-messages that P_1 received from P_2 in all the activations of π until now ($\overline{m}_1$ is initially empty). P_1 runs the code of π on its input list $\overline{x}$, messages $\overline{m}_1$, and the random tape r_1 (as generated above).

(c) *Outgoing message transmission.* For any outgoing message m that π instructs P_1 to send to P_2, P_1 sends (**CP-prover**,sid_1,(m, r_1^2, $\overline{m}_1$)) to $\mathcal{F}_{CP}^1$ where the relation R_π for $\mathcal{F}_{CP}^1$ is defined as follows:

$$R_\pi = \{((m, r_1^2, \overline{m}_1), (\overline{x}, r_1^1)) \mid m = \pi(\overline{x}, r_1^1 \oplus r_1^2, \overline{m}_1)\}.$$

In this step, P_1 proves that m is truly the correct message generated by π with the input list $\overline{x}$, the random tape $r_1 = r_1^1 \oplus r_1^2$, and the series of incoming π-messages $\overline{m}_1$.

3. **Activation due to incoming message.** When activated with an incoming message (**CP-proof**, sid_2, (m, r_2^1, $\overline{m}_2$)) from $\mathcal{F}_{CP}^2$, P_1 first verifies that the following conditions hold ($\mathcal{F}_{CP}^2$ is parameterized by the same relation R_π as $\mathcal{F}_{CP}^1$):

(a) r_2^1 is the string that P_1 sent to P_2 in the step of 1-(b)-i above.

(b) $\overline{m}_2$ equals the series of π-messages received by P_2 from P_1 in all the activations until now.

If the conditions do not hold, then P_1 ignores the message. Otherwise, P_1 appends m to its list of incoming π-messages $\overline{m}_1$ and proceeds as in the steps 2-(b) and 2-(c).

4. **Output.** Whenever π generates an output, $\mathrm{Comp}(\pi)$ generates the same output.

In the UC framework, if a protocol π has UC security against semi-honest adversaries, the compiled protocol $\mathrm{Comp}(\pi)$-$\mathcal{F}_{CP}$ is proved to be secure against malicious adversaries. The proof can be shown by the following steps: First, let $\mathcal{A}$ be a malicious adversary against the compiled protocol $\mathrm{Comp}(\pi)$ in the $\mathcal{F}_{CP}$-hybrid model and let $\mathcal{A}'$ be a semi-honest adversary against the plain protocol π, then $\mathcal{A}'$ runs a simulated copy of $\mathcal{A}$ internally and interacts with π. Recall that $\mathcal{A}'$ follows the specification of the protocol π, on the other hand, $\mathcal{A}$ is allowed to behave arbitrarily. However, the malicious adversary $\mathcal{A}$ cannot cheat since each message sent throughout the protocol is verified by $\mathcal{F}_{CP}$ so that it has no choice but to behave in a semi-honest manner. For this reason, the semi-honest adversary $\mathcal{A}'$ can simulate the behavior of $\mathcal{A}$ by delivering a message only when $\mathcal{A}$ sends a correct message. In other words, from the view of the environment $\mathcal{Z}$, it is impossible to distinguish whether it is interacting with $\mathrm{Comp}(\pi)$ and $\mathcal{A}$ in the $\mathcal{F}_{CP}$-hybrid model, or with the plain protocol π and $\mathcal{A}'$. In the circumstances, it is shown that the compiled protocol UC-realizes the target functionality in the malicious model.

3.2   A Compiler in the LUC Framework

To utilize the compiler $\mathrm{Comp}(\cdot)$ in the LUC framework, we need to similarly complete the simulation mentioned in the previous subsection even in the LUC framework. However, we cannot do that without any modification on the existing process. The reason of this impossibility lies in the difference between the models of UC and LUC. In the UC model, communications between parties are mediated by the centralized adversary and it directly delivers a message to recipients. In contrast, in the LUC model, plural adversaries mediate communications and messages are supposed to go through the environment in the process. That means the environment $\mathcal{Z}$ can tell whether parties communicate each other through the ideal functionality, since if messages were delivered by the ideal functionality, they would not go through the environment. Therefore, by focusing on this point, the simulation will be distinguishable.

Based on the above point, we consider switching the interacting process of an original protocol π from the one totally controlled by the environment to the one which uses a subroutine so that the problem does not occur. Specifically, we consider a message transmission functionality, denoted by $\hat{\mathcal{F}}_{MT}$, below. Note that this functionality can be realized in the LUC framework, though the functionality is originally considered in the UC framework [6] (Fig. 2).

Subsequently, we show an adjusting point in regards to the commit-and-prove functionality required for constructing the protocol compiler. In the UC model, we can use the ideal functionality $\mathcal{F}_{CP}$ since it has been proved that there exists a protocol which UC-realizes it. However, to use such a functionality in the LUC model, we first need to show an existence of a protocol which LUC-realizes it. In this paper, we adopt the notion of the merger functionality in [8].

Functionality $\hat{\mathcal{F}}_{MT}$

$\hat{\mathcal{F}}_{MT}$, which is running with parties $P_1, \ldots, P_n$ and adversaries $\mathcal{S}_{(i,j)}$, where $(i,j) \in \mathcal{P}^2$ $(i \neq j)$ and $\mathcal{P}$ is the set of identities, proceeds as follows:
- Upon receiving a message (**Send**, sid, m, P_j) from a party P_i, send a public delayed output (**Send**, sid, m, P_i) to the party P_j.
- Upon receiving a message (**Deliver**, m, (l,k)) from an adversary $\mathcal{S}_{(i,j)}$, where $l,k,i,j \in \mathcal{P}$, send the message (**Delivered**, m, (i,j)) to the adversary $\mathcal{S}_{(l,k)}$.

Fig. 2. The message transmission functionality in the LUC model.

In short, we modify the functionality of [7] (in the UC model) artificially so that the protocol will LUC-realize the resulting functionality. Unfortunately, this modification allows adversaries to communicate freely when the modified ideal functionality, denoted by $\hat{\mathcal{F}}_{CP}$, is used as a subroutine of other protocols (hybrid model). Concerning this point, if we demand collusion-freeness for designing protocols, we cannot adopt this method. However, such a property is not needed in this work. Generally, in two-party protocols, if both parties are corrupted by the corresponding adversaries respectively and they coordinate their actions, the mechanism of protocol compiler seems to be totally unnecessary. Considering that, we should focus on the case where both adversaries are not cooperative. (The situation either P_1 or P_2 is corrupted by the corresponding adversary can be covered by the protocol compiler in the UC model, however, the situation both parties are corrupted by different adversaries cannot be covered except if we consider it in the LUC framework.) Therefore, the most important point is whether the simulation can be completed in this framework. First, we propose an ideal functionality $\hat{\mathcal{F}}_{CP}$ as follows (Fig. 3).

Then, we can show the following results.

Functionality $\hat{\mathcal{F}}_{CP}$

$\hat{\mathcal{F}}_{CP}$, which is running with a committer C, a receiver V and adversaries $\mathcal{S}_{(C,V)}$ and $\mathcal{S}_{(V,C)}$, and being parameterized by a value k and a relation R, proceeds as follows:
- **Commit phase.** Upon receiving a message (**commit**, sid,w) from C, where $w \in \{0,1\}^k$, append the value w to the list $\overline{w}$, and send a public delayed output (**receipt**, sid) to V. (Initially, the list $\overline{w}$ is empty.)
- **Prove phase.** Upon receiving a message (**CP-prover**, sid,x) from C where $x \in \{0,1\}^{poly(k)}$, compute $R(x,\overline{w})$; If $R(x,\overline{w}) = 1$, then send a public delayed output (**CP-proof**, sid,x) to V; Otherwise, ignore the message.
- Upon receiving a message (**Deliver**, m, (j,i)) from the adversary $\mathcal{S}_{(i,j)}$, if $\mathcal{S}_{(i,j)}, \mathcal{S}_{(j,i)} \in \{\mathcal{S}_{(C,V)}, \mathcal{S}_{(V,C)}\}$, send the message (**Delivered**, m, (i,j)) to the adversary $\mathcal{S}_{(j,i)}$.

Fig. 3. The commit-and-prove functionality in the LUC model.

Theorem 1. *Let π be a two-party protocol and let* $\mathrm{Comp}(\pi\text{-}\hat{\mathcal{F}}_{MT})$ *be the protocol obtained by applying the compiler to π in the $\hat{\mathcal{F}}_{MT}$-hybrid model. Then, for every malicious adversary $\mathcal{A}$ that interacts with* $\mathrm{Comp}(\pi\text{-}\hat{\mathcal{F}}_{MT})$ *in the $\hat{\mathcal{F}}_{CP}$-hybrid model there exists a semi-honest adversary $\mathcal{A}'$ that interacts with $\pi\text{-}\hat{\mathcal{F}}_{MT}$, such that for every environment $\mathcal{Z}$,*

$$\mathrm{LREAL}_{\pi\text{-}\hat{\mathcal{F}}_{MT},\mathcal{A}',\mathcal{Z}} \equiv \mathrm{LEXEC}^{\hat{\mathcal{F}}_{CP}}_{\mathrm{Comp}(\pi\text{-}\hat{\mathcal{F}}_{MT}),\mathcal{A},\mathcal{Z}}.$$

Proof. Let $\mathcal{A}'_{(1,2)}$ be an adversary for P_1's side and $\mathcal{A}'_{(2,1)}$ be an adversary for P_2's side. As in the UC case, $\mathcal{A}'_{(1,2)}$ and $\mathcal{A}'_{(2,1)}$ run a simulated copy of $\mathcal{A}_{(1,2)}$ and $\mathcal{A}_{(2,1)}$ respectively, and their actions are utilized as a guide for the interaction with $\pi\text{-}\hat{\mathcal{F}}_{MT}$ and $\mathcal{Z}$. We regard the communications of $\mathcal{A}'_{(1,2)}$ and $\mathcal{A}'_{(2,1)}$ with $\mathcal{Z}$ and $\pi\text{-}\hat{\mathcal{F}}_{MT}$ as an external communication, and the communications of $\mathcal{A}'_{(1,2)}$ and $\mathcal{A}'_{(2,1)}$ with the corresponding simulated $\mathcal{A}'_{(1,2)}$ or $\mathcal{A}'_{(2,1)}$ as an internal communication. $\mathcal{A}'_{(1,2)}$ and $\mathcal{A}'_{(2,1)}$ proceed as follows.

- **Simulating the communication with $\mathcal{Z}$.** Every input coming from $\mathcal{Z}$ is sent to the corresponding simulated adversary $\mathcal{A}_{(1,2)}$ or $\mathcal{A}_{(2,1)}$ as if coming from their own environment. In the same way, every output from internal adversaries is treated as an output of corresponding simulator.
- **Simulating the random tape generation phase.** We consider the following cases below.

1. **Both parties are honest:** We describe the simulation for the P_1's random tape generation (the simulation for P_2 is analogous). $\mathcal{A}'_{(1,2)}$ begins by passing the message (**receipt**, sid_1) to $\mathcal{A}_{(1,2)}$ as if coming from $\hat{\mathcal{F}}^1_{CP}$, and after $\mathcal{A}_{(1,2)}$ approved, $\mathcal{A}'_{(1,2)}$ delivers this message to $\mathcal{A}'_{(2,1)}$ using $\hat{\mathcal{F}}_{MT}$. Similarly, $\mathcal{A}'_{(2,1)}$ passes the message (**receipt**, sid_1) to $\mathcal{A}_{(2,1)}$ and if it approves, then chooses a random r^2_1 and passes the value to $\mathcal{A}_{(2,1)}$ as if coming from P_2. Furthermore, confirming that $\mathcal{A}_{(2,1)}$ delivers this value to P_1 using $\hat{\mathcal{F}}_{MT}$, $\mathcal{A}'_{(2,1)}$ actually delivers it to $\mathcal{A}'_{(1,2)}$ using $\hat{\mathcal{F}}_{MT}$. Finally, $\mathcal{A}'_{(1,2)}$ receives an approval from $\mathcal{A}_{(1,2)}$.
2. **P_1 is honest and P_2 is corrupted:** At first, we consider the generation of P_1's random tape. The simulation proceeds as in the case 1. $\mathcal{A}'_{(2,1)}$ receives the message (**receipt**, sid_1), then passes it to $\mathcal{A}_{(2,1)}$. If $\mathcal{A}_{(2,1)}$ delivers r^2_1 to P_2 using $\hat{\mathcal{F}}_{MT}$, $\mathcal{A}'_{(2,1)}$ actually delivers it to $\mathcal{A}'_{(1,2)}$ using $\hat{\mathcal{F}}_{MT}$. The rest of the process is the same as in the case 1. Next, we consider the generation of P_2's random tape. $\mathcal{A}'_{(2,1)}$ obtains the message (**commit**, sid_2, r^2_2) from $\mathcal{A}_{(2,1)}$ which sends it to $\hat{\mathcal{F}}^2_{CP}$ on behalf of P_2 in execution of $\mathrm{Comp}(\pi\text{-}\hat{\mathcal{F}}_{MT})$. Now, as the direction of this simulation, we must let the random tape of internal P_2 equal the random tape of P_2 in external execution of $\pi\text{-}\hat{\mathcal{F}}_{MT}$ so that $\mathcal{A}_{(2,1)}$ is forced to use the same randomness throughout the computation. For that reason, $\mathcal{A}'_{(2,1)}$ delivers the random tape of external P_2, denoted by r_2, to

$\mathcal{A}'_{(1,2)}$ using $\hat{\mathcal{F}}_{MT}$. Then $\mathcal{A}'_{(1,2)}$ simulates $\mathcal{A}_{(1,2)}$'s behavior as in the case 1, and then delivers $r_2^1 = r_2$ to $\mathcal{A}'_{(2,1)}$. Finally, $\mathcal{A}'_{(2,1)}$ sets $r_2^1 = r_2 \oplus r_2^2$ and passes it to $\mathcal{A}_{(2,1)}$.

3. **P_1 is corrupted and P_2 is honest:** This case can be simulated analogously to the previous one. That is, the random tape of internal P_1 corrupted by $\mathcal{A}_{(1,2)}$ becomes to be equal to that of external P_1 corrupted by $\mathcal{A}'_{(1,2)}$.

4. **Both parties are corrupted:** Similarly, this case can be simulated by applying simultaneously the simulators of the cases 2 and 3 above.

– **Simulating an activation due to a new input.** We describe the simulation from P_1's side (the simulation for P_2 is analogous).

1. **P_1 is not corrupted:** $\mathcal{A}'_{(1,2)}$ learns the fact that external P_1 is given a new input when it receives an approval request from $\hat{\mathcal{F}}_{MT}$. Then, $\mathcal{A}'_{(1,2)}$ passes the message (**receipt**, sid_1) to $\mathcal{A}_{(1,2)}$ as if coming from $\hat{\mathcal{F}}^1_{CP}$ and after receiving an approval from $\mathcal{A}_{(1,2)}$, $\mathcal{A}'_{(1,2)}$ delivers the same message to $\mathcal{A}'_{(2,1)}$ using $\hat{\mathcal{F}}_{MT}$. Subsequently, $\mathcal{A}'_{(2,1)}$ proceeds the rest process by checking whether $\mathcal{A}'_{(2,1)}$ approves or not.

2. **P_1 is corrupted:** $\mathcal{A}'_{(1,2)}$ receives the message (**commit**, sid_1, x) from $\mathcal{A}_{(1,2)}$. Then, $\mathcal{A}'_{(1,2)}$ adds x to its list $\overline{x}$ and passes (**receipt**, sid_1) to $\mathcal{A}_{(1,2)}$, as if coming from $\hat{\mathcal{F}}^1_{CP}$. After receiving an approval from $\mathcal{A}_{(1,2)}$, $\mathcal{A}'_{(1,2)}$ sets the input tape of external P_1 being equal to x (Note that a semi-honest adversary is allowed to modify the input values of corrupted parties, which is mentioned in [7] and this definition is due to the fact that there is no difference in terms of security between the case where the semi-honest adversary can modify a corrupted party's input value and the case where it cannot). Furthermore, $\mathcal{A}'_{(1,2)}$ delivers (**receipt**, sid_1) to $\mathcal{A}'_{(2,1)}$ using $\hat{\mathcal{F}}_{MT}$ and $\mathcal{A}'_{(2,1)}$ proceeds the rest process as in the case 1.

– **Dealing with π-$\hat{\mathcal{F}}_{MT}$ messages sent externally by uncorrupted parties.** When external P_1 who is not corrupted sends a message m to P_2 using $\hat{\mathcal{F}}_{MT}$, $\mathcal{A}'_{(1,2)}$ internally passes $\mathcal{A}_{(1,2)}$ the message (**CP-proof**, sid_1, $(m, r_1^2, \overline{m}_1)$) as $\mathcal{A}_{(1,2)}$ expects to receive from $\hat{\mathcal{F}}^1_{CP}$), where r_1^2 is the value used in the P_1's random tape generation phase, and $\overline{m}_1$ is the series of all messages P_1 received in the execution of π-$\hat{\mathcal{F}}_{MT}$ so far. Similarly, if P_2 sends a message m to P_1, $\mathcal{A}'_{(2,1)}$ would pass the message (**CP-proof**, sid_2, $(m, r_2^1, \overline{m}_2)$) to $\mathcal{A}_{(2,1)}$. Each simulator delivers a message to the recipients only when the internal adversary approves of the message delivery.

– **Dealing with $\text{Comp}(\pi$-$\hat{\mathcal{F}}_{MT})$ messages sent internally by corrupted parties.** Consider the case where P_1 is corrupted. When $\mathcal{A}_{(1,2)}$ sends the message (**CP-prover**, sid_1, $(m, r_1'^2, \overline{m}_1')$) to $\hat{\mathcal{F}}^1_{CP}$, $\mathcal{A}'_{(1,2)}$ can verify that $\overline{m}_1' = \overline{m}_1$ and $r_1^2 = r_1'^2$, besides, $m = \pi$-$\hat{\mathcal{F}}_{MT}(\overline{x}, r_1^1 \oplus r_1^2, \overline{m}_1)$, since P_1 is corrupted so that $\mathcal{A}'_{(1,2)}$ can obtain all the information needed for these checking. If no error is found, $\mathcal{A}'_{(1,2)}$ passes (**CP-proof**, sid_1, $(m, r_1'^2, \overline{m}_1')$) to $\mathcal{A}_{(1,2)}$ as if coming from $\hat{\mathcal{F}}_{CP}^1$. Then, when $\mathcal{A}_{(1,2)}$ approves of delivering this message, $\mathcal{A}'_{(1,2)}$ delivers m to $\mathcal{A}'_{(2,1)}$ using $\hat{\mathcal{F}}_{MT}$. After that, $\mathcal{A}'_{(2,1)}$ passes (**CP-proof**,...) message to $\mathcal{A}_{(2,1)}$, and regardless of whether P_2 is corrupted or not, $\mathcal{A}'_{(2,1)}$ approves of message delivering only when $\mathcal{A}_{(2,1)}$ approves. With this, the simulation is completed and the simulation for P_2 is analogous.   □

Corollary 1. *Let $\mathcal{F}$ be a two-party functionality and let π be a non-trivial protocol that LUC-realizes $\mathcal{F}$ in the presence of semi-honest adversaries. Then, $\mathrm{Comp}(\pi\text{-}\hat{\mathcal{F}}_{MT})$ is a non-trivial protocol that LUC-realizes $\mathcal{F}$ in the $\hat{\mathcal{F}}_{CP}$-hybrid model and in the presence of malicious adversaries.*

4   Oblivious Transfer with UC and Game-Theoretic Security

4.1   Oblivious Transfer in the UC Framework

The oblivious transfer [10,23] is a two-party cryptographic functionality implemented by a sender T who has input $x_1, x_2, \ldots, x_l$ and a receiver R who has input $i \in \{1, 2, \ldots, l\}$. When they follow the given specifications correctly, R receives the message x_i such that R cannot obtain any more information, while T obtains no information about the selection of R. We describe the ideal functionality $\mathcal{F}_{OT}$ in [7], and the protocol SOT (1-out-of-l) for the static and semi-honest adversarial model in [7,11,12] as follows (Figs. 4 and 5).

Functionality $\mathcal{F}_{OT}$

$\mathcal{F}_{OT}$, which is parameterized with an integer l, and running with an sender T, a receiver R and an adversary $\mathcal{S}$, proceeds as follows:

• Upon receiving a message (**sender**, sid, $x_1, \ldots, x_l$) from T, where $x_i \in \{0,1\}^m$, record the tuple $(x_1, \ldots, x_l)$.

• Upon receiving a message (**receiver**, sid, i) from R, where $i \in \{1, \ldots, l\}$, send (sid, x_i) to R and (sid) to $\mathcal{S}$, and halt.

Fig. 4. Functionality of oblivious transfer in the UC model.

4.2   Oblivious Transfer in the LUC Framework

For game-theoretic analysis, we consider realizing functionality of oblivious transfer in the LUC framework. To do so, we first investigate whether some modification will be needed in changing the framework from UC to LUC as follows: We consider the case that we use the previous ideal functionality $\mathcal{F}_{OT}$ for the protocol simulation in the LUC framework. If sender T is not corrupted, in the ideal process, the corresponding dummy party T passes its own input value to

---

Protocol SOT

Proceed with a security parameter k as follows.

• Given input (**sender**, sid, $x_1, \ldots, x_l$), the party T chooses a trapdoor permutation f over $\{0,1\}^k$, together with its inverse f^{-1}, and sends (sid, f) to the receiver R. (The permutation f is chosen uniformly from a given family of trapdoor permutations.)

• Given input (**receiver**, sid, i), and having received (sid, f) from T, the receiver R chooses $y_1, \ldots, y_{i-1}, r, y_{i+1}, \ldots, y_l \in_R \{0,1\}^k$, computes $y_i = f(r)$, and sends $(sid, y_1, \ldots, y_l)$ to T.

• Having received $(sid, y_1, \ldots, y_l)$ from R, the sender T sends $(sid, x_1 \oplus B(f^{-1}(y_1)), \ldots, x_l \oplus B(f^{-1}(y_l)))$ to R, where $B(\cdot)$ is a hard-core predicate for f.

• Having received $(sid, b_1, \ldots, b_l)$ from T, the receiver R outputs $(sid, b_i \oplus B(r))$.

Fig. 5. A static and semi-honest oblivious transfer protocol.

$\mathcal{F}_{OT}$ automatically at the first step. Then, after receiving the value from T, $\mathcal{F}_{OT}$ records it and enters a waiting state. Following that, the environment $\mathcal{Z}$ is activated next and it is supposed to activate the receiver R with an input value. On the other hand, in the real life protocol execution, the process of the first message delivery is as follows. At first, the sender T passes its input value to the corresponding adversary, denoted by $\mathcal{A}_{(T,R)}$, and then $\mathcal{A}_{(T,R)}$ delivers the value to the opponent adversary, denoted by $\mathcal{A}_{(R,T)}$, through the environment $\mathcal{Z}$. Finally, the receiver R receives the value from $\mathcal{A}_{(R,T)}$. Therefore, the environment $\mathcal{Z}$ can obviously tell whether it is facing the ideal process or the real life protocol execution, since there is great difference between the two situations.

For this reason, we need to modify the definition of the previous ideal functionality of OT by changing the framework so that the difference mentioned above does not arise. We describe the modified ideal functionality $\hat{\mathcal{F}}_{OT}$ as follows (Fig. 6).

Then, we show that the protocol SOT meets the following security.

Theorem 2. *Suppose that f in the protocol* SOT *is an enhanced trapdoor permutation[1]. Then,* SOT *LUC-realizes* $\hat{\mathcal{F}}_{OT}$ *in the presence of semi-honest and static adversaries.*

Proof. As in the UC case, $\mathcal{S}_{(T,R)}$ and $\mathcal{S}_{(R,T)}$ run a simulated copy of $\mathcal{A}_{(T,R)}$ and $\mathcal{A}_{(R,T)}$ respectively, and their actions are utilized as a guide for the interaction with $\hat{\mathcal{F}}_{OT}$ and $\mathcal{Z}$. $\mathcal{S}_{(T,R)}$ and $\mathcal{S}_{(R,T)}$ proceed as follows.

– **Simulating the communication with $\mathcal{Z}$.** Every input coming from $\mathcal{Z}$ is sent to the corresponding simulated adversary $\mathcal{A}_{(T,R)}$ or $\mathcal{A}_{(R,T)}$ as if coming

---

[1] The enhanced trapdoor permutation has the property that a random element generated by the domain sampler is hard to invert, even given the random coins used by the sampler. Note that any trapdoor permutation over $\{0,1\}^k$ is clearly enhanced, since this domain can be easily and directly sampled.

Functionality $\hat{\mathcal{F}}_{OT}$

$\hat{\mathcal{F}}_{OT}$, which is parameterized with an integer l and running with a sender T, a receiver R and adversaries $\mathcal{S}_{(T,R)}$ and $\mathcal{S}_{(R,T)}$, proceeds as follows:

● Upon receiving a message (**sender**, sid, $x_1, \ldots, x_l$) from T, where $x_i \in \{0,1\}^m$, record the tuple $(x_1, \ldots, x_l)$. If the message from R has already been recorded, then send a private delayed output (sid, x_i) to R, and halt. Otherwise, send a public delayed output (**receipt**, sid) to R.

● Upon receiving a message (**receiver**, sid, i) from R, where $i \in \{1, \ldots, l\}$, record the value i. If the message from T has already been recorded, then send a private delayed output (sid, x_i) to R, and halt. Otherwise, send a public delayed output (**receipt**, sid) to T.

● Upon receiving a message (**Deliver**, m, (j,i)) from the adversary $\mathcal{S}_{(i,j)}$, if $\mathcal{S}_{(i,j)}, \mathcal{S}_{(j,i)} \in \{\mathcal{S}_{(T,R)}, \mathcal{S}_{(R,T)}\}$, send the message (**Delivered**, m, (i,j)) to the adversary $\mathcal{S}_{(j,i)}$.

Fig. 6. Functionality of oblivious transfer in the LUC model.

from their own environment. In the same way, every output from internal adversaries is treated as an output of corresponding simulator.

- **Simulating the case where no party is corrupted.** At first, the simulator $\mathcal{S}_{(T,R)}$ is activated by receiving the message (**receipt**, sid) from $\hat{\mathcal{F}}_{OT}$ ($\mathcal{S}_{(T,R)}$ is demanded for approving of the message delivery). Then, $\mathcal{S}_{(T,R)}$ randomly chooses a trapdoor permutation f over $\{0,1\}^k$ with its inverse f^{-1} and passes (sid, f) to the simulated adversary $\mathcal{A}_{(T,R)}$. When $\mathcal{A}_{(T,R)}$ delivers the message to the environment $\mathcal{Z}$, $\mathcal{S}_{(T,R)}$ actually delivers it to the opponent simulator $\mathcal{S}_{(R,T)}$ through $\mathcal{Z}$. Following that, $\mathcal{S}_{(R,T)}$ activates $\mathcal{S}_{(T,R)}$ by using $\hat{\mathcal{F}}_{OT}$, and $\mathcal{S}_{(T,R)}$ approves of $\hat{\mathcal{F}}_{OT}$'s message delivery at this timing. Then, $\mathcal{S}_{(R,T)}$ is activated again with a request for an approval from $\hat{\mathcal{F}}_{OT}$. $\mathcal{S}_{(R,T)}$ approves after confirming that $\mathcal{A}_{(R,T)}$ delivers (sid, f) to R. Next, the dummy party R receives (**receiver**, sid, i) from $\mathcal{Z}$ as an input and sends it to $\hat{\mathcal{F}}_{OT}$. Then, $\mathcal{S}_{(T,R)}$ is activated with a request for an approval from $\hat{\mathcal{F}}_{OT}$. At this timing, $\mathcal{S}_{(T,R)}$ approves of $\hat{\mathcal{F}}_{OT}$'s message delivery. After that, $\mathcal{S}_{(R,T)}$ is activated with a request for an approval similar to the process of $\mathcal{S}_{(T,R)}$. Then, $\mathcal{S}_{(R,T)}$ chooses $y_1, \ldots, y_l \in \{0,1\}^k$ and passes these values to $\mathcal{A}_{(R,T)}$. After confirming that $\mathcal{A}_{(R,T)}$ delivers the message $(sid, y_1, \ldots, y_l)$ to $\mathcal{Z}$, $\mathcal{S}_{(R,T)}$ actually delivers it to $\mathcal{S}_{(T,R)}$ through $\mathcal{Z}$. Similarly, $\mathcal{S}_{(T,R)}$ simulates $\mathcal{A}_{(T,R)}$ delivering the message to T internally. Following that, $\mathcal{S}_{(T,R)}$ chooses $b_1, \ldots, b_l$ uniformly, and passes the message $(sid, b_1, \ldots, b_l)$ to $\mathcal{A}_{(T,R)}$. If $\mathcal{A}_{(T,R)}$ delivers the message correctly, then $\mathcal{S}_{(T,R)}$ actually delivers it to $\mathcal{S}_{(R,T)}$ through $\mathcal{Z}$. Finally, $\mathcal{S}_{(R,T)}$ concludes the simulation by confirming that $\mathcal{A}_{(R,T)}$ delivers the message to R and approving of $\hat{\mathcal{F}}_{OT}$'s message delivery.

- **Simulating the case where only the sender T is corrupted.** $\mathcal{S}_{(T,R)}$ begins by sending the message (**sender**, sid, $x_1, \ldots, x_l$) to $\hat{\mathcal{F}}_{OT}$ and receives a request for an approval of message delivery. Before approving, $\mathcal{S}_{(T,R)}$ activates $\mathcal{A}_{(T,R)}$ with an input value and receives (sid, f) that $\mathcal{A}_{(T,R)}$ is supposed to deliver R in a real life protocol execution. Furthermore, $\mathcal{S}_{(T,R)}$ delivers the message to $\mathcal{S}_{(R,T)}$ through $\mathcal{Z}$ as in the above case, and once activated next, it approves the message delivery of $\hat{\mathcal{F}}_{OT}$. The following process is analogous to the above case. Next, after R sends its input to $\hat{\mathcal{F}}_{OT}$, $\mathcal{S}_{(T,R)}$ is activated with a request for approving of $\hat{\mathcal{F}}_{OT}$'s message delivery. At the time when $\mathcal{S}_{(T,R)}$ receives $(sid, y_1, \ldots, y_l)$, it passes the message to $\mathcal{A}_{(T,R)}$. Subsequently, after receiving $(sid, b_1, \ldots, b_l)$ from $\mathcal{A}_{(T,R)}$, $\mathcal{S}_{(T,R)}$ delivers it to $\mathcal{S}_{(R,T)}$ through $\mathcal{Z}$. Finally, $\mathcal{S}_{(R,T)}$ concludes the simulation in the same way as in the above case.

- **Simulating the case where only the receiver R is corrupted.** The simulation proceeds similar to the case where no party is corrupted until $\mathcal{S}_{(R,T)}$, controlling R, is activated with an input (**receiver**, sid, i). Following that, $\mathcal{S}_{(R,T)}$ passes it to $\mathcal{A}_{(R,T)}$ and receives $(sid, y_1, \ldots, y_l)$. Furthermore, $\mathcal{S}_{(R,T)}$ delivers the message to $\mathcal{S}_{(T,R)}$ through $\mathcal{Z}$. After receiving that message, $\mathcal{S}_{(T,R)}$ delivers $(sid, b_1, \ldots, b_l)$ to $\mathcal{S}_{(R,T)}$ through $\mathcal{Z}$, and $\mathcal{S}_{(R,T)}$ obtains f^{-1} by using $\hat{\mathcal{F}}_{OT}$. Then, $\mathcal{S}_{(R,T)}$ sends (**receiver**, sid, i) to $\hat{\mathcal{F}}_{OT}$ and subsequently both simulators approve of the message delivery ($\mathcal{S}_{(R,T)}$ receives x_i). Next, $\mathcal{S}_{(R,T)}$ sets $b_i = x_i \oplus B(f^{-1}(y_i))$ and passes $(sid, b_1, \ldots, b_l)$ to $\mathcal{A}_{(R,T)}$. Finally, $\mathcal{S}_{(R,T)}$ concludes the simulation by outputting x_i when $\mathcal{A}_{(R,T)}$ does so.
- **Simulating the case where both parties are corrupted.** This case can be simulated by applying simultaneously the simulators of each case where only one of the parties is corrupted. □

4.3 Analysis of Game-Theoretic Security

Next, we consider the case where rational parties implement the protocol SOT in the case $l = 2$ as in [14], since 1-out-of-2 OT is simple and fundamental. As already mentioned, SOT is designed for the semi-honest adversarial model so that its security does not concern the behaviors of rational parties. We investigate whether SOT is game-theoretically secure, before and after being compiled, respectively.

First, we define utility functions for two-message OT protocols similar to the work of [14]. In [14], Higo et al. studied the game-theoretic concepts of two-message OT protocols with reasonable utility functions, so it seems to be appropriate to follow the previous definitions. For doing it, we consider sender's (i.e., T's) and receiver's (i.e., R's) preferences as follows:

- T does not prefer the receiver R to know the input bit x_{1-i}, where the index of the receiver's selection is $i \in \{0, 1\}$ (This explains the case where R obtains x_0 and x_1 simultaneously),
- T prefers to complete the protocol execution,
- T prefers to know the input index of the receiver's selection $i \in \{0, 1\}$; and
- R does not prefer the sender S to know its input index of the selection,
- R prefers to complete the protocol execution,
- R prefers to know the other sender's input bit x_{1-i}.

Then, a formal definition of utility functions is given as follows.

Definition 4 (Utility Functions). *Let π be an OT protocol having a sender T with inputs $x_0, x_1 \in \{0,1\}$ and a receiver R with an input $i \in \{0,1\}$. Let $\alpha_T, \beta_T, \gamma_T, \alpha_R, \beta_R, \gamma_R$ be positive constants. The utility functions U_T for T and U_R for R are defined by*

$$\begin{aligned}
U_T := &- \alpha_T \cdot (\Pr\{x' = x_{1-i} \mid guess_R(T(x_0, x_1), R(i)) = x'\} - 1/2) \\
&+ \beta_T \cdot (\Pr\{fin(T(x_0.x_1), R(i)) = 1\} - 1) \\
&+ \gamma_T \cdot (\Pr\{i' = i \mid guess_T(T(x_0, x_1), R(i)) = i'\} - 1/2), \\
U_R := &- \alpha_R \cdot (\Pr\{i' = i \mid guess_T(T(x_0, x_1), R(i)) = i'\} - 1/2) \\
&+ \beta_R \cdot (\Pr\{fin(T(x_0.x_1), R(i)) = 1\} - 1) \\
&+ \gamma_R \cdot (\Pr\{x' = x_{1-i} \mid guess_R(T(x_0, x_1), R(i)) = x'\} - 1/2),
\end{aligned}$$

where $guess_T(\cdot)$ and $guess_R(\cdot)$ mean guessing by T and R, respectively, for the opponent's private value, and $fin(\cdot)$ represents the completion of the protocol execution: $fin(\cdot) = 1$ if the protocol satisfies the specifications correctly; otherwise $fin(\cdot) = 0$.

In addition, as in the work of [4,14,15], we consider Nash equilibrium as the solution concept in terms of the game theory.

Definition 5 (Nash Equilibrium). *For a pair of utility functions (U_T, U_R), we say that a pair of strategies (σ_T, σ_R) is in Nash equilibrium, if for every pair of strategies (σ_T^*, σ_R^*), it holds that $U_T(\sigma_T, \sigma_R) \geq U_T(\sigma_T^*, \sigma_R) - negl(n)$ and $U_R(\sigma_T, \sigma_R) \geq U_R(\sigma_T, \sigma_R^*) - negl(n)$.*

Definition 6 (Game-Theoretic Security for OT). *Let π be an OT protocol having a sender T and a receiver R. Let σ_T and σ_R be strategies planned to follow all the specifications of π, respectively. We say that π is game-theoretically secure, if the pair of strategies (σ_T, σ_R) is in Nash equilibrium with respect to the pair of utility functions (U_T, U_R).*

Then, we can show the game-theoretic security of SOT before/after application of the compiler in the LUC model below.

Theorem 3. *The protocol* SOT *is not game-theoretically secure in the presence of rational parties, however, the compiled protocol* $\mathrm{Comp}(\mathrm{SOT}\text{-}\hat{\mathcal{F}}_{MT})$ *in the $\hat{\mathcal{F}}_{CP}$-hybrid model is game-theoretically secure in the presence of rational parties.*

Proof. First, we show that the plain protocol SOT is not secure. Once both parties are allowed to behave rationally, this protocol becomes quite imbalanced. If the receiver R attempts to enhance its own utility more than that of the case where it acts honestly, it takes action such as the following. In the step where R is supposed to choose $y_{1-i}, r \in_R \{0,1\}^k$ and computes $y_i = f(r)$, it also applies f for generating y_{1-i}. For this, R can obviously obtain the T's private value x_{1-i} in addition to x_i unless the sender T aborts the protocol execution. (Note that we take no account of the case where each party changes its own input value, since it seems reasonable to assume so. Furthermore even if that occurs, the result is not affected essentially.) In addition, since y_{1-i} and r are randomly chosen, R's dishonest behavior is not detectable. Thus, this results in increasing the value $\gamma_R \cdot (\Pr\{x' = x_{1-i} \mid guess_R(T(x_0, x_1), R(i)) = x'\} - 1/2)$.

On the sender T's side, he/she would think that R does wrong absolutely. However, T has only two choices, either following the specifications of the protocol or aborting, since T obtains no information from the received values y_0, y_1 and there is no way to benefit in the subsequent process. The selection depends on to which T gives much weight *the completion of the protocol* or *protecting the secret*. If T prefers the completion of the protocol, it results in decreasing the value $-\alpha_T \cdot (\Pr\{x' = x_{1-i} \mid guess_R(T(x_0, x_1), R(i)) = x'\} - 1/2)$ and increasing the value $\beta_T \cdot (\Pr\{fin(T(x_0.x_1), R(i)) = 1\} - 1)$ in comparison with the latter. On the contrary, if T prefers to protect the secret and chooses to abort, it results in increasing the value $-\alpha_T \cdot (\Pr\{x' = x_{1-i} \mid guess_R(T(x_0, x_1), R(i)) = x'\} - 1/2)$ and decreasing the value $\beta_T \cdot (\Pr\{fin(T(x_0.x_1), R(i)) = 1\} - 1)$ in comparison with the former. From the above discussion, at least the pair of strategies (σ_T, σ_R) is not in Nash equilibrium.

Next, we show that the compiled protocol Comp(SOT-$\hat{\mathcal{F}}_{MT}$)-$\hat{\mathcal{F}}_{CP}$ is secure. Regarding the dishonest actions of R mentioned above, R cannot enhance its own utility even if applying f for generating y_0 and y_1. Since the functionality $\hat{\mathcal{F}}_{CP}$ rejects incorrect messages, the protocol execution would never be completed. Therefore, it results in decreasing the value $\beta_R \cdot (\Pr\{fin(T(x_0.x_1), R(i)) = 1\} - 1)$ compared to that of the case where R follows the protocol specifications. On the T's side, there is no need to worry about the R's dishonest actions, and hence T can obtain the highest utility by following the protocol honestly. Similarly to the R's case, if T chooses to deviate from the protocol, T's total utility obviously decreases. Thus, the pair of strategies (σ_T, σ_R) is in Nash equilibrium.   □

5   Concluding Remarks

In this paper, we have proposed a compiler of two-party protocols in the LUC framework such that it transforms any two-party protocol secure against semi-honest adversaries into a protocol secure against malicious adversaries. Then, we have shown the application of our compiler to an oblivious transfer protocol to achieve a primitive with both UC and game-theoretic security. We emphasize that our main purpose was to address how protocols with security in the game-theoretic model can be composed to obtain an overall game-theoretically secure protocol. In this sense, our result is successful and the constructed protocol has desirable properties.

An interesting line for future work is to address whether this resulting protocol carries over to the general multi-party computation protocols as a building block in the game-theoretic setting.

Acknowledgments. We would like to thank anonymous referees for their helpful comments. This work was partially supported by JSPS KAKENHI Grant Number

15H02710, and it was partially conducted under the auspices of the MEXT Program for Promoting the Reform of National Universities.

References

1. Alwen, J., Katz, J., Lindell, Y., Persiano, G., Shelat, A., Visconti, I.: Collusion-free multiparty computation in the mediated model. In: Halevi, S. (ed.) CRYPTO 2009. LNCS, vol. 5677, pp. 524–540. Springer, Heidelberg (2009)
2. Alwen, J., Katz, J., Maurer, U., Zikas, V.: Collusion-preserving computation. In: Safavi-Naini, R., Canetti, R. (eds.) CRYPTO 2012. LNCS, vol. 7417, pp. 124–143. Springer, Heidelberg (2012)
3. Alwen, J., Shelat, A., Visconti, I.: Collusion-free protocols in the mediated model. In: Wagner, D. (ed.) CRYPTO 2008. LNCS, vol. 5157, pp. 497–514. Springer, Heidelberg (2008)
4. Asharov, G., Canetti, R., Hazay, C.: Towards a game theoretic view of secure computation. In: Paterson, K.G. (ed.) EUROCRYPT 2011. LNCS, vol. 6632, pp. 426–445. Springer, Heidelberg (2011)
5. Backes, M., Pfitzmann, B., Waidner, M.: A universally composable cryptographic library. In: IACR Cryptology ePrint Archive (2003)
6. Canetti, R.: Universally composable security: a new paradigm for cryptographic protocols. In: 42nd Annual Symposium on Foundations of Computer Science (FOCS 2001), pp. 136–145 (2001)
7. Canetti, R., Lindell, Y., Ostrovsky, R., Sahai, A.: Universally composable two-party and multi-party secure computation. In: 34th Annual ACM Symposium on Theory of Computing (STOC 2002), pp. 494–503 (2002)
8. Canetti, R., Vald, M.: Universally composable security with local adversaries. In: Visconti, I., De Prisco, R. (eds.) SCN 2012. LNCS, vol. 7485, pp. 281–301. Springer, Heidelberg (2012)
9. Dodis, Y., Halevi, S., Rabin, T.: A cryptographic solution to a game theoretic problem. In: Bellare, M. (ed.) CRYPTO 2000. LNCS, vol. 1880, pp. 112–130. Springer, Heidelberg (2000)
10. Even, S., Goldreich, O., Lempel, A.: A randomized protocol for signing contracts. Commun. ACM **28**(6), 637–647 (1985)
11. Goldreich, O.: The Foundations of Cryptography: Basic Applications, vol. 2. Cambridge University Press, New York (2004)
12. Goldreich, O., Micali, S., Wigderson, A.: How to play any mental game or a completeness theorem for protocols with honest majority. In: 19th Annual ACM Symposium on Theory of Computing (STOC 1987), pp. 218–229 (1987)
13. Gradwohl, R., Livne, N., Rosen, A.: Sequential rationality in cryptographic protocols. In: 51th Annual IEEE Symposium on Foundations of Computer Science (FOCS 2010), pp. 623–632 (2010)
14. Higo, H., Tanaka, K., Yamada, A., Yasunaga, K.: A game-theoretic perspective on oblivious transfer. In: Susilo, W., Mu, Y., Seberry, J. (eds.) ACISP 2012. LNCS, vol. 7372, pp. 29–42. Springer, Heidelberg (2012)
15. Higo, H., Tanaka, K., Yasunaga, K.: Game-theoretic security for bit commitment. In: Sakiyama, K., Terada, M. (eds.) IWSEC 2013. LNCS, vol. 8231, pp. 303–318. Springer, Heidelberg (2013)
16. Izmalkov, S., Lepinski, M., Micali, S.: Perfect implementation. Games Econ. Behav. **71**(1), 121–140 (2011)

17. Izmalkov, S., Micali, S., Lepinski, M.: Rational secure computation and ideal mechanism design. In: 46th Annual IEEE Symposium on Foundations of Computer Science (FOCS 2005), pp. 585–595 (2005)
18. Katz, J.: Bridging game theory and cryptography: recent results and future directions. In: Canetti, R. (ed.) TCC 2008. LNCS, vol. 4948, pp. 251–272. Springer, Heidelberg (2008)
19. Kol, G., Naor, M.: Cryptography and game theory: designing protocols for exchanging information. In: Canetti, R. (ed.) TCC 2008. LNCS, vol. 4948, pp. 320–339. Springer, Heidelberg (2008)
20. Lepinski, M., Micali, S., Shelat, A.: Collusion-free protocols. In: 37th Annual ACM Symposium on Theory of Computing (STOC 2005), pp. 543–552 (2005)
21. Maurer, U.: Constructive cryptography – a primer. In: Sion, R. (ed.) FC 2010. LNCS, vol. 6052, p. 1. Springer, Heidelberg (2010)
22. Maurer, U., Renner, R.: Abstract cryptography. In: Second Symposium on Innovations in Computer Science (ICS 2011), pp. 1–21 (2011)
23. Rabin, M.O.: How to exchange secrets with oblivious transfer. Technical report TR-81, Aiken Computation Lab., Harvard University (1981)

Zero-Knowledge Interactive Proof Systems for New Lattice Problems

Claude Crépeau and Raza Ali Kazmi[(✉)]

McGill University, Montreal, Canada
crepeau@cs.mcgill.ca, raza-ali.kazmi@mail.mcgill.ca

Abstract. In this work we introduce a new hard problem in lattices called **Isometric Lattice Problem** (**ILP**) and reduce **Linear Code Equivalence** over prime fields and **Graph Isomorphism** to this problem. We also show that this problem has an (efficient prover) perfect zero-knowledge interactive proof; this is the only hard problem in lattices that is known to have this property (with respect to malicious verifiers). Under the assumption that the polynomial hierarchy does not collapse, we also show that **ILP** cannot be **NP-complete**. We finally introduce a variant of **ILP** over the rationals radicands and provide similar results for this new problem.

1 Introduction

Zero-Knowledge interactive proof systems **ZKIP** [6] have numerous applications in cryptography such as Identification Schemes, Authentication Schemes, Multiparty Computations, etc. Appart from cryptographic applications these proof systems play an important part in the study of complexity theory. The first **IP** for lattice problems (**coGapCVP$_\gamma$**, **coGapSVP$_\gamma$**) was presented by Goldreich and Goldwasser [13]. However, these proofs are only *honest-verifier* Perfect Zero-Knowledge and known to have *inefficient provers*. Micciancio and Vadhan [10] presented Interactive Proofs for **GapCVP$_\gamma$** and **GapSVP$_\gamma$**. These proofs are Statistical Zero-Knowledge and have efficient provers[1] as well. In this paper we introduce a new hard problem called **ISOMETRIC LATTICE PROBLEM** (**ILP**). We present **IP** systems for the **ILP**. These proof systems are *Perfect* Zero-Knowledge and have *efficient* provers. We show that a variant of **ILP** over the integers is at least as hard as **Graph Isomorphism** (**GI**) [4,5] and **Linear Code Equivalence** (**LCE**) [5,7]. This is the only hard problem known in lattices that have a (*malicious*-verifier) Perfect Zero-Knowledge **IP** system with an *efficient* prover. We also show that **ILP** is unlikely to be **NP-complete**. Finally we also introduce another variant of **ILP** over the rational-radicands and provide similar results for this problem.

C. Crépeau and R.A. Kazmi—Supported in part by Québec's FRQNT, Canada's NSERC and CIFAR.

[1] The Prover runs in probabilistic polynomial time given a certificate for the input string.

© Springer International Publishing Switzerland 2015
J. Groth (Ed.): IMACC 2015, LNCS 9496, pp. 152–169, 2015.
DOI: 10.1007/978-3-319-27239-9_9

2   Notations

For any matrix $\mathbf{A}$, we denote its transpose by $\mathbf{A}^t$. Let $O(n, \mathbb{R}) = \{Q \in \mathbb{R}^{n \times n} : Q \cdot Q^t = \mathbf{I}\}$ denote the group of $n \times n$ orthogonal matrices over $\mathbb{R}$. Let $GL_k(\mathbb{Z})$ denote the group of $k \times k$ invertible (unimodular) matrices over $\mathbb{Z}$. Let $GL_k(\mathbb{F}_q)$ denote the set of $k \times k$ invertible matrices over the finite field $\mathbb{F}_q$. Let $\mathcal{P}_n$ denote the set of $n \times n$ permutation matrices. Let σ_n be the set of all permutations of $\{1, \ldots, n\}$. For $\pi \in \sigma_n$, we denote P_π the corresponding $n \times n$ permutation matrix. $\mathcal{P}(n, \mathbb{F}_q)$ denotes the set of $n \times n$ monomial matrices (there is exactly one nonzero entry in each row and each column) over $\mathbb{F}_q$. $\mathcal{D}_{\epsilon_n}$ is the set of diagonal matrices $D_\epsilon = diag(\epsilon_1, \ldots, \epsilon_n)$, $\epsilon_i = \pm 1$ for $i = 1, \ldots, n$. For a real vector $\mathbf{v} = (v_1, \ldots, v_n)$ we denote its Euclidean norm by $\|\mathbf{v}\| = \sqrt{v_1^2 + \cdots + v_n^2}$ and max-norm $\|\mathbf{v}\|_\infty = max_{i=1}^n |v_i|$ and for any matrix $\mathbf{B} = [\mathbf{b}_1 | \mathbf{b}_2 | \ldots | \mathbf{b}_k] \in \mathbb{R}^{n \times k}$ we define its norm by $\|\mathbf{B}\| = max_{i=1}^n \|\mathbf{b}_i\|$. For any ordered set of linearly independent vectors $\{\mathbf{b}_1, \mathbf{b}_2, \ldots, \mathbf{b}_k\}$, we denote $\{\tilde{\mathbf{b}}_1, \tilde{\mathbf{b}}_2, \ldots, \tilde{\mathbf{b}}_k\}$, its Gram-Schmidt orthogonalization.

2.1   Lattices

Let $\mathbb{R}^n$ be an n-dimensional Euclidean space and let $\mathbf{B} \in \mathbb{R}^{n \times k}$ be a matrix of rank k. A lattice $\mathcal{L}(\mathbf{B})$ is the set of all vectors

$$\mathcal{L}(\mathbf{B}) = \{\mathbf{B}\mathbf{x} : \mathbf{x} \in \mathbb{Z}^k\}.$$

The integer n and k are called the dimension and rank of $\mathcal{L}(\mathbf{B})$. A lattice is called full dimensional if $k = n$. Two lattices $\mathcal{L}(\mathbf{B}_1)$ and $\mathcal{L}(\mathbf{B}_2)$ are equivalent if and only if there exists a unimodular matrix $\mathbf{U} \in \mathbb{Z}^{k \times k}$ such that $\mathbf{B}_1 = \mathbf{U}\mathbf{B}_2$.

2.2 q-ary Lattices

A lattice $\mathcal{L}$ is called q-ary, if it satisfies $q\mathbb{Z}^n \subseteq \mathcal{L} \subseteq \mathbb{Z}^n$ for a positive integer q. In other words, the membership of a vector $\mathbf{v} \in \mathcal{L}$ is given by $\mathbf{v} \mod q$. Let $\mathbf{G} = [\mathbf{g}_1 | \ldots | \mathbf{g}_k] \in \mathbb{Z}_q^{n \times k}$ be a $n \times k$ matrix of rank k over $\mathbb{Z}_q^{n \times k}$. We define below two important families of q-ary lattices used in cryptography

$$\Lambda_q(\mathbf{G}) = \{\mathbf{y} \in \mathbb{Z}^n : \mathbf{y} \equiv \mathbf{G} \cdot \mathbf{s} \pmod{q}, \text{ for some vector } \mathbf{s} \in \mathbb{Z}^k\}$$

$$\Lambda_q^\top(\mathbf{G}) = \{\mathbf{y} \in \mathbb{Z}^n : \mathbf{y} \cdot \mathbf{G} \equiv 0 \pmod{q}\}.$$

A basis $\mathbf{B}$ of $\Lambda_q(\mathbf{G})$ is

$$\mathbf{B} = [\mathbf{g}_1 | \ldots | \mathbf{g}_k | \mathbf{b}_{k+1} | \ldots | \mathbf{b}_n] \in \mathbb{Z}^{n \times n}$$

where $\mathbf{b}_j = (0, \ldots, q, \ldots, 0) \in \mathbb{Z}^n$ is a vector with its j-th coordinate equal to q and all other coordinates are 0, $k + 1 \leq j \leq n$. A basis of $\Lambda_q^\top$ is given by $q \cdot (\mathbf{B}^{-1})^t$.

2.3   Discrete Gaussian Distribution on Lattices

For any $s > 0$, $\mathbf{c} \in \mathbb{R}^n$, we define a Gaussian function on $\mathbb{R}^n$ centered at $\mathbf{c}$ with parameter s.

$$\forall \mathbf{x} \in \mathbb{R}^n, \ \rho_{s,\mathbf{c}}(\mathbf{x}) = e^{\frac{-\pi \|\mathbf{x} - \mathbf{c}\|}{s^2}}.$$

Let $\mathcal{L}$ be any n dimensional lattice and $\rho_{s,\mathbf{c}}(\mathcal{L}) = \sum_{y \in \mathcal{L}} \rho_{s,\mathbf{c}}(y)$. We define a *Discrete Gaussian* distribution on $\mathcal{L}$

$$\forall \mathbf{x} \in \mathcal{L}, \ D_{s,\mathbf{c},\mathcal{L}}(\mathbf{x}) = \frac{\rho_{s,\mathbf{c}}(\mathbf{x})}{\rho_{s,\mathbf{c}}(\mathcal{L})}.$$

Theorem 1. *Given a basis* $\mathbf{B} = [\mathbf{b}_1 | \dots | \mathbf{b}_k] \in \mathbb{R}^{n \times k}$ *of an n-dimensional lattice* $\mathcal{L}$, *a parameter* $s \geq \|\tilde{\mathbf{B}}\| \cdot \omega(\sqrt{\log n})$ *and a center* $\mathbf{c} \in \mathbb{R}^n$, *the algorithm* SampleD *([2], Sect. 4.2, p. 14) outputs a sample from a distribution that is statistically close to* $D_{s,\mathbf{c},\mathcal{L}}$.

Theorem 2. *There is a deterministic polynomial-time algorithm that, given an arbitrary basis* $\{\mathbf{b}_1, \dots, \mathbf{b}_k\}$ *of an n-dimensional lattice* $\mathcal{L}$ *and a set of linearly independent lattice vectors* $\mathbf{S} = [\mathbf{s}_1 | \mathbf{s}_2 \dots | \mathbf{s}_k] \in \mathcal{L}$ *with ordering* $\|\mathbf{s}_1\| \leq \|\mathbf{s}_2\| \leq \dots \leq \|\mathbf{s}_k\|$, *outputs a basis* $\{\mathbf{r}_1 \dots \mathbf{r}_k\}$ *of* $\mathcal{L}$ *such that* $\|\tilde{\mathbf{r}}_i\| \leq \|\tilde{\mathbf{s}}_i\|$ *for* $1 \leq i \leq k$.

2.4   Orthogonal Matrices and Givens Rotations

A Givens rotation is an orthogonal $n \times n$ matrix of the form

$$G_{(i,j,\theta)} = \begin{bmatrix} 1 & \cdots & 0 & \cdots & 0 & \cdots & 0 \\ \vdots & \ddots & \vdots & & \vdots & & \vdots \\ 0 & \cdots & c & \cdots & -s & \cdots & 0 \\ \vdots & & \vdots & \ddots & \vdots & & \vdots \\ 0 & \cdots & s & \cdots & c & \cdots & 0 \\ \vdots & & \vdots & & \vdots & \ddots & \vdots \\ 0 & \cdots & 0 & \cdots & 0 & \cdots & 1 \end{bmatrix}, i \neq j$$

The non-zero elements of a Givens matrix $G_{(i,j,\theta)}$ are given by

$$g_{k,k} = 1 \text{ for } k \neq i, j \text{ and } g_{i,i} = g_{j,j} = c$$

$$g_{i,j} = s = -g_{j,i} \text{ for } i < j$$

where $c = \cos(\theta)$ and $s = \sin(\theta)$.

The product $G_{(i,j,\theta)} \cdot \mathbf{v}$ represents a counter-clockwise rotation of the vector $\mathbf{v}$ in the (i,j) plane by angle θ. Moreover, only the i-th and j-th entries of $\mathbf{v}$ are affected and the rest remains unchanged. Any orthogonal matrix $Q \in \mathbb{R}^{n \times n}$ can be written as a product of $\frac{n(n-1)}{2}$ Givens matrices and a diagonal matrix $D_\epsilon \in \mathcal{D}_{\epsilon_n}$

$$Q = D_\epsilon \left(G_{(1,2,\theta_{1,2})} \cdots G_{(1,n,\theta_{1,n})} \right) \cdot \left(G_{(2,3,\theta_{2,3})} \cdots G_{(2,n,\theta_{2,n})} \right) \cdots \left(G_{(n-1,n,\theta_{n-1,n})} \right).$$

The angles $\theta_{i,j} \in [0, 2\pi]$, $1 \leq i < j \leq n$ are called angles of rotation.

2.5   Properties of Givens Matrices

1. *Additivity:* For angles $\theta, \phi \in [0, 2\pi]$ and any vector $\mathbf{v} \in \mathbb{R}^n$

$$G_{(i,j,\phi)} \cdot G_{(i,j,\theta)} \mathbf{v} = G_{(i,j,\phi+\theta)} \mathbf{v}.$$

2. *Commutativity:* For angles $\theta_{i,j}, \theta_{j,i}, \theta_{y,z} \in [0, 2\pi]$ and $\{i,j\} \cap \{y,z\} = \emptyset$ or $\{i,j\} = \{y,z\}$.

$$G_{(i,j,\theta_{i,j})} \cdot G_{(j,i,\theta_{j,i})} \mathbf{v} = G_{(j,i,\theta_{j,i})} \cdot G_{(i,j,\theta_{i,j})} \mathbf{v}$$

$$G_{(i,j,\theta_{i,j})} \cdot G_{(y,z,\theta_{y,z})} \mathbf{v} = G_{(y,z,\theta_{y,z})} \cdot G_{(i,j,\theta_{i,j})} \mathbf{v}.$$

3. *Linearity:* For any Givens matrix $G_{(i,j,\theta_{i,j})}$, any vector $\mathbf{v} \in \mathbb{R}^n$ and any permutation $\pi \in \sigma_n$

$$G_{\left(\pi(i),\pi(j),\theta_{\pi(i),\pi(j)}\right)} P_\pi \cdot \mathbf{v} = P_\pi G_{(i,j,\theta_{i,j})} \cdot \mathbf{v}$$

P_π is the corresponding permutation matrix of π.

2.6   The Set $\mathcal{R}$

Computationally it is not possible to work over arbitrary real numbers as they require infinite precision. However, there are reals that can be represented finitely and one can add and multiply them without losing any precision. For example we can represent numbers $\sqrt{7}$ and $\sqrt[4]{5}$ as $< 2, 7 >$ and $< 4, 5 >$. In, general, a real number r that has the following form

$$r = a_1 \sqrt[n_{11}]{x_{11} + \sqrt[n_{21}]{x_{21} + \cdots + \sqrt[n_{k1}]{x_{k1}}}} + a_2 \sqrt[n_{12}]{x_{12} + \sqrt[n_{22}]{x_{22} + \cdots + \sqrt[n_{k2}]{x_{k2}}}}$$

$$+ \cdots + a_l \sqrt[n_{1l}]{x_{1l} + \sqrt[n_{2l}]{x_{2l} + \cdots + \sqrt[n_{kl}]{x_{kl}}}}.$$

where $a_j{'}s, n_{ij}{'}s \in \mathbb{Q}, x_{i_j}{'}s \in \mathbb{Q}^+ \cup \{0\}$ and $l, k_1 \cdots k_l \in \mathbb{N}$; can be represented as

$$r = a_1 < n_{11}, x_{11} + < n_{21}, x_{21} + \cdots + < n_{k1}, x_{k1} >> \cdots >$$
$$+ a_2 < n_{12}, x_{12} + < n_{22}, x_{22} + \cdots + < n_{k2}, x_{k2} >> \cdots >$$
$$+ \cdots + a_l < n_{1l}, x_{1l} + < n_{2l}, x_{2l} + \cdots + < n_{kl}, x_{kl} >> \cdots > .$$

We call such numbers *rational radicands* and denote the set of all rational radicands $\mathcal{R}$.[2]

2.7   The Set $\overline{O(n, \mathcal{R})}$

Let $O(n, \mathcal{R})$ denote a set of $n \times n$ orthogonal matrices over $\mathcal{R}$. In this sub-section we will define a subset $\overline{O(n, \mathcal{R})} \subset O(n, \mathcal{R})$ that has the following properties:

---

[2] In this notation any rational number x can be represented as $\pm < 1, x >$.

- Any orthogonal matrix $Q \in \overline{O(n, \mathcal{R})}$ has finite representation.
- If $Q \in \overline{O(n, \mathcal{R})}$, then $Q^t \in \overline{O(n, \mathcal{R})}$.
- $\overline{O(n, \mathcal{R})}$ is a finite set.

Let $\mathcal{P}$ be any desired publicly known positive polynomial in the size of the input bases $\mathbf{B}_1, \mathbf{B}_2 \in O(n, \mathcal{R})$ and $\delta = \frac{\pi}{2^{\mathcal{P}}}$. We denote the set of angles $C = \{0, \delta, 2\delta, \ldots, \theta, \ldots, 2\pi - \delta\}$. We denote $\overline{O(n, \mathcal{R})}$ to be the set of $n \times n$ orthogonal matrices corresponding to C that can be written as a product of commuting Givens rotations. More, precisely

$$\overline{O(n, \mathcal{R})} = \{G_{(1,2,\theta_1)} \cdot G_{(3,4,\theta_2)} \cdots G_{(x-1,x,\theta_{x/2})} : \theta_i \in C, 1 \leq i \leq x\}.$$

where $x = n$ if n is even, otherwise $x = n - 1$. Clearly $\overline{O(n, \mathcal{R})}$ is a finite set, since C is a finite set. Furthermore for any integer $\mathcal{P} \geq 2$,

$$\sin\left(\frac{\pi}{2^{\mathcal{P}}}\right) = \frac{1}{2} \underbrace{< 2, 2- < 2, 2 + \cdots + < 2, 2 >> \cdots >}_{\mathcal{P}-1}$$

$$\cos\left(\frac{\pi}{2^{\mathcal{P}}}\right) = \frac{1}{2} \underbrace{< 2, 2+ < 2, 2 + \cdots + < 2, 2 >> \cdots >}_{\mathcal{P}-1}.$$

For any integer $0 \leq N \leq 2^{\mathcal{P}+1}$ $\sin(\frac{N\pi}{2^{\mathcal{P}}})$ and $\cos(\frac{N\pi}{2^{\mathcal{P}}})$ can be computed in $O(\mathcal{P})$ time (see Appendix A). Let $Q \in \overline{O(n, \mathcal{R})}$,

$$Q = G_{(1,2,\theta_1)} \cdot G_{(3,4,\theta_2)} \cdots G_{(x-1,x,\theta_{x/2})} \text{ for some } \theta_1, \ldots, \theta_i, \ldots, \theta_{x/2} \in C.$$

We will show that $Q^t \in \overline{O(n, \mathcal{R})}$. Let

$$Q' = G_{(1,2,2\pi-\theta_1)} \cdot G_{(3,4,2\pi-\theta_2)} \cdots G_{(x-1,x,2\pi-\theta_{x/2})}.$$

Clearly if $\theta_i \in C$, then $2\pi - \theta_i \in C$. Therefore, it follows that $Q' \in \overline{O(n, \mathcal{R})}$.

$$Q \cdot Q' = \left(G_{(1,2,\theta_1)}G_{(1,2,2\pi-\theta_1)}\right) \cdot \left(G_{(3,4,\theta_2)}G_{(3,4,2\pi-\theta_2)}\right) \cdots$$
$$\left(G_{(x-1,x,\theta_{x/2})}G_{(x-1,x,2\pi-\theta_{x/2})}\right)$$
$$= G_{(1,2,\theta_1+2\pi-\theta_1)} \cdot G_{(3,4,\theta_2+2\pi-\theta_2)} \cdots G_{(x-1,x,\theta_{x/2}+2\pi-\theta_{x/2})}$$
$$= G_{(1,2,2\pi)} \cdot G_{(3,4,2\pi)} \cdots G_{(x-1,x,2\pi)}$$

but $G_{(i,j,2\pi)} = \mathbf{I}$ therefore $G_{(1,2,2\pi)} \cdot G_{(3,4,2\pi)} \cdots G_{(x-1,x,2\pi)} = \mathbf{I}$.

3   Isometric Lattices

Definition 1. *Let $\mathbf{B}_1, \mathbf{B}_2 \in \mathbb{R}^{n \times k}$ be two bases of rank k. We say that two lattices $\mathcal{L}(\mathbf{B}_1) \cong \mathcal{L}(\mathbf{B}_2)$ are isometric if there exists a matrix $U \in GL_k(\mathbb{Z})$ and a matrix $Q \in O(n, \mathbb{R})$ such that $\mathbf{B}_2 = Q\mathbf{B}_1 U$.*

Decision Problem ILP: Given two matrices $\mathbf{B}_1, \mathbf{B}_2 \in \mathbb{R}^{n \times k}$, decide whether $\mathcal{L}(\mathbf{B}_1) \cong \mathcal{L}(\mathbf{B}_2)$.

3.1   Variants of ILP

Let $\mathbf{S}_{(\mathbf{B}_1, \mathbf{B}_2)} = \{\mathbf{B} \in \mathbb{R}^{n \times k} : \mathcal{L}(\mathbf{B}) \cong \mathcal{L}(\mathbf{B}_1) \cong \mathcal{L}(\mathbf{B}_2)\}$ be the set of bases that are isometric to $\mathbf{B}_1$ and $\mathbf{B}_2$. The **ILP** seems to be very similar to **LCE** [5,7]. Therefore, it is natural to ask if one can obtain a **PZKIP** for **ILP** by mimicking the **LCE** proof system.[3] However, if we try to mimic the proof system for **LCE** we are faced with following problems. Recall that a proof system is zero-knowledge if there exists a probabilistic polynomial time simulator that can forge transcripts that are distributed identically (or statistically close to) real transcripts.

- In the **LCE** proof system the prover picks uniformly and independently invertible matrices from $\mathbb{F}_q^{k \times k}$. In comparison the corresponding set $(GL_k(\mathbb{Z}))$ in **ILP** is countably infinite. Therefore there exists no uniform distribution on $GL_k(\mathbb{Z})$.
- Computationally it is not possible to work over reals as they required infinite precision and almost all elements in $O(n, \mathbb{R})$, have infinite representation. Whereas in **LCE** every element in the corresponding set $\mathcal{P}(n, \mathbb{F}_q)$ can be represented with $O(n^2 \log q)$ bits. Note that in theory the uniform distribution exists on $O(n, \mathbb{R})$ [14–16], but computationally it is not possible to pick uniformly from $O(n, \mathbb{R})$ as this would require infinite computational power.

A natural solution would be to define some finite subsets $\overline{GL_k(\mathbb{Z})}$, $\overline{O(n, \mathbb{R})}$ of $GL_k(\mathbb{Z})$, $O(n, \mathbb{R})$ and pick uniformly from $\overline{GL_k(\mathbb{Z})}$ and $\overline{O(n, \mathbb{R})}$. However, this solution may not preserve the zero-knowledge property of the proof system. To see this let $\mathbf{B}_2 = \overline{Q} \mathbf{B}_1 \overline{U}$, be two isometric bases that can be represented finitely, where $\overline{Q} \in \overline{O(n, \mathbb{R})}$ and $\overline{U} \in \overline{GL_k(\mathbb{Z})}$.

$$[\mathbf{B}_1] = \left\{ \overline{Q'} \mathbf{B}_1 \overline{U'} : \overline{Q'} \in \overline{O(n, \mathbb{R})} \text{ and } \overline{U'} \in \overline{GL_k(\mathbb{Z})} \right\}$$

$$[\mathbf{B}_2] = \left\{ \overline{Q'} \mathbf{B}_2 \overline{U'} : \overline{Q'} \in \overline{O(n, \mathbb{R})} \text{ and } \overline{U'} \in \overline{GL_k(\mathbb{Z})} \right\}.$$

1. The prover picks uniformly $i \in \{1, 2\}$.
2. The prover picks uniformly $\mathbf{B} \in [\mathbf{B}_i]$ and sends $\mathbf{B}$ to the receiver.
3. The verifier uniformly picks $j \in \{1, 2\}$ and sends j to the prover.

Note that the zero-knowledge property requires that from $\mathbf{B}$ the verifier should not be able to learn i except with probability $\frac{1}{2}$ (for perfect zero-knowledge) or $\frac{1}{2} + negl$ (for statistical zero-knowledge). This implies that $[\mathbf{B}_1] = [\mathbf{B}_2]$ (for perfect zero-knowledge) or $|[\mathbf{B}_1] \cup [\mathbf{B}_2]| - |[\mathbf{B}_1] \cap [\mathbf{B}_2]| = negl$ (for statistical zero-knowledge). Note that any $\mathbf{B} \in [\mathbf{B}_1]$ can only be in $[\mathbf{B}_2]$ if and only if $\overline{Q'} \cdot \overline{Q}^{\mathrm{t}} \in \overline{O(n, \mathbb{R})}$ and $\overline{U}^{-1} \cdot \overline{U'} \in \overline{GL_k(\mathbb{Z})}$. Similarly, any $\mathbf{B} \in [\mathbf{B}_2]$ can only be in $[\mathbf{B}_1]$ if and only if $\overline{Q'} \cdot \overline{Q} \in \overline{O(n, \mathbb{R})}$ and $\overline{U} \cdot \overline{U'}^{-1} \in \overline{GL_k(\mathbb{Z})}$. Therefore sets $\overline{O(n, \mathbb{R})}$ and $\overline{GL_k(\mathbb{Z})}$ must be a group under multiplication. But this seems unlikely to happen in general. To see this lets try to construct a finite subgroup $\overline{O(n, \mathbb{Q})} \leq O(n, \mathbb{Q})$.

---

[3] The **IP** for **LCE** is **PZKIP** with an efficient prover see [5].

– Let $Q \in O(n, \mathbb{Q})$. We add Q in $\overline{O(n, \mathbb{Q})}$, therefore $\overline{O(n, \mathbb{Q})} \leftarrow \overline{O(n, \mathbb{Q})} \cup \{Q\}$.
– Since $\overline{O(n, \mathbb{Q})}$ has to be a multiplicative group, we must add $Q \cdot Q$ and Q^t to it. Hence $\overline{O(n, \mathbb{Q})} \leftarrow \overline{O(n, \mathbb{Q})} \cup \{Q \cdot Q\} \cup \{Q^t\}$.
– By the same argument $Q \cdot Q \cdot Q$ and $Q^t \cdot Q^t$ must also be added to $\overline{O(n, \mathbb{Q})}$. Hence, this process may never end and $\overline{O(n, \mathbb{Q})}$ will become an infinite set. Similarly if we try to construct a finite subgroup $\overline{GL_k(\mathbb{Z})} \leq GL_k(\mathbb{Z})$ we will face the same problem.

In order to deal with these issues we will present two variants of isometric lattice problems. We will show that one of the variant are at least hard as **GI** and **LCE**. We further show that both variants are unlikely to be **NP-complete** unless the polynomial hierarchy collapses [18,19].

3.2   Isometric Lattices over $\mathbb{Z}$

Definition 2. *Let* $\mathbf{B}_1, \mathbf{B}_2 \in \mathbb{Z}^{n \times k}$ *be two bases of rank k. We say that two lattices $\mathcal{L}(\mathbf{B}_1) \cong_{\mathbb{Z}} \mathcal{L}(\mathbf{B}_2)$ are isometric over integers if there exists a matrix $U \in GL_k(\mathbb{Z})$ and a matrix $Q \in O(n, \mathbb{Z})$ such that $\mathbf{B}_2 = Q\mathbf{B}_1 U$.*

Decision Problem ILP$_{\mathbb{Z}}$: Given two matrices $\mathbf{B}_1, \mathbf{B}_2 \in \mathbb{Z}^{n \times k}$, decide whether $\mathcal{L}(\mathbf{B}_1) \cong_{\mathbb{Z}} \mathcal{L}(\mathbf{B}_2)$.

3.3   Isometric Lattices over $\mathcal{R} \subset \mathbb{R}$

Definition 3. *Let* $\mathbf{B}_1, \mathbf{B}_2 \in \mathcal{R}^{n \times k}$ *be two bases of rank k. We say that two lattices $\mathcal{L}(\mathbf{B}_1) \cong_{\mathcal{R}} \mathcal{L}(\mathbf{B}_2)$ are isometric over $\mathcal{R}$ if there exists a matrix $U \in GL_k(\mathbb{Z})$ and a matrix $Q \in \overline{O(n, \mathcal{R})}$ such that $\mathbf{B}_2 = Q\mathbf{B}_1 U$.*

Decision Problem ILP$_{\mathcal{R}}$: Given two matrices $\mathbf{B}_1, \mathbf{B}_2 \in \mathcal{R}^{n \times k}$, decide whether $\mathcal{L}(\mathbf{B}_1) \cong_{\mathcal{R}} \mathcal{L}(\mathbf{B}_2)$.

4   Interactive Proof System for ILP$_{\mathbb{Z}}$

The set of $n \times n$ orthogonal matrices over integers $O(n, \mathbb{Z})$ is finite and of cardinality $2^n \cdot n!$. In fact the set $O(n, \mathbb{Z})$ is exactly equal to the set of $n \times n$ signed permutation matrices. Therefore, any element $Q \in O(n, \mathbb{Z})$ can be written as a product $Q = D \cdot P$ for some $D \in \mathcal{D}_{\epsilon_n}$ and $P \in \mathcal{P}_n$. Furthermore, for any matrix $\mathbf{B} \in \mathbb{Z}^{k \times n}$ the Hermite normal form $\mathbf{HNF}(\mathbf{B})$ only depends on the lattice $\mathcal{L}(\mathbf{B})$ generated by $\mathbf{B}$ and not on a particular lattice basis. Moreover, one can compute $\mathbf{HNF}(\mathbf{B}')$ from any basis $\mathbf{B}'$ of $\mathcal{L}$ in polynomial time [17]. Since $\mathbf{HNF}(\mathbf{B}) = \mathbf{HNF}(\mathbf{B}')$, the Hermite normal form does not give any information about the input basis. This will completely bypass the need for picking random elements from the set $GL_k(\mathbb{Z})$.

An Interactive Proof for ILP$_\mathbb{Z}$

- Input $\mathbf{B}_1, \mathbf{B}_2 \in \mathbb{Z}^{n \times k}$.
  1. Repeat for $l := poly(\|\mathbf{B}_1\| + \|\mathbf{B}_2\|)$ rounds.
     (a) Prover picks uniformly an orthogonal matrix $Q' \in O(n, \mathbb{Z})$.
     (b) Prover computes $\mathbf{H} \leftarrow \mathbf{HNF}(Q'\mathbf{B}_1)$ and sends it to the verifier.
     (c) Verifier randomly picks $c \in \{1, 2\}$ and sends it to the prover.
     (d) Prover sends the verifier an orthogonal matrix $P \in O(n, \mathbb{Z})$.
         i. if $c = 1$ then $P = Q'$.
         ii. if $c = 2$ then $P = Q'Q^t$.
  2. Verifier will accept the proof if for all l rounds $\mathbf{H} = \mathbf{HNF}(PB_c)$.

Theorem 3. *The proof system for* **ILP$_\mathbb{Z}$** *is a malicious verifier perfect-zero knowledge interactive proof with an efficient prover.*

Proof:

Completeness: Clearly, if $\mathcal{L}(\mathbf{B}_1)$ and $\mathcal{L}(\mathbf{B}_2)$ are isometric lattices over the integers, then the prover will never fail convincing the verifier.

Soundness: If $\mathcal{L}(\mathbf{B}_1)$ and $\mathcal{L}(\mathbf{B}_2)$ are not isometric over integers, then the only way for the prover to cheat is to guess c correctly in each round. Since, c is chosen uniformly and independently from $\{1, 2\}$, the probability of prover guessing c in all round is 2^{-l}. Note that verifier's computations are done in polynomial time.

Efficient Prover: The steps $1a$ and $1d$ can be done efficiently. The Hermite normal forms can be computed in polynomial time using the algorithm presented in [17]. Therefore the expected running time of the prover is polynomial.

Zero-Knowledge: Let V^* be any probabilistic polynomial time (possibly malicious) verifier. Let $\mathcal{T}(V^*)$ denote the set of all possible transcripts that could be produced as a result of the prover P and V^* carrying out the interactive proof with a yes instance $(\mathbf{B}_1, \mathbf{B}_2)$ of **ILP$_\mathbb{Z}$**. Let S denote the simulator, which will produce the possible set of forged transcripts $\mathcal{T}(S)$. We denote $\mathbf{Pr}_{V^*}(\mathcal{T})$ the probability distribution on $\mathcal{T}(V^*)$ and we denote $\mathbf{Pr}_S(\mathcal{T})$ the probability distribution on $\mathcal{T}(S)$.

We will show that:

1. The expected running time of S is polynomial.
2. $\mathbf{Pr}_{V^*}(\mathcal{T}) = \mathbf{Pr}_S(\mathcal{T})$ i.e. the two distributions are identical.

Input: $\mathbf{B}_1, \mathbf{B}_2 \in \mathbb{Z}^{n \times k}$ such that $\mathcal{L}(\mathbf{B}_1) \cong_{\mathbb{Z}} \mathcal{L}(\mathbf{B}_2)$.

1. $T = (\mathbf{B}_1, \mathbf{B}_2)$.
2. **for** $j = 1$ **to** $l = poly(\|\mathbf{B}_1\| + \|\mathbf{B}_2\|)$ **do**
   (a) old state $\leftarrow$ state(V^*)
   (b) repeat
      i. Pick uniformly $i \in \{1, 2\}$.
      ii. Pick uniformly Q'_j from $O(n, \mathbb{Z})$.
      iii. Compute $\mathbf{H}'_j \leftarrow \mathbf{HNF}(Q'_j \mathbf{B}_i)$.
      iv. Call V^* with input $\mathbf{H}'_j$ and obtain c'.
      v. **if** $i = c'$ **then**
         – Concatenate $(\mathbf{H}'_j, i, Q'_j)$ to the end of T.
         else
         – Set state$(V^*) \leftarrow$ old state.
      vi. until $i = c'$

Simulator S for ILP$_{\mathbb{Z}}$.

Since V^* runs in polynomial time and that the probability $i = c'$ is $1/2$, on average S will generate two triples $(\mathbf{H}'_j, i, Q'_j)$ for every triple it concatenates to the transcript T and hence, the average running time of S is polynomial.

Using induction we will show that $\mathbf{Pr}_{V^*}(T) = \mathbf{Pr}_S(T)$. Let $\mathbf{Pr}_{V^*}(T_j)$ and $\mathbf{Pr}_S(T_j)$ denote the probability distributions on the partial set of transcripts that could occur at the end of the j-th round.

Base Case: If $j = 0$, then in both case $T = (\mathbf{H}_1, \mathbf{H}_2)$, hence both probabilities are identical.

Inductive Step: Suppose both distributions $\mathbf{Pr}_{V^*}(T_{j-1})$ and $\mathbf{Pr}_S(T_{j-1})$ are identical for some $j \geq 1$.

Now let's go back and see what happens at the j-th round of our interactive proof for ILP$_{\mathbb{Z}}$. The probability that at this round V^* picks $c = 1$ is some number $0 \leq p \leq 1$ and the probability that $c = 2$ is $1 - p$. Moreover, the prover picks an orthogonal matrix Q' with probability $\frac{1}{2^n n!}$. This probability is independent of how the verifier picks $c \in \{1, 2\}$. Therefore the probability that at the j-th round $(\mathbf{H}'_j, i, Q'_j)$ is on the transcript of the **IP** if $c = 1$ is $\frac{p}{2^n n!}$ and if $c = 2$ is $\frac{1-p}{2^n n!}$

The simulator S in any round will pick an orthogonal matrix Q'_j with probability $\frac{1}{2^n n!}$. The probability that $i = 1$ and $c' = 1$ is $\frac{p}{2}$ and the probability $i = 2$ and $c' = 2$ is $\frac{1-p}{2}$.

In both cases the corresponding triple $(\mathbf{H}'_j, i, Q'_j)$ will be written to the transcript. Note with probability $1/2$ nothing is added to the transcript. The probability that $(\mathbf{H}'_j, 1, Q'_j)$ is written on the transcript in j-th round during the m-th iteration of the **repeat** loop is $\frac{p}{2^m \times (2^n n!)}$. Therefore the total probability that $(\mathbf{H}'_j, 1, Q'_j)$ is written on the transcript in the j-th round is

$$\frac{p}{2 \times (2^n n!)} + \frac{p}{2^2 \times (2^n n!)} + \dots + \frac{p}{2^m \times (2^n n!)} + \dots$$

$$= \frac{p}{2 \times (2^n n!)} \left(1 + \frac{1}{2} + \frac{1}{4} + \dots + \frac{1}{2^{m-1}} + \dots \right) = \frac{p}{2^n n!}.$$

Similarly the total probability that $(\mathbf{H}'_j, 2, Q'_j)$ is written on the transcript in the j-th round is $\frac{1-p}{2^n n!}$. Hence, by induction, the two probability distributions are identical $\mathbf{Pr}_{V^*}(\mathcal{T}) = \mathbf{Pr}_S(\mathcal{T})$.

5   Sampling a Lattice Basis in Zero-Knowledge and ILP$_\mathcal{R}$

Suppose $\mathbf{B} \in \mathbb{R}^{n \times k}$ is a basis of some lattice $\mathcal{L}(\mathbf{B})$. Recall that $\mathbf{B}'$ is a basis of $\mathcal{L}(\mathbf{B})$ if and only if $\mathbf{B}' \in \{\mathbf{B}U : U \in GL_k(\mathbb{Z})\}$ and that the algorithm SampleD [2] takes an input basis $\mathbf{B} = [\mathbf{b}_1 | \mathbf{b}_2 | \dots | \mathbf{b}_k] \in \mathbb{R}^{n \times k}$, an appropriate parameters $s \in \mathbb{R}$ and $\mathbf{c} \in \mathbb{R}^n$ and outputs a lattices point $\mathbf{v} \in \mathcal{L}(\mathbf{B})$ that is distributed according to the discrete Gaussian distribution $D_{s,\mathbf{c},\mathcal{L}}$ [2]. SampleD is zero-knowledge in a sense that the output point $\mathbf{v}$ leaks almost no information about the input basis $\mathbf{B}$ except the bound s with overwhelming probability [2]. Furthermore, for an n dimensional $\mathcal{L}$ if we pick $\mathbf{V} = \{\mathbf{v}_1, \mathbf{v}_2, \dots, \mathbf{v}_{n^2}\}$ lattice points independently according to $D_{s,\mathcal{L}}$, then $\mathbf{V}$ contain a subset of k linearly independent vectors, except with $negl(n)$ probability ([12], Corollary 3.16).

Let $\mathbf{B} = \{\mathbf{b}_1, \dots, \mathbf{b}_k\}$ be a basis of a lattice $\mathcal{L}$ and suppose $\mathbf{S} = \{\mathbf{s}_1, \dots, \mathbf{s}_k\}$ is a set of linearly independent vectors that belong to $\mathcal{L}$. There exists a deterministic polynomial time algorithm that will output a basis $\mathbf{T} = \{\mathbf{t}_1, \dots, \mathbf{t}_k\}$ of $\mathcal{L}$ such that $||\mathbf{t}_i||_2 \le ||\mathbf{s}_i||_2$ for $1 \le i \le k$ ([1], p. 129).

Using the above two algorithms we will present a probabilistic polynomial time algorithm Sample$\mathcal{L}$ that will take an input basis $\mathbf{B} = \{\mathbf{b}_1, \dots, \mathbf{b}_k\}$ of some lattice $\mathcal{L}$, $\mathbf{c} \in \mathbb{R}^n$, a parameter $s \ge \omega(\sqrt{\log n}) \cdot ||\widetilde{\mathbf{B}}||$ and outputs a basis $\mathbf{T}$, such that $\mathbf{T}$ leaks no information about the basis $\mathbf{B}$, except s (the bound on the norm of $\mathbf{B}$) with overwhelming probability.

Protocol 1. Sample$\mathcal{L}$

Input $(\mathbf{B} \in \mathbb{R}_\mathcal{R}^{n \times k}, k, n, s)$

1. Sample $\mathbf{V} = \{\mathbf{v}_1, \mathbf{v}_2, \dots, \mathbf{v}_{n^2}\}$ points independently using the algorithm SampleD$(\mathbf{B}, 0, s)$).
2. Pick $\mathbf{S} = \{\mathbf{s}_1, \mathbf{s}_2, \dots, \mathbf{s}_k\} \subset \mathbf{V}$, such that $\mathbf{S}$ is a set of linearly independent vectors.
3. Using the deterministic algorithm output the basis $\mathbf{T}$, such that $\mathcal{L}(\mathbf{T}) = \mathcal{L}(\mathbf{B})$.

It is easy to see that if $\mathbf{B} \in \mathbb{R}_\mathcal{R}^{n \times k}$ then so $\mathbf{T} \in \mathbb{R}_\mathcal{R}^{n \times k}$. Since $\mathbf{T}$ and $\mathbf{B}$ are bases of the same lattice, there exists a $U \in GL_k(\mathbb{Z})$ such that

$$\mathbf{T} = \mathbf{B}U.$$

6   An Interactive Proof for ILP$_\mathcal{R}$

- Input $\mathbf{B}_1, \mathbf{B}_2 \in \mathcal{R}^{n \times k}$ such that $\mathcal{L}(\mathbf{B}_1) \cong_\mathcal{R} \mathcal{L}(\mathbf{B}_2)$.
  1. Prover set $s = \log n \cdot max\{\|\widetilde{\mathbf{B}}_1\|, \|\widetilde{\mathbf{B}}_2\|\}$.
  2. for $i = 1$ to $l = poly(\|\mathbf{B}_1\| + \|\mathbf{B}_2\|)$ rounds do.
     (a) Prover picks uniformly an orthogonal matrix $Q'_j \leftarrow \overline{O(n, \mathcal{R})}$.
     (b) Prover picks $\mathbf{B}'_j \leftarrow \mathrm{Sample}\mathcal{L}\left(Q'_j \mathbf{B}_1, k, n, s\right)$.
     (c) Prover sends the basis $\mathbf{B}'_j$ to the verifier.
     (d) Verifier randomly picks $c_j \in \{1, 2\}$ and sends it to the prover.
     (e) Prover sends the verifier an orthogonal matrix $P_j \in \overline{O(n, \mathcal{R})}$.
         i. if $c_j = 1$, then $P_j = Q'_j$.
         ii. if $c_j = 2$ then $P_j = Q'_j Q^\mathsf{t}$, where $Q \in \overline{O(n, \mathcal{R})}$ is such that $\mathcal{L}(\mathbf{B}_2) = \mathcal{L}(Q \mathbf{B}_1)$.
  3. Verifier will accept the proof if for all l rounds $\mathcal{L}(\mathbf{B}) = \mathcal{L}(P_j \mathbf{B}_{c_j})$.

Theorem 4. *The proof system for* **ILP**$_\mathcal{R}$ *is a statistical zero-knowledge interactive proof with an efficient prover.*

Proof:

Completeness: If $\mathcal{L}(\mathbf{B}_1)$ and $\mathcal{L}(\mathbf{B}_2)$ are isometric lattices, then $\mathbf{B}_2 = Q\mathbf{B}_1 U$ for some $Q \in \overline{O(n, \mathcal{R})}$ and $U \in GL_k(\mathbb{Z})$. Clearly,

$$\mathcal{L}(Q'_j \mathbf{B}_1) = \mathcal{L}(\mathbf{B}) = \mathcal{L}(Q'_j Q^\mathsf{t} \mathbf{B}_2)$$

since $\mathbf{B}'_j = Q'_j \mathbf{B}_1 U'_j$ and $\mathbf{B}'_j = Q'_j Q^\mathsf{t} \mathbf{B}_2 U U'_j$ for some $U'_j \in GL_k(\mathbb{Z})$. Therefore, the prover will always be able to convince the verifier.

Soundness: If $\mathcal{L}(\mathbf{B}_1)$ and $\mathcal{L}(\mathbf{B}_2)$ are not isometric over $\mathcal{R}$, then the only way for the prover to deceive the verifier is for him to guess correctly c_j in each round. Since c_j is chosen uniformly from $\{1, 2\}$, the probability of the prover guessing c_j in all rounds is 2^{-l}. Hence, the protocol is sound.

Efficient Prover: Clearly the prover can perform steps 1, 2a, 2c and 2e in expected polynomial-time. In step 2b the prover picks a lattice basis using Sample$\mathcal{L}$, which runs in expected polynomial time. Hence the total expected running time of the prover is polynomial.

Zero-Knowledge: Let V^* be any probabilistic polynomial time (possibly malicious) verifier. Let $\mathcal{T}(V^*)$ denote the set of all possible transcripts that could be produced as a result of P and V^* carrying out the interactive proof on a **yes** instance $(\mathbf{B}_1, \mathbf{B}_2)$ of **ILP**$_\mathcal{R}$. Let $S_\mathcal{R}$ denote the simulator, which will produce the possible set of forged transcripts $\mathcal{T}(S_\mathcal{R})$. We denote $\mathbf{Pr}_{V^*}(\mathcal{T})$ the probability distribution on $\mathcal{T}(V^*)$ and we denote $\mathbf{Pr}_{S_\mathcal{R}}(\mathcal{T})$ the probability distribution on $\mathcal{T}(S_\mathcal{R})$. We will prove that:

1. $S_{\mathcal{R}}$ is polynomial.
2. $\mathbf{Pr}_{V^*}(\mathcal{T}) \sim \mathbf{Pr}_{S_{\mathcal{R}}}(\mathcal{T})$ i.e. the two distributions are statistically close.

Input: $\mathbf{B}_1, \mathbf{B}_2 \in \mathcal{R}^{n \times k}$ such that $\mathcal{L}(\mathbf{B}_1) \cong_{\mathcal{R}} \mathcal{L}(\mathbf{B}_2)$.

1. Set $s = \log n \cdot max\{\|\widetilde{\mathbf{B}}_1\|, \|\widetilde{\mathbf{B}}_2\|\}$.
2. $T = (\mathbf{B}_1, \mathbf{B}_2)$.
3. **for** $j = 1$ **to** $l = poly(\|\mathbf{B}_1\| + \|\mathbf{B}_2\|)$ **do**
   (a) old state $\leftarrow$ state(V^*)
   (b) repeat
       i. Pick uniformly $i_j \in \{1, 2\}$.
       ii. Pick uniformly Q'_j from $\in \overline{O(n, \mathcal{R})}$.
       iii. Compute $\mathbf{H}'_j \leftarrow \text{Sample}\mathcal{L}\left(Q'_j \mathbf{B}_{i_j}, k, n, s\right)$.
       iv. Call V^* with $\mathbf{H}'_j$ and obtain i'.
       v. **if** $i_j = i'$ **then**
          – Concatenate $\left(\mathbf{H}'_j, i_j, Q'_j\right)$ to the end of T.
          else
          – Set state(V^*) $\leftarrow$ old state.
       vi. until $i_j = i'$.

Simulator $S_{\mathcal{R}}$ for ILP$_{\mathcal{R}}$.

Running Time of the Simulator: What is the probability that $i_j = i'$? In other words, on average how many triples $\left(\mathbf{H}'_j, i_j, Q'_j\right)$ will the simulator $S_{\mathcal{R}}$ generate for every triple it concatenates to T? We note that $Q'Q^t$ and Q' are uniformly distributed over $\overline{O(n, \mathcal{R})}$, and $\mathcal{L}(Q'\mathbf{B}_1) = \mathcal{L}(Q'Q^t\mathbf{B}_2)$ therefore the probability that the lattice $\mathcal{L}(\mathbf{H}'_j)$ is obtained by rotating the lattice $\mathcal{L}(\mathbf{B}_1)$ is equal to the probability that it is obtain by rotating $\mathcal{L}(\mathbf{B}_2)$. Furthermore the algorithm Sample$\mathcal{L}$ ensures that as far as the parameters are chosen appropriately, $\mathbf{H}'_j$ will leak almost no information (apart from the bound s) about the input basis except with negligible probability. Hence, on the average the simulator will generate roughly 2 triples for every triple it adds to T. Therefore the expected running time of $S_{\mathcal{R}}$ is roughly twice the running time of V^*. By definition V^* runs in probabilistic polynomial time. Hence the running time of $S_{\mathcal{R}}$ is also expected polynomial time.

We will prove that the two probability distributions $\mathbf{Pr}_{V^*}(\mathcal{T})$ and $\mathbf{Pr}_{S_{\mathcal{R}}}(\mathcal{T})$ are statistically close as follows. We first prove that the two distributions are statistically close for one round ($l = 1$). Then we will invoke the sequential composition Lemma 4.3.11 on page 216 of [9], which implies that an interactive proof which is zero-knowledge for one round remains zero-knowledge for polynomially many rounds.

Case $l = 1$: Let $(\mathbf{B}'_1, c_1, P'_1)$ denote a transcript produced as a result of an interactive proof and $(\mathbf{H}'_1, i_1, Q'_1)$ denote a transcript produced by the simulator. In the interactive proof P picks uniformly P'_1 over $\overline{O(n, \mathcal{R})}$ and $S_{\mathcal{R}}$ also picks Q'_1

uniformly over $\overline{O(n, \mathcal{R})}$. Hence both P_1' and Q_1' are identically distributed. Also $\mathbf{B}_1'$ and $\mathbf{H}_1'$ are computed by Sample$\mathcal{L}$. Therefore they are almost identically distributed to $D_{s,c,\mathcal{L}}$ and thus to each other.

Let p be the probability that V^* picks $c_1 = 1$ and $1 - p$ be the probability that it picks $c_1 = 2$ in the interactive proof. The probability may depend on the state of V^*. The simulator picks $i_1 \in \{1, 2\}$ uniformly and independent of how V^* picks i'. Also given $\mathbf{H}_1'$, the probability that V^* can guess the index i_1 is at most $\frac{1}{2} + negl.$ Therefore probability that V^* picks $i' = 1$ is nearly p and $i' = 2$ is nearly $1 - p$ respectively. This means that i_1 and c_1 have nearly the same distributions.

Therefore, it follows that $(\mathbf{B}_1', c_1, P_1')$ and $(\mathbf{H}_1', i_1, Q_1')$ are statistically close. Hence for one round the two distributions are statistically close. Hence, by Lemma 4.3.11 for any polynomially many rounds we have $\mathbf{Pr}_{V^*}(\mathcal{T}) \sim \mathbf{Pr}_{S_{\mathcal{R}}}(\mathcal{T})$.

7   Isometric Lattice Problem Is Not Easy

In this section we will show that $\mathbf{ILP}_{\mathbb{Z}}$ is at least as hard as Linear Code Equivalence problem over prime fields $\mathbb{F}_p$ and Graph Isomorphism.

Theorem 5. $\mathbf{ILP}_{\mathbb{Z}}$ *is at least as hard as* $\mathbf{LCE}$ (*Linear Code Equivalence problem*) *over prime fields* $\mathbb{F}_p$.

Proof: Let $\mathbf{G} = [\mathbf{g}_1 | \dots | \mathbf{g}_k] \in \mathbb{F}_p^{n \times k}$ be a basis of some $[k, n]$ linear code C

$$\psi : C \longrightarrow \Lambda_2(\mathbf{G}); \quad \mathbf{G} \longrightarrow \mathbf{B}$$

where $\Lambda_2(\mathbf{G})$ be the corresponding p-ary lattice. Recall from Sect. 2 that $\mathbf{B} = [\mathbf{g}_1 | \dots | \mathbf{g}_k | \mathbf{b}_{k+1} | \dots | \mathbf{b}_n] \in \mathbb{Z}^{n \times n}$ is a basis of $\Lambda_p(\mathbf{G})$. Where $\mathbf{b}_j = (0, ..., p, ..., 0) \in \mathbb{Z}^n$ and the j-th coordinate is equal to p, for $k + 1 \leq j \leq n$. Clearly the map ψ can be computed in polynomial time. Let $\mathbf{G}_1 = [\mathbf{g}_{11} | \dots | \mathbf{g}_{1k}] \in \mathbb{F}_p^{n \times k}$ and $\mathbf{G}_2 = [\mathbf{g}_{21} | \dots | \mathbf{g}_{2k}] \in \mathbb{F}_p^{n \times k}$ be two code generators.

$\Longrightarrow$ Suppose $\mathbf{G}_1$ and $\mathbf{G}_2$ generate linearly equivalent codes i.e. $\mathbf{G}_2 = P\mathbf{G}_1 M$ for $M \in GL_k(\mathbb{F}_p)$ and monomial matrix $P' \in \mathcal{P}(n, \mathbb{F}_q)$. Note that we can write P' as a product of a permutation matrix $P \in \mathcal{P}_n$ and an invertible diagonal matrix $D \in \mathbb{F}_p^{n \times k}$. Write $\mathbf{G}_2 = P\mathbf{G}_1' M$, where $\mathbf{G}_1' = D\mathbf{G}_1$ and let $\Lambda_p(\mathbf{G}_1')$ and $\Lambda_p(\mathbf{G}_2)$ be corresponding lattices.

$$\text{For any } \mathbf{v} \in \Lambda_p(\mathbf{G}_2) \Longleftrightarrow \mathbf{v} \equiv \mathbf{G}_2 \cdot \mathbf{s} \pmod{p}, \text{ for some } \mathbf{s} \in \mathbb{Z}^k$$
$$\Longrightarrow \mathbf{v} \equiv P\mathbf{G}_1' M \cdot \mathbf{s} \pmod{p} \equiv P\mathbf{G}_1' \cdot \mathbf{s}' \pmod{p}, \mathbf{s}' = M\mathbf{s} \in \mathbb{Z}^k$$
$$\Longrightarrow \mathbf{v} \in \Lambda_p(P\mathbf{G}_1')$$

Hence, $\Lambda_p(\mathbf{G}_2) \subseteq \Lambda_p(P\mathbf{G}_1')$. Since, $P\mathbf{G}_1' = \mathbf{G}_2 M^{-1}$, by the same argument $\Lambda_p(P\mathbf{G}_1') \subseteq \Lambda_p(\mathbf{G}_2)$, we have $\Lambda_p(P\mathbf{G}_1') = \Lambda_p(\mathbf{G}_2)$. Therefore, there exists a $U \in GL_k(\mathbb{Z})$ such that

$$\psi(\mathbf{G}_2) = \psi(P\mathbf{G}_1')U = P\psi(\mathbf{G}_1')U$$

$\Longleftarrow$ Now suppose $\mathbf{G}_1$ and $\mathbf{G}_2$ are not linearly equivalent and suppose $\psi(\mathbf{G}_2) = Q\psi(\mathbf{G}_1)U$ for $Q \in O(n, \mathbb{Z})$ and $U \in GL_k(\mathbb{Z})$. Note we can write any $Q \in O(n, \mathbb{Z})$ as $Q = PD_\epsilon$, for some $D_\epsilon \in \mathcal{D}_{\epsilon_n}$ and $P \in \mathcal{P}_n$. But $P' = PD_\epsilon \pmod{p}$ is a monomial matrix. Further U is also non-singular over $\mathbb{F}_p$. Therefore, $\psi(\mathbf{G}_2) = Q\psi(\mathbf{G}_1)U$, which implies

$$\mathbf{G}_2 = P'(\mathbf{G}_1)M \quad \bmod p \text{for some} M \in GL_k(\mathbb{F}_p) \text{ and } M \equiv U \pmod{p}$$

This contradicts the assumption that $\mathbf{G}_1$ and $\mathbf{G}_2$ are not linearly equivalent. Therefore $\mathbf{ILP}_\mathbb{Z}$ is at least as hard as $\mathbf{LCE}$.

Theorem 6. $\mathbf{ILP}_\mathbb{Z}$ *is at least as hard as the* $\mathbf{GI}$ *(Graph Isomorphism) problem.*

Proof: Petrank and Roth [20] reduced $\mathbf{GI}$ to $\mathbf{PCE}$ (Permutation Code Equivalence). More precisely they provided a polynomial time mapping ϕ from the set of all graphs to the set of generator matrices over $\mathbb{F}_2$ such that two graphs G_1 and G_2 are isomorphic if and only if $\phi(G_1)$ and $\phi(G_2)$ are permutation equivalent codes. We will prove that $\mathbf{ILP}$ is at least as hard as $\mathbf{GI}$, by reducing the $\mathbf{PCE}$ over $\mathbb{F}_2$ to $\mathbf{ILP}$. Let $\mathbf{G} = [\mathbf{g}_1|\ldots|\mathbf{g}_k] \in \mathbb{F}_2^{n \times k}$ be a basis of some $[k, n]$ linear code C

$$\psi : C \longrightarrow \Lambda_2(\mathbf{G}); \quad \mathbf{G} \longrightarrow \mathbf{B}$$

where $\Lambda_2(\mathbf{G})$ is the corresponding 2-ary lattice. Recall from Sect. 2 that $\mathbf{B} = [\mathbf{g}_1|\ldots|\mathbf{g}_k|\mathbf{b}_{k+1}|\ldots|\mathbf{b}_n] \in \mathbb{Z}^{n \times n}$ is a basis of $\Lambda_2(\mathbf{G})$. Where $\mathbf{b}_j = (0, ..., 2, ..., 0) \in \mathbb{Z}^n$ and the j-th coordinate is equal to2, for $k + 1 \le j \le n$. Clearly the map ψ can be computed in polynomial time. Let $\mathbf{G}_1 = [\mathbf{g}_{11}|\ldots|\mathbf{g}_{1k}] \in \mathbb{F}_2^{n \times k}$ and $\mathbf{G}_2 = [\mathbf{g}_{21}|\ldots|\mathbf{g}_{2k}] \in \mathbb{F}_2^{n \times k}$ be two code generators and $\Lambda_2(\mathbf{G}_1)$ and $\Lambda_2(\mathbf{G}_2)$ be corresponding lattices.

$\Longrightarrow$) Suppose $\mathbf{G}_1$ and $\mathbf{G}_2$ are permutation equivalent i.e. $\mathbf{G}_2 = P\mathbf{G}_1M$ for $M \in GL_k(\mathbb{F}_2)$ and $P \in \mathcal{P}_n$. Let $\mathbf{G}_1' = P\mathbf{G}_1$. Therefore we can write $\mathbf{G}_2 = \mathbf{G}_1'M$. By definition for any $\mathbf{v} \in \Lambda_2(\mathbf{G}_2)$, there exists an $\mathbf{s} \in \mathbb{Z}^k$ such that

$$\mathbf{v} \equiv \mathbf{G}_2 \cdot \mathbf{s} \equiv \mathbf{G}_1'M \cdot \mathbf{s} \pmod{2}.$$
$$\Longrightarrow \mathbf{v} \equiv P\mathbf{G}_1 \cdot \mathbf{s}' \pmod{2}, \text{ where } \mathbf{s}' = M \cdot \mathbf{s} \in \mathbb{Z}^k \Longrightarrow \mathbf{v} \in \Lambda_2(P\mathbf{G}_1).$$

Hence, $\Lambda_2(\mathbf{G}_2) \subseteq \Lambda_2(P\mathbf{G}_1)$. Since, $P^t\mathbf{G}_2M^{-1} = \mathbf{G}_1$ by the same argument $\Lambda_2(P\mathbf{G}_1) \subseteq \Lambda_2(\mathbf{G}_2)$. Hence, there exist a $U \in GL_k(\mathbb{Z})$ such that

$$\psi(\mathbf{G}_2) = \psi(P\mathbf{G}_1)U \Longrightarrow \mathbf{B}_2 = P\mathbf{B}_1U$$

$\Longleftarrow$) Now suppose $\mathbf{G}_1$ and $\mathbf{G}_2$ are not permutation equivalent and suppose $\psi(\mathbf{G}_2) = Q\psi(\mathbf{G}_1)U$ for $Q \in O(n, \mathbb{Z})$ and $U \in GL_k(\mathbb{Z})$. Note that $Q \equiv P \pmod{2}$, for some $P \in \mathcal{P}_n$. For every $\mathbf{v} \in \Lambda_2(\mathbf{G}_2)$ we have

$$\mathbf{v} \equiv \mathbf{G}_2\mathbf{u} \pmod{2} \text{ for some } \mathbf{u} \in \mathbb{Z}^k.$$

Since, $\Lambda_2(Q\mathbf{G}_1) = \Lambda_2(\mathbf{G}_2)$, we also have $\mathbf{v} \equiv (Q\mathbf{G}_1)\mathbf{u} \equiv (P\mathbf{G}_1)\mathbf{u} \pmod{2}$ for some $\mathbf{u} \in \mathbb{Z}^k$. This means that $P\mathbf{G}_1$ and $\mathbf{G}_2$ have the same span over $\mathbb{F}_2$. This contradicts the assumption that $\mathbf{G}_1$ and $\mathbf{G}_2$ are not permutation equivalent. This proves that $\mathbf{ILP}$ is at least as hard as $\mathbf{GI}$.

7.1   ILP is unlikely to be NP-complete

In this sub-section we show that $\textbf{ILP}_\mathbb{S}$ is unlikely to be NP-complete (where $\mathbb{S} = \mathbb{Z}$ or $\mathbb{S} = \mathcal{R}$ see Sect. 3.1). We do this by constructing a constant round interactive proof for the **Non-Isometric Lattice problem (co-ILP$_\mathbb{S}$)**, i.e. the complementary problem of $\textbf{ILP}_\mathbb{S}$. Then we invoke results from the field of complexity theory, implying that if the complement of a problem Π has a constant round interactive proof and Π is NP-complete then the polynomial hierarchy collapses [18,19]. It is widely believed that the polynomial hierarchy does not collapse, therefore we end up with the conclusion that $\textbf{ILP}$ is unlikely to be NP-complete.

Constant Round IP for co-ILP$_\mathbb{S}$

- Input $\mathbf{B}_1, \mathbf{B}_2 \in \mathbb{S}^{n \times k}$ bases such that $\mathcal{L}(\mathbf{B}_1) \not\cong_\mathbb{S} \mathcal{L}(\mathbf{B}_2)$.
  1. Verifier sets $l = poly(|\mathbf{B}_1| + |\mathbf{B}_2|)$.
  2. Verifier picks uniformly $j_1, \ldots, j_l \in \{1, 2\}$.
  3. If $\mathbb{S} = \mathbb{Z}$ then the verifier picks independent random orthogonal matrices
  $$Q_1, \ldots, Q_l \in O(n, \mathbb{Z}).$$
  Else verifier picks independently random orthogonal matrices
  $$Q_1, \ldots, Q_l \in \overline{O(n, \mathcal{R})}.$$
  4. For $1 \leq i \leq l$, verifier computes a basis $\mathbf{H}_i'$ for the lattice $\mathcal{L}(Q_i \mathbf{B}_{j_i})$. If $\mathbb{S} = \mathbb{Z}$, then $\mathbf{H}_i' \leftarrow \mathbf{HNF}(Q_i \mathbf{B}_{j_i})$, otherwise $\mathbf{H}_i'$ is computed using algorithm Sample$\mathcal{L}$ from Sect. 5.
  5. For $1 \leq i \leq l$, the all-powerful prover computes and sends j_i' such that $\mathbf{H}_i'$ and $\mathbf{B}_{j_i'}$ are isometric.
  6. Verifier accepts the proof if $j_i = j_i'$ for all $1 \leq i \leq l$.

Completeness: Clearly, if $\mathcal{L}(\mathbf{B}_1)$ and $\mathcal{L}(\mathbf{B}_2)$ are non-isometric lattices then the prover will never fail convincing the verifier.

Soundness: Suppose $\mathcal{L}(\mathbf{B}_1)$ and $\mathcal{L}(\mathbf{B}_2)$ are isometric lattices. The probability that prover can guess $(i_1, ..., i_l)$ given $(\mathbf{H}_1', ..., \mathbf{H}_l')$ is 2^{-l} if $\mathbb{S} = \mathbb{Z}$ and $2^{-l} + negl$ if $\mathbb{S} = \mathcal{R}$.

8   Conclusion and Acknowledgement

We conclude with an open problem related to our work. Construct a Malicious verifier statistical zero-knowledge proof system with an efficient prover for the Isometric Lattice Problem over rationals $\textbf{ILP}_\mathbb{Q}$. We would also like to thank Professor Chris Peikert, for his help and patience, who always took time out of his busy schedule to answer our questions.

A   Computing Sine and Cosine Efficiently

Let $p(n)$ be any desired publicly known positive polynomial. Recall that

$$\sin\left(\frac{\pi}{2^{p(n)}}\right) = \frac{1}{2} \underbrace{< \frac{1}{2}, 2 - < \frac{1}{2}, 2 + \cdots + < \frac{1}{2}, 2 >> \cdots >}_{p(n)-1}$$

$$\cos\left(\frac{\pi}{2^{p(n)}}\right) = \frac{1}{2} \underbrace{< \frac{1}{2}, 2 + < \frac{1}{2}, 2 + \cdots + < \frac{1}{2}, 2 >> \cdots >}_{p(n)-1}.$$

Suppose we have to compute $\sin\left(\frac{l \cdot \pi}{2^{p(n)}}\right)$ for some $0 \leq l \leq 2^{p(n)}$.

$$\sin(\alpha + \beta) = \sin(\alpha)\cos(\beta) + \sin(\beta)\cos(\alpha)$$
$$\cos(\alpha + \beta) = \cos(\alpha)\cos(\beta) - \sin(\alpha)\sin(\beta)$$

Write $l = \sum_{i=0}^{k} x_i \cdot 2^i$, $x_i \in \{0,1\}$ and $k \leq p(n)$. WLOG we can assume that l is not even.

$$\sin\left(\frac{l \cdot \pi}{2^{p(n)}}\right) = \sin\left(\frac{\pi}{2^{p(n)-k}} + \cdots + \frac{\pi}{2^{p(n)}}\right)$$

$$= \sin\left(\frac{\pi}{2^{p(n)-k}}\right)\cos\left(\frac{\left[\sum_{i=0}^{k-1} x_i 2^i\right]\pi}{2^{p(n)}}\right)$$

$$+ \sin\left(\frac{\left[\sum_{i=0}^{k-1} x_i 2^i\right]\pi}{2^{p(n)}}\right)\cos\left(\frac{\pi}{2^{p(n)-k}}\right).$$

Note that $\sin\left(\frac{\pi}{2^{p(n)-k}}\right)$ and $\cos\left(\frac{\pi}{2^{p(n)-k}}\right)$ can be computed directly. Now we can recursively compute $\cos\left(\frac{\left[\sum_{i=0}^{k-1} x_i 2^i\right]\pi}{2^{p(n)}}\right)$ and $\sin\left(\frac{\left[\sum_{i=0}^{k-1} x_i 2^i\right]\pi}{2^{p(n)}}\right)$. But since $\sin(\theta)^2 = 1 - \cos^2(\theta)$, in recursion we will only have to compute either $\cos\left(\frac{\left[\sum_{i=0}^{k-1} x_i 2^i\right]\pi}{2^{p(n)}}\right)$ or $\sin\left(\frac{\left[\sum_{i=0}^{k-1} x_i 2^i\right]\pi}{2^{p(n)}}\right)$.

Clearly depth of the recursion is $k \leq p(n)$ and for each recursive step we will have four values, with each value is of size $O(p(n))$. Hence in total running time is at most $O(p(n))$ operations. Similarly, one can show that $\cos\left(\frac{l \cdot \pi}{2^{p(n)}}\right)$ for any $0 \leq l \leq 2^{p(n)}$, can be computed in polynomial time as well.

References

1. Micciancio, D., Goldwasser, S.: Complexity of Lattice Problems: A Cryptographic Perspective. Springer International Series in Engineering and ComputerScience, vol. 671. Springer, USA (2002)

2. Gentry, C., Peikertm, C., Vaikuntanathan, V.: How to use a short basis: trapdoors for hard lattices and new cryptographic constructions. In: STOC, pp. 197–206 (2008)

3. Cash, D., Hofheinz, D., Kiltz, E., Peikert, C.: Bonsai trees, or how to delegate a lattice basis. In: Gilbert, H. (ed.) EUROCRYPT 2010. LNCS, vol. 6110, pp. 523–552. Springer, Heidelberg (2010)

4. Garey, M.R., Johnson, D.S.: Computers and Intractability: A Guide to the Theory of NP-Completeness. W. H. Freeman & Company, New York (1990)

5. Kazmi, R.A.: Cryptography from Post-Quantum Assumptions. Ph.D. Thesis, School of Computer Science, McGill University, 2015. Supervised by Claude Crépeau. https://eprint.iacr.org/2015/376

6. Goldwasser, S., Micali, S., Rackoff, C.: The knowledge complexity of interactive proof systems. SIAM J. Comput. **18**(1), 186–208 (1989)

7. Sendrier, N., Simos, D.E.: The hardness of code equivalence over $\mathbb{F}_q$ and its application to code-based cryptography. In: Gaborit, P. (ed.) PQCrypto 2013. LNCS, vol. 7932, pp. 203–216. Springer, Heidelberg (2013)

8. Goldreich, O., Micali, S., Wigderson, A.: How to play any mental game or a completeness theorem for protocols with honest majority. In: STOC, pp. 218–229 (1987)

9. Goldreich, O.: Foundations of Cryptography, vol. I & II. Cambridge University Press (2001–2004)

10. Micciancio, D., Vadhan, S.P.: Statistical zero-knowledge proofs with efficient provers: lattice problems and more. In: Boneh, D. (ed.) CRYPTO 2003. LNCS, vol. 2729, pp. 282–298. Springer, Heidelberg (2003)

11. Peikert, C., Vaikuntanathan, V.: Noninteractive statistical zero-knowledge proofs for lattice problems. In: Wagner, D. (ed.) CRYPTO 2008. LNCS, vol. 5157, pp. 536–553. Springer, Heidelberg (2008)

12. Regev, O.: On lattices, learning with errors, random linear codes, and cryptography. In: STOC, pp. 84–93 (2005)

13. Goldreich, O., Goldwasser, S.: On the limits of non-approximability of lattice problems. In: STOC, pp. 23–26 (1998)

14. Stewart, G.W.: The efficient generation of random orthogonal matrices with an application to condition estimators. SIAM J. Numer. Anal. **17**(3), 403–409 (1980)

15. Marsaglia, G.: Choosing a point from the surface of a sphere. Ann. Math. Stat. **43**(2), 645–647 (1972)

16. Schmutz, E.: Rational points on the unit sphere. Cent. Eur. J. Math. **6**(3), 482–487 (2008)

17. Pernet, C., Stein, W.: Fast computation of hermite normal forms of random integer matrices. J. Number Theor. **130**(7), 1675–1683 (2010)

18. Goldwasser, S., Sipser, M.: Private coins versus public coins in interactive proof systems. In: STOC, pp. 59–68 (1986)

19. Boppana, R.B., Håastad, J., Zachos, S.: Does co-NP have short interactive proofs? J. Inf. Process. Lett. **25**(2), 127–132 (1987)

20. Petrank, E., Roth, R.M.: Is code equivalence easy to decide? IEEE Trans. Inf. Theor. **43**(5), 1602–1604 (1997)

21. Micciancio, D., Peikert, C.: Trapdoors for lattices: simpler, tighter, faster, smaller. In: Pointcheval, D., Johansson, T. (eds.) EUROCRYPT 2012. LNCS, vol. 7237, pp. 700–718. Springer, Heidelberg (2012)

22. Goldreich, O., Sahai, A., Vadhan, S.: Honest-verifier statistical zero-knowledge equals general statistical zero-knowledge. In: STOC, pp. 399–408 (1998)

23. Tanner, M.A., Thisted, R.A.: Appl. Stat. **31**, 199–206 (1982)
24. Liebeck, H.: Osborne, anthony: the generation of all rational orthogonal matrices. Am. Math. Monthly **98**(2), 131–133 (1991)
25. Bernstein, D.J., Buchmann, J.A., Dahmen, E.: Post-Quantum Cryptography. Number Theory and Discrete Mathematics. Springer, Heidelberg (2008). ISBN 978-3-540-88701-0
26. Garey, M., Johnson, D.S.: Computers and Intractability: A Guide to the Theory of NP-Completeness. W. H. Freeman & Co., New York (1990). ISBN 0-7167-1045-5
27. McEliece, R.J.: A public-key cryptosystem based on algebraic coding theory. Technical memo, California Institute of Technology (1978)

Codes

Soft Distance Metric Decoding of Polar Codes

Monica C. Liberatori[1], Leonardo J. Arnone[2], Jorge Castiñeira Moreira[3]([✉]),
and Patrick G. Farrell[4]

[1] Communications Laboratory, ICYTE and Electronics Departament,
Engineering School, Mar del Plata University, Mar del Plata, Argentina
[2] Components Laboratory, ICYTE and Electronics Departament,
Engineering School, Mar del Plata University, Mar del Plata, Argentina
[3] Communications Laboratory, ICYTE and Electronics Departament, Engineering
School, Mar del Plata University and CONICET, Mar del Plata, Argentina
casti@fi.mdp.edu.ar
[4] FREng, FIMA, FIET, FIEEE, Deal, UK

Abstract. In this paper, we implement the Successive Cancellation (SC) decoding algorithm for Polar Codes by using Euclidean distance estimates as the metric of the algorithm. This implies conversion of the classic statistical recursive expressions of the SC decoder into a suitable form, adapting them to the proposed metric, and properly expressing the initialization values for this metric. This leads to a simplified version of the logarithmic SC decoder, which offers the advantage that the algorithm can be directly initialised with the values of the received channel samples. Simulations of the BER performance of the SC decoder, using both the classic statistical metrics, and the proposed Euclidean distance metric, show that there is no significant loss in BER performance for the proposed method in comparison with the classic implementation. Calculations are simplified at the initialization step of the algorithm, since neither is there a need to know the noise power variance of the channel, nor to perform complex and costly mathematical operations like exponentiations, quotients and products at that step. This complexity reduction is especially important for practical implementations of the SC decoding algorithm in programmable logic technology like Field Programmable Gate Arrays (FPGAs).

Keywords: Soft distance · Successive cancellation decoding · Polar codes

1 Introduction

Ever since Claude Shannon proved that there is a maximum rate at which information can be reliably transmitted over a channel, a parameter that is known as the channel capacity [1], researchers have looked for efficient error-control techniques that can approach capacity. However, most of this research only addressed heuristic methods. The design of structural capacity-achieving codes had to wait until the appearance of Polar Codes, an error-control coding technique proposed

© Springer International Publishing Switzerland 2015
J. Groth (Ed.): IMACC 2015, LNCS 9496, pp. 173–183, 2015.
DOI: 10.1007/978-3-319-27239-9_10

by Arikan [2], who proved that polar codes can asymptotically achieve the capacity of Binary-input discrete memoryless channels (B-DMCs).

Subsequent research also proved that Polar Codes achieve the capacity of any discrete or continuous input alphabets in memoryless channels [3]. The construction of optimized Polar Codes depends on the characteristics of the channel. A first method of constructing capacity-achieving codes for B-DMCs was presented in [2], for the Binary Erasure Channel (BEC). The extension to channels like the AWGN channel or the Rayleigh channel requires a more elaborate design. Arikan proposed a heuristic approach in [4]. Another method, making use of density evolution, was proposed in [5,6]. This last approach requires large amounts of memory and involves a high computation complexity that increases with the code length.

An alternative method, used here in this paper, was presented in [8], where Bhattacharyya parameters, involved in the determination of the best sub-channels in the polarization process, are evaluated using a Gaussian approximation (GA). This method leads to the construction of Polar Codes for the AWGN channel.

Arikan showed that these codes can be efficiently encoded and decoded with complexity $O\left(N log N\right)$ where N is the code block length [2]. The decoding method is called Successive Cancellation (SC) decoding, because in essence it sequentially provides estimates of the received bits, in the order in which they are input to the decoder. This process can be greatly simplified by taking advantage of the recursive structure of polar codes. The metric used in the algorithm is the Likelihood Ratio (LR), and a further simplification is obtained by using calculations in the logarithmic domain, operating with Logarithmic Likelihood Ratios (LLR).

Polar Codes need to operate with very long block lengths, in order to effectively approach the capacity of the channel, which results in highly complex SC decoders. Many simplification procedures have been applied to the SC decoding algorithm in order to reduce its complexity, particularly for practical implementations that make use of programmable technology, such as in the case of FPGA devices. Examples of such techniques include semi-parallel decoder implementations [7], and decomposition of Polar Codes into its constituent sub-codes which are easier to decode [9].

In this paper we explore successive cancellation decoding of Polar Codes using the non-statistical Euclidean Soft Distance (SD) metric. This metric, when applied to the sum-product decoding algorithm for Low-Density Parity-Check (LDPC) codes over AWGN and Rayleigh channels, provides a performance very close to that obtained with the LLR metric, but with a significant reduction in complexity in its logarithmic computational form [10]. In this case, the computational complexity reduction is given at the initialization step, and also for the iterative part of the algorithm. This Soft Decision (SD) decoding algorithm does not require the calculation of LLRs, and so it is not necessary to measure or estimate the SNR at the receiver before decoding. The SD can be applied directly to the recursive structure of the SC decoder of a polar code. Our investigation is based on the notation and description of the construction and SC decoding of Polar Codes presented in [2,7].

2   Brief Description of Polar Codes

Polar Codes are linear block codes of length $N = 2^n$ that are constructed by forming their generator matrix with K rows of the nth Kronecker power of the root matrix $F = \begin{bmatrix} 1 & 0 \\ 1 & 1 \end{bmatrix}$. The following is an example of the Kronecker matrix for $n = 3$:

$$F^{\otimes 3} = \begin{bmatrix} 1&0&0&0&0&0&0&0 \\ 1&1&0&0&0&0&0&0 \\ 1&0&1&0&0&0&0&0 \\ 1&1&1&1&0&0&0&0 \\ 1&0&0&0&1&0&0&0 \\ 1&1&0&0&1&1&0&0 \\ 1&0&1&0&1&0&1&0 \\ 1&1&1&1&1&1&1&1 \end{bmatrix} \tag{1}$$

The corresponding encoder is seen in Fig. 1.

For this example, the input (information-bit) vector is denoted as $\mathbf{u} = u_0^7$, whereas $x = x_0^7$ represents the output (codeword bit) vector to be sent through the channel. In this notation u_a^b describes the sequence of bits $u_a, \ldots, u_b$, of the vector $\mathbf{u}$. Input bits are encoded by the Kroenecker operation, resulting in the codeword vector $\mathbf{x}$. Each pair of input u_i and output x_i bits, where $i = 0, 1, \ldots, N - 1$, corresponds to a sub-channel in the channel polarisation scheme. The transmitted word is a serial version of the output bits.

Polar Codes are usually decoded by the procedure called SC decoding, which operates using a recursive butterfly-based SC decoder [2,7], to estimate an error probability for each input bit $\hat{u}_i$. As proved in [2], there is a phenomenon called channel polarization, which for a large enough N means that some of the N sub-channels transmit bits with error probability close to zero, whereas other sub-channels transmit bits with error probability close to 0.5; that is, some sub-channels become very reliable, whereas others become highly noisy. As indicated by Arikan in [2], we can take advantage of this phenomenon, and devise an error control coding technique, known as polar coding, by selecting the most reliable sub-channels to transmit K input information bits, and by setting the remaining $N - K$ input bits to a known value, typically equal to 0. These bits are usually known as the frozen bits.

Following transmission, which will be affected by noise and interference, a sampling process at the receiver converts the transmitted codeword $\mathbf{x}$ into a received vector $\mathbf{y}$. Based on the received vector, a SC decoder generates estimates of bits u_0 to u_{N-1}.

Bit u_i requires, in order to be estimated, that bits u_0^{i-1} have been already decoded. The bit to be decoded can be a frozen bit or not. If the bit is a frozen bit, estimation does not take place, and the decoded value is set to be that of the frozen bit. If the bit is not a frozen bit, its estimate is calculated as

$$\hat{u}_l = \begin{cases} 0 \ if \ \frac{P\left(\mathbf{y}, u_0^{i-1} u_i = 0\right)}{P\left(\mathbf{y}, u_0^{i-1} u_i = 1\right)} \\ 1 \quad otherwise \end{cases} \tag{2}$$

where $P\left(y, u_0^{i-1} u_i = b\right)$ is the probability of receiving the vector y with the condition that the previous decoded bits are u_0^{i-1}, and that the current bit is in either the state $b = 0$ or $b = 1$.

The selection of the best channels is done by determining the Bhattacharyya coefficients [2], which are a measure of the transformation of the sub-channels to be noise-free or completely noisy. In our example of the polar code with $K = 4$ and $N = 8$, the best sub-channels are those numbered 3, 5, 6 and 7, and channels 0, 1, 2 and 4 are the sub-channels for the frozen bits.

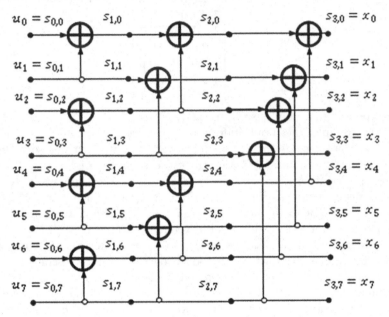

Fig. 1. Encoder of a polar code with $K = 4$, $N = 8$ [7]

Thus, rows 3, 5, 6 and 7 of the Kronecker matrix $F^{\otimes 3}$ form the generator matrix of the polar code, as they have the higher Bhattacharyya parameter values.

In general, the calculation of the Bhattacharyya parameters depends on the characteristics of the channel. One method we will use is the construction method of Arikan for the Binary Erasure Channel (BEC), where Bhattacharyya parameters are determined by using the following recursive expression [2,7]:

$$Z\left(W_{2N}^{(2i-1)}\right) = 2Z\left(W_N^{(i)}\right) - Z\left(W_N^{(i)}\right)^2$$
$$Z\left(W_{2N}^{(2i)}\right) = Z\left(W_N^{(i)}\right)^2 \tag{3}$$

The initial value in this recursion is relevant to the calculation, and we have set it to be 0.5. For comparison purposes we will apply this construction, labelled A-BEC, to a polar code transmitting over both the Gaussian and the Rayleigh channels.

Another construction method is presented in [8] for the Gaussian channel, which is shown to improve on the A-BEC construction. This method is based on a Gaussian approximation for determining the Bhattacharyya parameters. We have also adopted this method here, and we refer it to as the GA construction. We have selected the two above described construction methods for polar codes in order to evaluate the proposed decoding algorithm under different scenarios of construction methods and of channel models.

In the analysis performed in [11] for determining a suitable recursion expression for generating the Bhattacharyya parameters, the conclusion in that paper is that for the Rayleigh channel, expression (3) is the most suitable among three possible recursions studied. Thus we have adopted this construction method for the Rayleigh channel. We also have simulated the GA construction over the Rayleigh channel, in order to show the dependence of the BER performance of the polar code with respect to the relationship between channel characteristics and the generator matrix of the code.

3   SC Decoding of Polar Codes Using Likelihood Ratio Metrics

Arikan [2] simplified the decoding of Polar Codes by proposing a successive cancellation (SC) decoder. This decoder is a butterfly-based decoder [7], and it operates over a Fourier-like structure or graph over which recursive likelihood ratio (LR) calculations are performed. The values passed in the decoder are LR values denoted as $L_{j,i}$, where j and i correspond to the graph stage index and the row index, respectively. At the message side of the graph the LR values are $L_{0,i} = L(\hat{u}_i)$, that is, the estimates of input bits, whereas $L_{n,i}$ are LR values calculated from the channel output side y_i.

The decoding procedure starts with the y_i values, and determines the input bit estimations by recursively using the following expressions [2, 7]:

$$L_{j,i} = \begin{cases} f\left(L_{j+1,i}; L_{j+1,i+2^j}\right) = f(a,b) & \text{if } B(j,i) = 0 \\ g\left(\hat{s}_{j,i-2^j}; L_{j+1,i-2^j}; L_{j+1,i}\right) = g(\hat{s},a,b) & \text{if } B(j,i) = 1 \end{cases} \tag{4}$$

Here, s is a modulo-2 partial sum of decoded bits, $B(j,i)$ is defined as $B(j,i) \triangleq \frac{i}{2^j} \bmod 2$. On the other hand, $0 \leq j < n$, and $0 \leq i < N$. In order to calculate the LR values $L_{j,i}$, functions f and g in Eq. (4) are calculated with the following expressions:

$$L_{j,i} = \begin{cases} f(a,b) & \frac{1+ab}{a+b} \\ g(\hat{s},a,b) & a^{1-2\hat{s}}b \end{cases} \tag{5}$$

Figure 2 shows the Butterfly graph for SC decoding [7].

4   A Successive Cancellation Decoder Based on the SD Metric

As proposed in [10], iterative decoding algorithms like the sum-product (SP) algorithm, which are used for decoding LDPC codes, can be implemented by

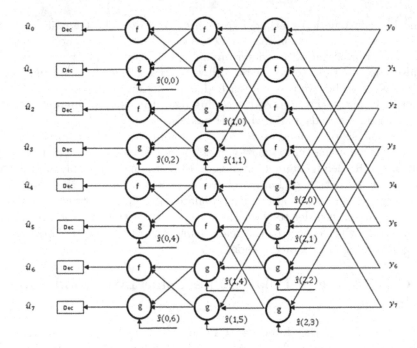

Fig. 2. Butterfly graph for SC decoding

using SD estimates as the decoder metric, instead of using classic probability or LR estimates. As shown in that paper, this brings both a complexity reduction in the decoding algorithm, and also a quite lower complexity initialization step, with the additional advantage of not requiring knowledge of the received signal-to-noise ratio, thus avoiding the need for an SNR estimation process and the possible significant loss of decoding performance due to inaccurate estimations [12]. The same metric can be used for SC decoding of polar codes.

In order to implement this modification of the SC decoder, we first return to the concept of LR values calculated in the logarithmic domain, and then analyse how they propagate by using expressions (5) in this domain. This will allow us to use SD values in this logarithmic version of the SC decoder. This resembles the concept of the anti-log sum operation used in [10] for SP decoding of LDPC codes.

The expressions for function f in Eq. (4) can be modified by using the following exponential expression, where we first set $a = e^{\lambda_a}, b = e^{\lambda_b}$, so:

$$f(a,b) = \frac{1 + e^{\lambda_a} e^{\lambda_b}}{e^{\lambda_a} + e^{\lambda_b}} = \frac{e^0 + e^{\lambda_a + \lambda_b}}{e^{\lambda_a} + e^{\lambda_b}} \tag{6}$$

Here, λ_a and λ_b are logarithmic versions of LR values; that is, they are Log-Likelihood Ratios (LLRs). We propose to replace these LLR values with the difference between the squared distances of the channel samples to the two possible received values $x_i = \pm 1$, as will be shown in Sect. 5 below.

First we calculate the LR ratios in the logarithmic domain:

$$f\left(\lambda_a, \lambda_b\right) = \ln\left(f\left(a, b\right)\right) = \ln\left(e^0 + e^{\lambda_a + \lambda_b}\right) - \ln\left(e^{\lambda_a} + e^{\lambda_b}\right) \tag{7}$$

For a logarithm of a sum, we have:

$$e^\gamma = e^\alpha + e^\beta$$
$$\gamma = \ln\left(e^\alpha + e^\beta\right) = \max\left(\alpha, \beta\right) + \ln\left(1 + e^{-|\beta - \alpha|}\right) \tag{8}$$

then

$$f\left(\lambda_a, \lambda_b\right) = \max\left(0, \lambda_a + \lambda_b\right) + \ln\left(1 + e^{-|\lambda_a + \lambda_b - 0|}\right)$$
$$- \left[\max\left(\lambda_a, \lambda_b\right) + \ln\left(1 + e^{-|\lambda_b - \lambda_a|}\right)\right]$$

By doing the following approximation, that discards the terms $\ln\left(1 + e^{-|\lambda_a + \lambda_b - 0|}\right)$ and $-\ln\left(1 + e^{-|\lambda_b - \lambda_a|}\right)$, we get:

$$f\left(\lambda_a, \lambda_b\right) \simeq \max\left(0, \lambda_a + \lambda_b\right) - \max\left(\lambda_a, \lambda_b\right) \tag{9}$$
$$\text{error} = \ln\left(1 + e^{-|\lambda_a + \lambda_b|}\right) - \ln\left(1 + e^{-|\lambda_b - \lambda_a|}\right)$$

This error term can be discarded without a significant loss in BER performance, a fact which we have verified in our BER performance simulations.

For the values of the g function in the logarithmic domain of Eq. (4) we calculate $\ln\left(g\left(\hat{s}, a, b\right)\right)$:

$$g\left(\hat{s}, a, b\right) = a^{1 - 2\hat{s}}b$$
$$g\left(\hat{s}, \lambda_a, \lambda_b\right) = \ln\left(g\left(\hat{s}, a, b\right)\right) = \left(1 - 2\hat{s}\right)\ln\left(a\right) + ln\left(b\right) \tag{10}$$

But since $a = e^{\lambda_a}$, $b = e^{\lambda_b}$, then we have $\lambda_a = \ln\left(a\right)$, $\lambda_b = \ln\left(b\right)$, so that:

$$g\left(\hat{s}, \lambda_a, \lambda_b\right) = \left(1 - 2\hat{s}\right)\lambda_a + \lambda_b \tag{11}$$

Summarising:

$$L_{j,i} = \left\{ \begin{array}{l} f\left(\lambda_a, \lambda_b\right) \simeq \max\left(0, \lambda_a + \lambda_b\right) - \max\left(\lambda_a, \lambda_b\right) \\ g\left(\hat{s}, \lambda_a, \lambda_b\right) = \left(1 - 2\hat{s}\right)\lambda_a + \lambda_b \end{array} \right\} \tag{12}$$

These expressions are equivalent to Eqs. (16) and (17) in [7], repeated here for clarity:

$$L_{j,i} = \left\{ \begin{array}{l} f\left(\lambda_a, \lambda_b\right) \simeq \text{sign}\left(\lambda_a\right)\text{sign}\left(\lambda_b\right)\min\left(|\lambda_a|, |\lambda_b|\right) \\ g\left(\hat{s}, \lambda_a, \lambda_b\right) = \left(-1\right)^{\hat{s}}\lambda_a + \lambda_b \end{array} \right\}$$

The propagation of these estimates can now happen as in the case of the logarithmic SC decoder. We now analyse the effect of initializing the logarithmic SC decoder with SD values, to give form to a SD-metric-based SC decoder for polar codes.

5   SC Decoding of Polar Codes Over the AWGN Channel Using Soft Distance Metrics

If we assume that transmitted values adopt a normalized polar format with amplitudes in the range $x_i = \pm 1$, LR values for the AWGN channel using Arikan's SC decoder [2] can be determined by using the following expressions:

$$W\left(y_i/0\right) = \frac{1}{\sqrt{2\pi}\sigma}e^{\frac{-(y_i+1)^2}{2\sigma^2}}$$

$$W\left(y_i/1\right) = \frac{1}{\sqrt{2\pi}\sigma}e^{\frac{-(y_i-1)^2}{2\sigma^2}}$$

$$L_i = \frac{W\left(y_i/0\right)}{W\left(y_i/1\right)} \tag{13}$$

Then the decision rule based on calculation of the LR value is:

$$u = \begin{cases} 0 \text{ if } L > 1 \\ 1 \text{ if } L < 1 \end{cases}$$

However, as indicated in the Introduction and Sect. 4 above, it is also possible to decode Polar Codes by using soft distance metrics on the same butterfly decoder structure as that for SC decoding using likelihood ratios.

Simplifying and re-labelling expression (13):

$$L_i = \frac{e^{\frac{-(y_i+1)^2}{2\sigma^2}}}{e^{\frac{-(y_i-1)^2}{2\sigma^2}}} = e^{\left[\frac{-(y_i+1)^2}{2\sigma^2} + \frac{(y_i-1)^2}{2\sigma^2}\right]} = e^{\frac{(y_i-1)^2 - (y_i+1)^2}{2\sigma^2}} \tag{14}$$

An additional simplification comes from the use of logarithmic domain calculation of L_i. We can determine the values $d_{0,i}^2 = (y_i+1)^2$ and $d_{1,i}^2 = (y_i-1)^2$, which are the squared soft distances from the channel information values to the two possible transmitted values $x_i = \pm 1$. Then:

$$L_i = e^{\frac{d_{1,i}^2 - d_{0,i}^2}{2\sigma^2}}$$

$$\ln\left(L_i\right) = \frac{d_{1,i}^2 - d_{0,i}^2}{2\sigma^2} \tag{15}$$

$$\hat{u} = \begin{cases} 0 \text{ if } \ln\left(L_i\right) > 0 \\ 1 \text{ if } \ln\left(L_i\right) < 0 \end{cases} \quad \text{or} \quad \hat{u} = \begin{cases} 0 \quad \text{if } d_{1,i}^2 > d_{0,i}^2 \\ 1 \quad \text{if } d_{1,i}^2 < d_{0,i}^2 \end{cases} \tag{16}$$

If we now have as the metric the parameter:

$$d_{1,i}^2 - d_{0,i}^2 = (y_i-1)^2 - (y_i+1)^2 = -4y_i \tag{17}$$

then

$$u = \begin{cases} 0 \text{ if } -4y_i > 0 \\ 1 \text{ if } -4y_i < 0 \end{cases} \tag{18}$$

The idea is to apply Eq. (12) where the logarithmic values are the differences of squared distances $d_{1,i}^2 - d_{0,i}^2$ by simply setting the initialization values $L_{n,i}$ to be equal to $d_{1,i}^2 - d_{0,i}^2 = -4y_i$. This simplifies calculations of the initialization values and avoids the need to estimate the noise level in the channel. On the other hand, the determination of the initialization values of the classic SC decoder implies squaring and quotient operations on the noise dispersion σ. These operations are usually quite costly in terms of hardware implementations, but they can be avoided by using the SD metric, without significant BER performance loss, as we show below.

6   Simulation Results

Simulations of the BER performance of a polar code were done using Arikan's SC decoder [2,7] with the LLR metric, and the same polar code decoded using the SD metric. These simulations were performed for both the AWGN channel and the Rayleigh channel, and are intended to measure any possible loss in BER performance as a result of the use of the SD metric. Since the construction of the corresponding generator matrix is relevant to the BER performance, we compare the classic construction of Arikan for Polar Codes over the BEC (A-BEC), and the construction using the GA approach (GA) to determine the Bhattacharyya parameters.

Simulation results for the BER performance of a polar code with parameters $(N, K) = (256, 128)$ over the AWGN channel are shown in Fig. 3. The simulated transmission over the AWGN channel involves transmitting 100,000 messages

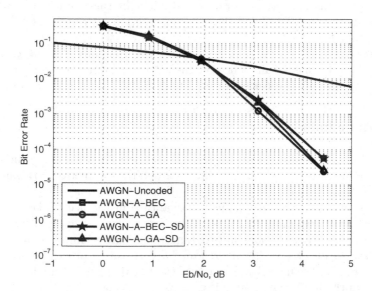

Fig. 3. BER performance of a $(256, 128)$ polar code for the AWGN channel

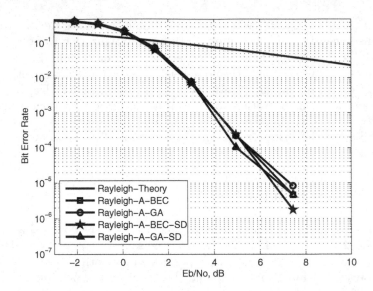

Fig. 4. BER performance of a $(256, 128)$ polar code for the Rayleigh channel

of 128 bits each, and is done for both the A-BEC and the GA code construction. For each construction both decoders, the classic SC decoder and the same algorithm structure using the SD metric, are applied.

Simulation results of the BER performance of the same polar code (256,128) over the Rayleigh channel are shown in Fig. 4. The simulated transmission over the Rayleigh channel involves transmitting 50,000 messages of 128 bits each, and is done for both the A-BEC and the GA construction, and both the SC and SD decoders, as before.

Simulations have shown that removal of the constant 4 in the expression $d_{1,i}^2 - d_{0,i}^2 = -4y_i$ has no influence on the BER performance, so we can set it equal to 1 to simplify the decision rule calculations.

7 Conclusions

The loss in BER performance using the SD metric is very small for the AWGN channel, and in the case of the Rayleigh channel, the SD metric performs better than the classic LLR metric, but again with small differences. Given that this loss is very small, there is a significant advantage in the use of the SD metric, because it avoids calculations of products, quotients and exponentials (costly operations for implementations in programmable logic technology like FPGAs) in the initialization step of the algorithm. Thus the main processing complexity improvement is in the initialization step of the SD algorithm. An additional simplification in the initialization step is that we can initialize the SC decoder with simple SD metrics that are directly the channel samples $-y_i$ or $+y_i$, without any significant loss in BER performance. Another major improvement in complexity arises because there is no need to estimate the variance of the channel noise, which also means that any loss of performance due to inaccurate estimation is avoided.

As expected, the GA polar code construction performs best over the Gaussian channel. However, we note that the A-BEC construction performs slightly better than the GA construction over the Rayleigh channel. This confirms the dependence of polar code BER performance on the code construction method as well as on the channel characteristics.

Finally, we point out that the simplified SC decoding methods proposed in [7, 9] can also make use of the SD metric, with all the advantages that this metric gives.

References

1. Shannon, C.E.: A mathematical theory of communication. Bell Syst. Tech. **27**, 379–423 (1948)
2. Arikan, E.: Channel polarization: a method for constructing capacity-achieving codes for symetric binary-input memoryless channels. IEEE Trans. Inf. Theor. **55**, 3051–3073 (2009)
3. Sasoglu, E., Telatar, E., Arikan, E.: Polarization for arbitrary discrete memoryless channels. In: Proceedings of IEEE Information Theory Workshop (ITW), pp. 144–148 (2009)
4. Arikan, E.: A performance comparison of Polar Codes and Reed-Muller codes. IEEE Commun. Lett. **12**(6), 447–449 (2008)
5. Mori, R., Tanaka, T.: Performance and construction of Polar Codes on symmetric binary-input memoryless channels. In: ISIT (2009)
6. Mori, R., Tanaka, T.: Performance of Polar Codes with the construction using density evolution. IEEE Commun. Lett. **13**(7), 519–521 (2009)
7. Leroux, C., Raymond, A., Sarkis, G., Gross, W.: A semi-parallel successive-cancellation decoder for polar codes. IEEE Trans. Sign. Process. **61**(3), 289–299 (2013)
8. Li, H., Yuan, J.: A practical construction method for Polar Codes in AWGN channels. In: 2013 TENCON Spring Conference, pp. 17–19. IEEE, Sydney, Australia (2013). ISBN: 978-1-4673-6347-1
9. Alamdar-Yazdi, A., Kschischang, F.R.: A simplified successive-cancellation decoder for polar codes. IEEE Commun. Lett. **15**(12), 1378–1380 (2011)
10. Farrell, P.G., Arnone, L.J., Castiñeira Moreira, J.: Euclidean distance soft-input soft-output decoding algorithm for low-density parity-check codes. IET Commun. **5**(16), 2364–2370 (2011)
11. Shi, P., Zhao, S., Wang, B., Zhou, L.: Performance of Polar Codes on wireless communications Channels (2012). (Submitted to GlobeCom)
12. Saeedi, H., Banihashemi, A.H.: Performance of belief propagation for decoding LDPC codes in the presence of channel estimation error. IEEE Trans. Commun. **55**(1), 83–89 (2007)

On the Doubly Sparse Compressed Sensing Problem

Grigory Kabatiansky[1]([⊠]), Serge Vlăduţ[1,2], and Cedric Tavernier[3]

[1] Institute for Information Transmission Problems, Russian Academy of Sciences,
Moscow, Russia
{kaba,vl}@iitp.ru

[2] Institut de Mathématiques de Marseille, Aix-Marseille Université, IML,
Marseille, France

[3] Assystem AEOS, Saint-Quentin en Yvelines, France
tavernier.cedric@gmail.com

Abstract. A new variant of the Compressed Sensing problem is investigated when the number of measurements corrupted by errors is upper bounded by some value l but there are no more restrictions on errors. We prove that in this case it is enough to make $2(t+l)$ measurements, where t is the sparsity of original data. Moreover for this case a rather simple recovery algorithm is proposed. An analog of the Singleton bound from coding theory is derived what proves optimality of the corresponding measurement matrices.

1 Introduction and Definitions

A vector $x = (x_1, \ldots, x_n) \in \mathbb{R}^n$ in n-dimensional vector space $\mathbb{R}^n$ called t-sparse if its Hamming weight $wt(x)$ or equivalently its l_0 norm $||x||_0$ is at most t, where by the definition $wt(x) = ||x||_0 = |\{i : x_i \neq 0\}|$. Let us recall that the Compressed Sensing (CS) Problem [1,2] is a problem of reconstructing of an n-dimensional t-sparse vector x by a few (r) linear measurements $s_i = \langle h^{(i)}, x \rangle$ (i.e. inner product of vectors x and $h^{(i)}$), assuming that measurements $(h^{(i)}, x)$ are known with some errors e_i, for $i = 1, \ldots, r$. Saying in other words, one needs to construct an $r \times n$ matrix H with minimal number of rows $h^{(1)}, \ldots, h^{(r)}$, such that the following equation

$$\hat{s} = Hx^T + e, \tag{1}$$

has either a unique t-sparse solution or all such solutions are "almost equal". The compressed sensing problem was mainly investigated under the assumption that the vector $e = (e_1, \ldots, e_r)$, is called the error vector, has relatively small

The work of Grigory Kabatiansky and Serge Vlăduţ was carried out at the Institute for Information Transmission Problems of the Russian Academy of Sciences at the expense of the Russian Science Foundation, project no. 14-50-00150. The work of Cedric Tavernier was partially supported by SCISSOR ICT project no. 644425, funded by the European Commission Information & Communication Technologies H2020 Framework Program.

© Springer International Publishing Switzerland 2015
J. Groth (Ed.): IMACC 2015, LNCS 9496, pp. 184–189, 2015.
DOI: 10.1007/978-3-319-27239-9_11

Euclidean norm (length) $||e||_2$. We consider another problem's statement assuming that the error vector e is also sparse but its Euclidean norm can be arbitrary large. In other words, we consider the doubly sparse CS problem when $||x||_0 \leq t$ and $||e||_0 \leq l$. The assumption $||e||_0 \leq l$ was first time considered in [3] as a proper replacement for discrete version of CS-problem of usual assumption that an error vector e has relatively small Euclidean norm.

Definition 1. *A real $r \times n$ matrix H called a (t, l)-compressed sensing (CS) matrix if*

$$||Hx^T - Hy^T||_0 \geq 2l + 1 \tag{2}$$

for any two distinct vectors $x, y \in \mathbb{R}^n$ such that $||x||_0 \leq t$ and $||y||_0 \leq t$.

This definition immediately leads (see [3]) to the following

Proposition 1. *A real $r \times n$ matrix H is a (t, l)-CS matrix iff*

$$||Hz^T||_0 \geq 2l + 1 \tag{3}$$

for any nonzero vector $z \in \mathbb{R}^n$ such that $||z||_0 \leq 2t$.

Our main result is an explicit and simple construction of (t, l)-CS matrices with $r = 2(t + l)$ for any n. We show this value of r is the minimal possible for (t, l)-CS matrices by proving an analog of the well-known in coding theory Singleton bound for the compressed sensing problem. Besides that we propose an efficient recovery (decoding) algorithm for the considered double sparse CS-problem.

2   Optimal Matrices for Doubly Sparse Compressed Sensing Problem

We start with constructing of (t, l)-CS matrices. Let a real $\tilde{r} \times n$ matrix $\tilde{H}$ be a parity-check matrix of an $(n, n - \tilde{r})$-code code over $\mathbb{R}$, correcting t errors, i.e. any $2t$ columns $\tilde{h}_{i_1}, \ldots, \tilde{h}_{i_{2t}}$ of $\tilde{H}$ are linearly independent. And let G be a generator matrix of an $(r, \tilde{r})$-code over $\mathbb{R}$ of length r, correcting l errors. Let matrix H consists of the columns $h_1, \ldots, h_n$, where

$$h_j^T = \tilde{h}_j^T G \tag{4}$$

and transposition ^{T} means, that vectors h_j and $\tilde{h}_j$ are considered in (4) as row vectors, i.e.

$$H = G^T \tilde{H} \tag{5}$$

In other words, we encode columns of parity-check matrix $\tilde{H}$, which is already capable to correct t errors, by a code, correcting l errors, in order to restore correctly the syndrom of $\tilde{H}$.

Theorem 1. *Matrix $H = G^T \tilde{H}$ is a (t, l)-CS matrix.*

Proof. According to Proposition 1 it is enough to prove that $||Hz^T||_0 \geq 2l + 1$ for any nonzero vector $z \in \mathbb{R}^n$ such that $||z||_0 \leq 2t$. Indeed, $u = \tilde{H}z^T \neq 0$ since any $2t$ columns of $\tilde{H}$ are linear independent. Then $Hz^T = G^T \tilde{H}z^T = G^T u = (u^T G)^T$ and $u^T G$ is a nonzero vector of a code over $\mathbb{R}$, correcting l errors. Hence $||Hz^T||_0 = ||u^T G||_0 \geq 2l + 1$. ☐

Now let us choose the well known Reed-Solomon (RS) codes (which are a particular case of evaluation codes construction) as both constituent codes. The length of the RS-code is restricted by the number of elements in the field so in the case of $\mathbb{R}$ the length of evaluation code can be arbitrary large. Indeed, consider the corresponding evaluation code $\mathbb{RS}_{(n,k)} = \{(f(a_1), \ldots, f(a_n)) : \deg f(x) < k\}$, where $a_1, \ldots, a_n \in \mathbb{R}$ are n different real numbers. The distance of the $\mathbb{RS}_{(n,k)}$ code $d = n - k + 1$ since the number of roots of a polynomial cannot exceed its degree and hence $d \geq n - k + 1$, but, on the other hand, the Singleton bound states that $d \leq n - k + 1$ for any code, see [4]. Therefore the resulting matrix H is a (t, l)-CS matrix with $r = 2(t + l)$. The next result, which is a generalization of the Singleton bound for the doubly sparse CS problem, shows these matrices are optimal in the sense having the minimal possible number r of linear measurements.

Theorem 2. *For any (t, l)-CS $r \times n$-matrix*

$$r \geq 2(t + l). \tag{6}$$

Proof. Let H be any (t, l)-CS matrix of size $r \times n$, i.e., $||Hz^T||_0 \geq 2l + 1$ for any nonzero vector $z \in \mathbb{R}^n$ such that $||z||_0 \leq 2t$. And let H_{2t-1} be the $(2t - 1) \times n$ matrix consisting of the first $2t - 1$ rows of H. There exists a nonzero vector $\hat{z} = (\hat{z}_1, \ldots, \hat{z}_{2t}, 0, 0, \ldots, 0) \in \mathbb{R}^n$ such that $H\hat{z}^T = 0$ (a system of linear homogenious equations with the number of unknown variables larger than the number of equations has a nontrivial solution). Then $||H\hat{z}^T||_0 \leq r - (2t - 1)$ and finally $r \geq 2t + 2l$ since $||H\hat{z}^T||_0 \geq 2l + 1$. ☐

3   Recovery Algorithm for Doubly Sparse Compressed Sensing Problem

Let us start from a simple remark that for $e = 0$ recovering of the original sparse vector x, i.e., solving the Eq. (1), is the same as syndrome decoding of some code (over $\mathbb{R}$) defined by matrix H as a parity-check matrix. In general, syndrome $s = Hx^T$ is known with some error, namely, as $\hat{s} = s + e$ and therefore we additionally encoded columns of H by some error-correcting code in order to recover the original syndrome s and then apply usual syndrome decoding algorithm. Therefore recovering, i.e., decoding algorithm for constructed in previous chapter optimal matrices is in some sense a "concatenation" of decoding algorithms of constituent codes.

Namely, first we decode vector $\hat{s} = s + e$ by a decoding algorithm of the code with generator matrix G. Since $||e||_0 \leq l$ this algorithm outputs the correct syndrome s. After that we form the syndrome $\tilde{s}$ by selecting first $\tilde{r}$ coordinates of s and then apply syndrome decoding algorithm (of the first code with parity-check matrix $\tilde{H}$) for the following syndrom equation

$$\tilde{s} = \tilde{H}x^T. \tag{7}$$

Now let us discuss a right choice of constituent codes. It is very convenient to use the class of Reed-Solomon codes over $\mathbb{R}$. There are well known algorithms of their decoding up to half of the code distance (bounded distance decoding, see [4]), for instance, Berlekamp-Massey algorithm, which in our case (codes over $\mathbb{R}$) is known also as Trench algorithm, see [5]. Hence the total decoding complexity does not exceed $O(n^2)$ operations over real numbers. Moreover we can even decode these codes over their half distances by application of Guruswami-Sudan list decoding algorithm [6].

It is well known that encoding-decoding procedures of Reed-Solomon codes become more simple in the case of cyclic codes, when the set $a_1, \ldots, a_n$ is a cyclic group under multiplication. In order to do it let us consider $a_1, \ldots, a_n$ as complex roots of degree n and define our codes through their "roots", i.e. our codes consist of *polynomials* $f(x)$ over $\mathbb{R}$ such that $f(e^{2\pi i \frac{m}{n}}) = 0$ for $m \in \{-s, \ldots, -1, 0, +1, \ldots, +s\}$ with $s = t$ for the first constituent code and $s = l$ for the second. It easy to check that such codes achieve the Singleton bound with $d = 2s + 2$, so the corresponding doubly sparse code has redundancy $r = 2(t + l + 1)$ what is slightly larger than the corresponding Singleton bound, but in return these codes can be decoded via FFT.

4   Discussion - No Small Errors Case and Slightly Beyond

Let us note that the initial papers on Compressed Sensing especially stated that this new technique (application of l_1 minimization instead of l_0) allows to recover information vector $x \in \mathbb{R}^n$ in case when not many coordinates of x were affected by errors. For instance, "one can introduce errors of arbitrary large sizes and still recover the input vector exactly by solving a convenient linear program...", see in [7]. To achieve such performance some special restriction on matrix H was placed, called Restricted Isometry Property (RIP), as follows

$$(1 - \delta_D)||x||_2 \leq ||Hx^T||_2 \leq (1 + \delta_D)||x||_2, \tag{8}$$

for any vector $x \in \mathbb{R}^n : ||x||_0 \leq D$, where $0 < \delta_D < 1$. The smallest possible δ_D called the isometry constant.

Then typical result in [7] (Theorem 1.1) is of the following form "if $\delta_{3t} + 3\delta_{4t} < 2$ *then the solution of linear programming problem is unique and equal to* x".

Let us note that the condition $\delta_{3t} + 3\delta_{4t} < 2$ implies $\delta_{4t} < 2/3$ (of course, it implies that $\delta_{4t} < 1/2$, but for us enough to have $\delta_{4t} < 1$). Hence $Hx^T \neq \mathbf{0}$ for any nonzero x with $wt(x) \leq 4t$, or in other words, an error-correcting code (over reals) corresponding to such parity-check matrix H has the minimal distance at least $4t + 1$ and can correct $2t$ errors (instead of t). So we lost twice in error-correction capability but maybe linear programming provides more easier way for decoding? In fact, NOT, since it is well known in coding theory that such problem can be solved rather easily (in complexity) over *any infinite field* by usage of the corresponding Reed-Solomon codes and known decoding algorithms. In case of real number or complex number fields one can use just an RS code with Fourier parity-check matrix, namely, $h_{j,p} = exp(2\pi i \frac{jp}{n})$, where for complex numbers $p \in \{1, 2, ..., n\}$, $j = a, a+d, a+2d, ..., a+(r-1)d$, and "reversible" RS-matrix H for real numbers, where $p \in \{1, 2, ..., n\}$, $j \in \{-f, -f+1, ..., 0, 1, ..., f\}$ and $r = 2f + 1$.

Fortunately, matrices with the RIP property allow to correct not only sparse errors but also additional errors with arbitrary support but relatively small (up to ε) Euclidean norm. Again, the RIP property is good for linear programming decoding but is too strong in general. Namely, it is enough to have the following property

$$\lambda_{2t}||x||_2 \leq ||Hx^T||_2, \tag{9}$$

for any vector $x \in \mathbb{R}^n : ||x||_0 \leq 2t$, where $\lambda_{2t} > 0$ and the largest such value we call *extension constant*. Indeed, then for any two solutions x and $\tilde{x}$ of the Eq. (1) we have that

$$||x - \tilde{x}||_2 \leq 2\lambda_{2t}^{-1}\varepsilon, \tag{10}$$

where $||e||_2, ||\tilde{e}||_2 \leq \varepsilon$. Hence (10) shows that all solutions of the Eq. (1) are "almost equal" if λ_{2t} is large enough. Let us note that for RS-matrices any r columns are linear independent and hence $\lambda_{2t} > 0$, but λ_{2t} tends to zero when n grows and code rate is fixed. To find better class of codes over the field of real (or complex) numbers is an open problem.

5   Conclusion

In this paper we extends technique, which was developed in [8] for error correction with errors in both the channel and syndrome, to the Compressed Sensing problem. We hope this approach will help to find limits of the unrestricted (i.e. without LP usage) compressed sensing.

References

1. Donoho, D.L.: Compressed sensing. IEEE Trans. Inf. Theor. **52**(4), 1289–1306 (2006)
2. Candes, E.J., Tao, T.: Near-optimal signal recovery from random projections: universal encoding strategies? IEEE Trans. Inf. Theor. **52**(12), 5406–5425 (2006)
3. Kabatiansky, G., Vladuts, S.: What to do if syndromes are corrupted also. In: Proceedings of the International Workshop Optimal Codes, Albena, Bulgaria (2013)
4. MacWilliams, F.J., Sloane, N.J.A.: The Theory of Error-Correcting Codes. North-Holland, Amsterdam (1991)
5. Trench, W.R.: An algorithm for the inversion of finite Toeplitz matrices. J. Soc. Indust. Appl. Math. **12**, 515–522 (1964)
6. Guruswami, V., Sudan, M.: Improved decoding of Reed-Solomon and algebraic-geometry codes. IEEE Trans. Inf. Theor. **45**, 1757–1767 (1999)

7. Candes, E.J., Tao, T.: Decoding by linear programming. IEEE Trans. Inf. Theor. **51**(12), 4203–4215 (2005)
8. Vlăduţ, S., Kabatiansky, G., Lomakov, V.: On error correction with errors in both the channel and syndrome. Probl. Inf. Transm. **51**(2), 50–57 (2015)

Codes of Length 2 Correcting Single Errors of Limited Size

Torleiv Kløve$^{(\boxtimes)}$

University of Bergen, 5020 Bergen, Norway
Torleiv.Klove@ii.uib.no

Abstract. Linear codes over $\mathbb{Z}_q$ of length 2, correcting single errors of size at most k, are considered. It is determined for which q such codes exists and explicit code constructions are given for those q. One case remains open, namely $q = (k+1)(k+2)$, where $k+1$ is a prime power. For this case we conjecture that no such codes exist.

Keywords: Error correcting codes · Single errors · Limited size errors

1   Introduction

Flash memories are non-volatile, high density and low cost memories. Flash memories find wide applications in cell phones, digital cameras, embedded systems, etc. and it is a major type of Non-Volatile Memory (NVM).

In order to improve the density of flash memories, multi- level (q-level) memory cells are used so that each cell stores $\log_2 q$ bits. Even though multi-level cells increase the storage density compared to single-level cells, they also impose two important challenges. The first one is that the voltage difference between the states is narrowed since the maximum voltage is limited. A natural consequence is that reliability issues such as low data retention and read/write disturbs become more significant. The errors in such cases are typically of limited magnitude.

The second major challenge in flash memory systems is that the writing mechanism is relatively very time consuming. A cell can be programmed from a lower level to a higher level by injecting additional amount of electrons in the floating gate. However, in order to program a cell from a higher level to lower level, an entire block of cells needs to be erased to zero and then using many iterations electrons are carefully injected to the floating gates of each and every cell to achieve the desired levels. Thus, rewriting a cell from the higher voltage level to a lower voltage level is quite expensive. The amount of time required for write operation can be reduced by using error correcting codes. The overshoot of voltage level while writing can be considered as asymmetric error of limited magnitude. Using codes capable of correcting limited magnitude asymmetric errors, the overshoot errors can be corrected. Because of this we do not need to be very precise about achieving the desired voltage level and so, the number of iterations required for charging the floating gates can be reduced, which in turn will reduce the write operations time.

J. Groth (Ed.): IMACC 2015, LNCS 9496, pp. 190–201, 2015.
DOI: 10.1007/978-3-319-27239-9_12

2   Notations

We denote the set of integers by $\mathbb{Z}$. For $a, b \in \mathbb{Z}$, $a \leq b$, we let

$$[a, b] = \{a, a + 1, a + 2, \ldots, b\}.$$

For integers $q > 0$ and a, we let $(a \bmod q)$ denotes the main residue of a modulo q, that is, the least non-negative integer r such that q divides $a - r$.

Assume q is given. We denote modular addition of two integers a, b by $a \oplus b$, that is, $a \oplus b = ((a + b) \bmod q)$. Similarly, we define modular subtraction by $a \ominus b = ((a - b) \bmod q)$ and modular multiplication by $a \otimes b = (ab \bmod q)$.

We define the channel more precisely. Let q and k be integers, where $1 \leq k < q$. The alphabet is $\mathbb{Z}_q = [0, q - 1]$. A symbol a in the alphabet $\mathbb{Z}_q$ may be modified during transmission into another symbol $a \oplus e \in \mathbb{Z}_q$ where e is an integer such that $|e| \leq k$. Error correcting codes for this channel have been considered in e.g. in [1, 5, 6]. Most of these are linear and single error correcting. The simplest non-trivial case are codes of length two. We consider this case in detail in this note.

3   General Description of the Codes

We define the codes and the problem precisely. Let $(a, b) \in \mathbb{Z}^2$. The corresponding code is

$$C = C_{a,b} = \{(u, v) \in \mathbb{Z}_q^2 \mid (a \otimes u) \oplus (b \otimes v) = 0\}.$$

When (u, v) is transmitted and (u', v') is received, the corresponding *syndrom* is $(a \otimes u') \oplus (b \otimes v')$. We see that if $(u', v') = (u \oplus e, v)$, the syndrom is

$$(a \otimes (u \oplus e)) \oplus (b \otimes v) = (a \otimes u) \oplus (a \otimes e) \oplus (b \otimes v) = a \otimes e.$$

Similarly, if $(u', v') = (u, v \oplus e)$, the syndrom is $b \otimes e$. Therefore, the code can correct a single error of size at most k if and only if the $1 + 4k$ syndroms

$$\{0\} \cup \{a \otimes e \mid e \in [-k, -1] \cup [1, k]\} \cup \{b \otimes e \mid e \in [-k, -1] \cup [1, k]\} \quad (1)$$

are all distinct. If this is the case, we say that (a, b) is a (q, k) *check pair* or just a check pair if the values of q and k are clear from the context.

Our problem can now be precisely formulated as follows:

For which q and k does a (q, k) check pair exist?

At first glance, this may seem to be a rather trivial problem. However, this appears not to be the case for all q and k. When a check pair exists, we also want describe the corresponding code and its encoding and decoding.

The following reformulation will be usefull.

Proposition 1. *For given q, k, $(a, b) \in \mathbb{Z}^2$ is a check pair if and only if all the following conditions are satisfied:*

1. $a \otimes e \neq b \otimes \varepsilon$ for $e, \varepsilon \in [-k, -1] \cup [1, k]$,
2. $\gcd(a, q) < q/(2k)$,
3. $\gcd(b, q) < q/(2k)$.

Proof. By the definitions, (a, b) is a check pair if and only if all the syndroms are distinct, that is, all the following conditions are satisfied:

1. $a \otimes e \neq b \otimes \varepsilon$ for $e, \varepsilon \in [-k, -1] \cup [1, k]$,
2. $a \otimes e \neq a \otimes \varepsilon$ for $-k \leq \varepsilon < e \leq k$,
3. $b \otimes e \neq b \otimes \varepsilon$ for $-k \leq \varepsilon < e \leq k$.

We will show that second of these conditions is equivalent to the second condition of the proposition and similarly for the third conditions. Let $d = \gcd(a, q)$. Then $\gcd(a/d, q/d) = 1$. Putting $z = e - \varepsilon$ we get the following chain of equivalent conditions:

$$
\begin{aligned}
(2) &\Leftrightarrow a \otimes z \not\equiv 0 \pmod{q} \text{ for all } z \in [1, 2k] \\
&\Leftrightarrow (a/d) \otimes z \not\equiv 0 \pmod{(q/d)} \text{ for all } z \in [1, 2k] \\
&\Leftrightarrow z \not\equiv 0 \pmod{(q/d)} \text{ for all } z \in [1, 2k] \\
&\Leftrightarrow 2k < q/d \\
&\Leftrightarrow d < q/(2k).
\end{aligned}
$$

Similarly for the third condition.

Lemma 1. *Let (a, b) be a (q, k) check pair. Then*

1. *(b, a) is a check pair.*
2. *$(a, -b)$, $(-a, b)$, and $(-a, -b)$ are check pairs.*
3. *If $z \in \mathbb{Z}$ such that $\gcd(q, z) = 1$, then (za, zb) is a check pair.*

Proof. The syndroms of (b, a) are clearly the same as the syndroms of (a, b). This proves case *1*. Also for case *2* the syndroms are the same.

Now, consider case *3*. Let $z' \otimes z = 1$. Multiplying by z', we see that

$$(za) \otimes e = (zb) \otimes \varepsilon \text{ if and only if } a \otimes e = b \otimes \varepsilon$$

for $e, \varepsilon \in [-k, -1] \cup [1, k]$. Further, $\gcd(za, q) = \gcd(a, q)$ and so

$$\gcd(za, q) < q/(2k) \text{ if and only if } \gcd(a, q) < q/(2k).$$

4   The Case $q \leq (k+1)^2$

In [5], the following result was shown.

Theorem 1. *If $k \geq 1$ and $q \leq (k+1)^2$, then there are no (q, k) check pairs.*

It was also shown that $(1, k+1)$ *is* a $((k+1)^2 + 1, k)$ check pair. In this paper, we consider all $q > (k+1)^2$. We split the presentation into two parts:

The case $q \geq (k+1)^2 + 1$, $q \neq (k+1)(k+2)$. For this case we show in Sect. 5 that there exists a simple check pair.

The case $q = (k+1)(k+2)$. This is the hardest case. A check pair exists for some k, but not all. We discuss this case in Sect. 6.

5   The Case $q \geq (k+1)^2 + 1, q \neq (k+1)(k+2)$

5.1   Check Pairs

We will give explicit check pairs for all q in this case.

First, consider the pair $(1, k+1)$. The corresponding syndrom set is

$$[0, k] \cup [q-k, q-1] \cup \{(k+1)e \mid e \in [1, k]\} \cup \{q - (k+1)x \mid x \in [1, k]\}.$$

If $q - k(k+1) > k(k+1)$, that is, $q \geq 2k(k+1) + 1$, then clearly all the syndroms are distinct and so $(1, k+1)$ is a check pair.

Similarly, if $q \in [(k+1)^2 + 1, 2k(k+1) - 1]$ but $q \not\equiv 0 \pmod{k+1}$, then again all the syndroms are distinct.

It remains to consider $q \in \{x(k+1) \mid x \in [k+3, 2k]\}$. For these q we have $q \not\equiv 0 \bmod (k+2)$. By an argument similar to the one above, we see that that $(1, k+2)$ is a check pair.

We summarize these results in a theorem.

Theorem 2. *We have the following cases.*

1. *If $q \geq 2k(k+1) + 1$, then $(1, k+1)$ is a check pair.*
2. *If $q \in [(k+1)^2 + 1, 2k(k+1) - 1]$ but $q \not\equiv 0 \pmod{k+1}$, then $(1, k+1)$ is a check pair.*
3. *If $q \in \{x(k+1) \mid x \in [k+3, 2k]\}$, then $(1, k+2)$ is a check pair.*

5.2   The Corresponding Codes

We take a closer look at the codes corresponding to check pairs in the second case. The other cases are very similar. The code is

$$\begin{aligned} C_{1,k+1} &= \{(u, v) \mid u, v \in \mathbb{Z}_q, u \oplus ((k+1) \otimes v) = 0\} \\ &= \{((-(k+1)) \otimes v, v) \mid v \in \mathbb{Z}_q\}. \end{aligned}$$

The most natural encoding for the information $m \in \mathbb{Z}_q$ is to encode it into $((-(k+1)) \otimes m, m))$. In particular, this gives a systematic encoding.

For decoding, we assume that (u', v') is received and that at most one of the elements are in error, and by an amount e of size at most k. From this we want to recover the sent information. We look at the possible syndroms.

- If there are no errors, the syndrom is 0.
- If $u' = u \oplus e$ where $e \in [1, k]$, then the syndrom is $s = e$. In this case the second part is error free and so $m = v' = v$.
- If $u' = u \oplus e$ where $e \in [-k, -1]$, then the syndrom is $s = q + e$. Also in this case $m = v' = v$.
- If $v' = v \oplus e$ where $e \in [1, k]$, then the syndrom is $s = (k+1)e$ and so $e = s/(k+1)$. In this case $m = v' \ominus e = v' \ominus s/(k+1)$.
- If $v' = v \oplus e$ where $e \in [-k, -1]$, then the syndrom is $s = q + (k+1)e$ and so $e = (s-q)/(k+1)$ and $m = v' \ominus (s-q)/(k+1)$.

This gives the following decoding algorithm:

- if $s \in [0, k]$ or $s \in [q - k, q - 1]$, then $m = v'$,
- else if $(s \bmod (k + 1)) = 0$, then $m = v' \ominus s/(k + 1)$,
- else if $((s - q) \bmod (k + 1)) = 0$, then $m = v' \ominus (s - q)/(k + 1)$.

This gives a correct answer for all errors of the type we consider. Of course, if other types of errors have occurred, the decoding algorithm will either give a wrong answer or no answer at all (when none of the conditions are satisfied).

For codes corresponding to the first and third cases in Theorem 2, we get a similar decoding algorithm.

6   The Case $q = (k + 1)(k + 2)$

This is the main case.

6.1   An Existence Result

Theorem 3. Let $k \geq 1$ and $q = (k + 1)(k + 2)$. For each integer a, $1 \leq a \leq q$, we have

$$\gcd(a, q) > k$$

or there exists integers $x \in [1, k]$ and $y \in [-k, -1] \cup [1, k]$ such that

$$y = a \otimes x. \tag{2}$$

Remark. We see that (2) is equivalent to

$$ax - tq = y \tag{3}$$

for some integer t. We note that this implies that $\gcd(a, q)$ divides y. In particular, it implies that $\gcd(a, q) \leq |y| \leq k$.

We will use Farey-sequences in the proof. For a discussion of Farey-sequences, see e.g. [2, p. 23ff]. The Farey-sequence F_k is the sequence of fractions t/n, where $0 \leq t \leq n \leq k$ and $\gcd(t, n) = 1$, listed in increasing order. The size of F_k is $1 + \Phi_k$, where

$$\Phi = \Phi_k = \sum_{r=1}^{k} \varphi(r).$$

We denote the elements of F_k by t_i/n_i, where $t_0/n_0 = 0/1$ and $t_\Phi/n_\Phi = 1/1$.

Example 1. F_6 is

$$\frac{0}{1}, \frac{1}{6}, \frac{1}{5}, \frac{1}{4}, \frac{1}{3}, \frac{2}{5}, \frac{1}{2}, \frac{3}{5}, \frac{2}{3}, \frac{3}{4}, \frac{4}{5}, \frac{5}{6}, \frac{1}{1}.$$

We see that the elements are symmetric around $1/2$, and this is clearly a general property: we have

$$n_{\Phi-i} = n_i \text{ and } t_{\Phi-i} = n_i - t_i, \text{ that is, } t_{\Phi-i}/n_{\Phi-i} = 1 - t_i/n_i.$$

for $0 \leq i \leq \Phi$.

The following lemma contains Theorems 28 and 30 in [2].

Lemma 2. *Let t_i/n_i and t_{i+1}/n_{i+1} be consecutive elements in F_k. Then*

$$t_{i+1}n_i - t_i n_{i+1} = 1, \tag{4}$$

and

$$n_i + n_{i+1} \geq k + 1. \tag{5}$$

Let

$$s_i = \frac{t_i}{n_i} + \frac{1}{n_i(n_i + n_{i+1})}.$$

Lemma 3.

$$s_i = \frac{t_i}{n_i} + \frac{1}{n_i(n_i + n_{i+1})} = \frac{t_{i+1}}{n_{i+1}} - \frac{1}{n_{i+1}(n_i + n_{i+1})}. \tag{6}$$

Proof.

$$\left(\frac{t_{i+1}}{n_{i+1}} - \frac{1}{n_{i+1}(n_i + n_{i+1})} \right) - \left(\frac{t_i}{n_i} + \frac{1}{n_i(n_i + n_{i+1})} \right)$$

$$= \left(\frac{t_{i+1}}{n_{i+1}} - \frac{t_i}{n_i} \right) - \left(\frac{1}{n_{i+1}(n_i + n_{i+1})} + \frac{1}{n_i(n_i + n_{i+1})} \right)$$

$$= \frac{t_{i+1}n_i - t_i n_{i+1}}{n_i n_{i+1}} - \frac{1}{n_i n_{i+1}} = 0.$$

It is easy to show that Theorem 3 is true for $k \leq 3$. Therefore, Theorem 3 is equivalent to the following lemma (note that (8) is equivalent to (3)).

Lemma 4. *Let $k \geq 4$ and $q = (k+1)(k+2)$. For each integer a, $1 \leq a \leq q$, such that $\gcd(a, q) \leq k$, there exists integers x, y, and t such that $1 \leq x \leq k$,*

$$1 \leq |y| \leq k, \tag{7}$$

and

$$\frac{a}{q} - \frac{t}{x} = \frac{y}{xq}. \tag{8}$$

We have

$$\frac{t_i}{n_i} \leq \frac{a}{q} < \frac{t_{i+1}}{n_{i+1}}$$

for some i. We split the proof into cases. We first consider the cases when

$$\frac{t_i}{n_i} \leq \frac{a}{q} \leq s_i.$$

Case I, $\frac{t_i}{n_i} = \frac{a}{q}$ or, equivalently, $n_i a = t_i q$. Since $\gcd(n_i, t_i) = 1$, n_i must divide q. Hence $a = t_i(q/n_i)$, and so

$$\gcd(a, q) \geq \frac{q}{n_i} \geq \frac{q}{k} = \frac{k^2 + 3k + 1}{k} > k + 3 > k.$$

Case II, $\frac{t_i}{n_i} < \frac{a}{q} \le \frac{t_i}{n_i} + \frac{1}{n_i(n_i + n_{i+1})}$. Then

$$0 < n_i a - t_i q \le \frac{q}{n_i + n_{i+1}}.$$

Subcase IIa, $n_i + n_{i+1} \ge k + 3$. Then

$$n_i a - t_i q \le \frac{k^2 + 3k + 2}{k + 3} = k + \frac{2}{k + 3} < k + 1$$

and so $0 < n_i a - t_i q \le k$.

Subcase IIb, $n_i + n_{i+1} = k + 2$. Then

$$n_i a - t_i q \le \frac{q}{k + 2} = k + 1.$$

Suppose that

$$n_i a - t_i q = k + 1. \tag{9}$$

Then

$$a = \frac{t_i(k + 2) + 1}{n_i}(k + 1). \tag{10}$$

From (4) we get

$$1 = t_{i+1} n_i - t_i n_{i+1} = t_{i+1} n_i - t_i(k + 2) + t_i n_i$$

and so $(t_i + t_{i+1})n_i = t_i(k + 2) + 1$. Hence $\gcd(t_i + t_{i+1}, k + 2) = 1$. Further, combining with (10) we get

$$a = (t_i + t_{i+1})(k + 1).$$

Hence, $\gcd(a, q) = k + 1 > k$.

Subcase IIc, $n_i + n_{i+1} = k + 1$ is similar. First, from (4) we get, in this case,

$$1 = t_{i+1} n_i - t_i n_{i+1} = t_{i+1} n_i - t_i(k + 1) + t_i n_i$$

and so

$$(t_i + t_{i+1})n_i = t_i(k + 1) + 1. \tag{11}$$

Hence $\gcd(t_i + t_{i+1}, k + 1) = 1$ and

$$\gcd(n_i, k + 1) = 1. \tag{12}$$

Further

$$n_i a - t_i q \le \frac{q}{k + 1} = k + 2.$$

Subcase IIc-1,

$$n_i a - t_i q = k + 2. \tag{13}$$

Then, by (11) and (13),

$$a = \frac{t_i(k + 1) + 1}{n_i}(k + 2) = (t_i + t_{i+1})(k + 2). \tag{14}$$

Hence, $\gcd(a, q) = k + 2 > k$.

Subcase IIc-2,

$$n_i a - t_i q = k + 1. \tag{15}$$

In this case,

$$n_i a = (t_i(k + 2) + 1)(k + 1),$$

and so, by (12), $n_i | (t_i(k + 2) + 1)$. Further, by (11), $n_i | (t_i(k + 1) + 1)$. Hence

$$n_i | ((t_i(k + 2) + 1) - (t_i(k + 1) + 1)) = t_i.$$

Since $\gcd(n_i, t_i) = 1$ and $t_i < n_i$, this is only possible if $n_i = 1$ and $t_i = 0$. Therefore, by (15), we must have $a = k + 1$ and so $\gcd(a, q) = k + 1 > k$.

Finally, we note that the cases where $s_i < \frac{a}{q} < \frac{t_{i+1}}{n_{i+1}}$ are similar. This completes the proof of Lemma 4 and so of Theorem 3.

Theorem 4. *Let* $q = (k + 1)(k + 2)$. *The pair* $(1, a)$ *is not a* (q, k) *check pair for any* a.

Proof. Suppose that $(1, a)$ is a check pair. By Proposition 1,

$$\gcd(a, q) < \frac{(k + 1)(k + 2)}{2k} < k.$$

By Theorem 3, there exist $e, \varepsilon \in [-k, -1] \cup [1, k]$ such that $e = u \otimes \varepsilon$. Hence, the syndroms are not all distinct. This contradicts our assumption that $(1, a)$ is a check pair.

Lemma 5. *Let* $q = (k + 1)(k + 2)$. *If* (a, b) *is a* (q, k) *check pair, then*

$$\gcd(a, q) > 1 \ and \ \gcd(b, q) > 1.$$

Proof. Suppose that $\gcd(a, q) = 1$. Let a' be defined by $a' \otimes a = 1$ and let $b' = a' \otimes b$. By Lemma 1 part 3, $(1, b')$ is a check pair. However this contradicts Theorem 4. Hence, $\gcd(a, q) > 1$. Similarly, $\gcd(b, q) > 1$.

In contrast to this lemma, we have the following lemma.

Lemma 6. *Let* $q = (k + 1)(k + 2)$. *If* (a, b) *is a* (q, k) *check pair, then*

$$gcd(a, b, q) = 1.$$

Proof. Suppose that $\gcd(a, b, q) = d > 1$. Then we see that $(a/d, b/d)$ is a $(q/d, k)$ check pair:

- If $(a/d) \otimes e \equiv (b/d) \otimes \varepsilon \pmod{q/d}$ where $e, \varepsilon \in [-k, -1] \cup [1, k]$, then $a \otimes e \equiv b \otimes \varepsilon \pmod{q}$, but this is not possible since (a, b) is a (q, k) check pair.
- We have

$$\gcd\left(\frac{a}{d}, \frac{q}{d}\right) = \frac{\gcd(a, q)}{d} < \frac{q/(2k)}{d} = \frac{q/d}{2k}.$$

– Similarly,

$$\gcd\left(\frac{b}{d}, \frac{q}{d}\right) < \frac{q/d}{2k}.$$

However,

$$\frac{q}{d} \le \frac{(k+1)(k+2)}{2} < (k+1)^2,$$

and so no $(q/d, k)$ check pair exists by Theorem 1, a contradiction.

6.2   Check Pairs When $k + 1$ Is Not a Prime Power

Theorem 5. *Let* $q = (k+1)(k+2)$. *If* $k + 1 = \sigma\rho$ *where* $\gcd(\sigma, \rho) = 1$, *then* $(\sigma, \rho(k+2-\sigma))$ *is a* (q, k) *check pair.*

Proof. Suppose that $k + 1 = \sigma\rho$ where $\gcd(\sigma, \rho) = 1$. Then

$$q = (k+1)(k+2) = \sigma\rho(k+2).$$

We break the proof up into three parts.

1. We have $\gcd(\sigma, q) \le \sigma = (k+1)/\rho < (k+2)/2 < q/(2k)$.
2. We have

$$\gcd(\rho(k+2-\sigma), q) = \rho\gcd(k+2-\sigma, \sigma(k+2)) = \rho d.$$

We will show that $d = 1$. Since $\sigma|(k+1)$, we have $\gcd(\sigma, k+2) = 1$. Hence,

$$\gcd(k+2-\sigma, \sigma) = \gcd(k+2, \sigma) = 1$$

and

$$\gcd(k+2-\sigma, k+2) = \gcd(-\sigma, k+2) = 1.$$

Therefore $d = 1$ and so

$$\gcd(\rho(k+2-\sigma), q) = \rho < (k+2)/2 < q/(2k).$$

3. Suppose that

$$\sigma e \equiv \rho(k+2-\sigma)\varepsilon \pmod{\sigma\rho(k+2)}, \tag{16}$$

where $e, \varepsilon \in [-k, -1] \cup [1, k]$. Without loss of generality, we can assume that $\varepsilon \in [1, k]$. From (16) we get $\sigma e \equiv 0 \pmod{\rho}$ and so $e \equiv 0 \pmod{\rho}$, that is $e = \rho e'$. Since $|e| \le k = \sigma\rho - 1$, we have $1 \le |e'| \le \sigma - 1$. Similarly, we get $\varepsilon = \sigma\varepsilon'$ where $1 \le \varepsilon' \le \rho - 1$. Substituting these in (16) we get

$$\sigma\rho e' \equiv \rho(k+2-\sigma)\sigma\varepsilon' \pmod{\sigma\rho(k+2)}$$

and so

$$e' \equiv (k+2-\sigma)\varepsilon' \pmod{k+2}.$$

This implies that

$$-e' \equiv \sigma\varepsilon' \pmod{k+2}. \tag{17}$$

However, since

$$(-e') \mod (k+2) \in [1, \sigma - 1] \cup [k+3-\sigma, k+1]$$

and

$$\sigma \le \sigma \varepsilon' \le \sigma(\rho - 1) = k + 1 - \sigma,$$

(17) is not possible. Hence, (16) is not possible.

6.3   Corresponding Codes

We look closer at the codes corresponding to the check pairs of Theorem 5 and their encoding and decoding. The code is

$$C = \{(u, v) \mid u, v \in [0, q-1], \sigma u \oplus \rho(k+2-\sigma)v = 0\}.$$

Lemma 7. *We have*

$$C = \{(\rho U, \sigma V) \mid U \in [0, \sigma(k+2)-1], V \in [0, \rho(k+2)-1], U+V \equiv 0 \ (\mathrm{mod} \ k+2)\}.$$

Proof. Since $\sigma u + \rho(k+2-\sigma)v \equiv 0 \ (\mathrm{mod} \ \sigma\rho(k+2))$, we get $\sigma u \equiv 0 \ (\mathrm{mod} \ \rho)$. Since $\gcd(\sigma, \rho) = 1$, this implies that $u \equiv 0 \ (\mathrm{mod} \ \rho)$. Hence $u = \rho U$ where $U \in [0, \sigma(k+2)-1]$.

Since $k+2 = \sigma\rho+1$, we similarly get $\rho v \equiv 0 \ (\mathrm{mod} \ \sigma)$ and so $v \equiv 0 \ (\mathrm{mod} \ \sigma)$ and $v = \sigma V$ where $V \in [0, \rho(k+2) - 1]$. Finally, $(\rho U, \sigma V) \in C$ if and only if

$$\sigma\rho U \oplus \rho(k+2-\sigma)\sigma V \equiv 0 \ (\mathrm{mod} \ \sigma\rho(k+2))$$

which is equivalent to

$$U + V \equiv 0 \ (\mathrm{mod} \ k+2). \tag{18}$$

Corollary 1. *We have* $|C| = q$.

Proof. Let $V \in [0, \rho(k+2) - 1]$. By (18), we have $(\rho U, \sigma V) \in C$ if and only if $U \equiv (-V) \ (\mathrm{mod} \ \sigma(k+2))$. Hence,

$$U \equiv (-V + z(k+2)) \ (\mathrm{mod} \ \sigma(k+2))$$

for some $z \in [0, \sigma - 1]$. Hence for each value of V there are σ possible values of U. Therefore, $|C| = \sigma\rho(k+2) = q$.

Theorem 4 showed that no systematic code exists in this case. However, also for the code given above there is an efficient bijection between $\mathbb{Z}_q$ and C.

The encoding (that is, the mapping from $\mathbb{Z}_q$ to C) can be done as follows: any integer $m \in [0, q-1]$ can be represented as

$$m = \sigma\mu + \nu \text{ where } \mu \in [0, \rho(k+2) - 1], \nu \in [0, \sigma - 1].$$

We encode m into $((\rho(-\mu + \nu(k+2)) \bmod q), \sigma\mu)$.

The information can easily be recovered from the representation $(\rho U, \sigma V)$. First, we let $\mu = V$. Then we know that

$$\rho(-\mu + \nu(k+2)) \equiv \rho U \pmod{\rho\sigma(k+2)},$$

and so

$$-\mu + \nu(k+2) \equiv U \pmod{\sigma(k+2)},$$

which in turn implies that $U + \mu \equiv 0 \pmod{(k+2)}$ and so

$$\nu = \left(\frac{U+\mu}{k+2} \bmod \sigma\right) \text{ and } m = \sigma V + \nu.$$

We next consider the correction of errors. A codeword is $(u, v) = (\rho U, \sigma V)$ where (18) is satisfied.

- If $u' = u + e$ where $e \in [0, k]$, then the syndrom is $s = \sigma e$ and so $e = s/\sigma$.
- If $u' = u + e$ where $e \in [-k, -1]$, then $s = q + \sigma e$ and so $e = (s - q)/\sigma$.
- If $v' = v + e$, where $e \in [-k, -1] \cup [1, k]$, then

$$s \equiv \rho(k + 2 - \sigma)e \pmod{\rho\sigma(k+2)}$$

and so ρ divides s and

$$\frac{s}{\rho} \equiv (k + 2 - \sigma)e \pmod{\sigma(k+2)}.$$

We see that $\gcd(k + 2 - \sigma, \sigma(k+2)) = 1$. Hence

$$e \equiv f \stackrel{\text{def}}{=} ((k + 2 - \sigma)^{-1}\frac{s}{\rho} \bmod \sigma(k+2)),$$

where the inverse is modulo $\sigma(k+2)$. If $f \leq k$, then $e = f$. If $f \geq \sigma(k+2) - k$, then $e = f - \sigma(k+2)$.

From this, we get the following decoding algorithm.

- if $s \equiv 0 \pmod{\sigma}$ and $s/\sigma \in [0, k]$, then decode into $(u \ominus (s/\sigma), v)$
- else if $s \equiv 0 \pmod{\sigma}$ and $s/\sigma \in [\rho(k+2) - k, \rho(k+2) - 1]$, then decode into $(u \ominus ((s-q)/\sigma), v)$
- else if $s \equiv 0 \pmod{\rho}$, let

$$f = ((k + 2 - \sigma)^{-1}\frac{s}{\rho} \bmod \sigma(k+2)),$$

- if $f \leq k$, then decode into $(u, v \ominus f)$,
- else decode into $(u, (v \ominus (f - \sigma(k+2)) \bmod q)$.

For $k \leq 100$ and $q = (k+1)(k+2)$, a complete search has shown that there are no check pairs when $k + 1$ a prime power. Possibly this is the case for all k and we formulate this a conjecture.

Conjecture 1. If $k+1$ a prime power, then there are no $((k+1)(k+2),k)$ check pairs.

When $k+1 \le 42$ is not a prime power, all the (q,k) check pairs are those given by Theorem 5, combined with Lemma 1. Possibly this is the case in general.

Conjecture 2. If $k+1$ a not prime power and $q = (k+1)(k+2)$, then all (q,k) check pairs are congruent $(c\sigma, c\rho(k+2-\sigma))$ or $(-c\sigma, c\rho(k+2-\sigma))$ modulo q, where $k+1 = \sigma\rho$, $\gcd(\sigma,\rho) = 1$, and $\gcd(c,q) = 1$.

7 Summary

In this paper we have considered linear codes of length two over the alphabet $\mathbb{Z}_q = \{0,1,\dots,q-1\}$, correcting single errors at size at most k. It was well known [5] that for $q \le (k+1)^2$ no such codes exist. For $q = (k+1)^2+1$ a simple code construction is known.

In this paper, we have studied the cases when $q \ge (k+1)^2+1$. In Sect. 5, we considered $q \ne (k+1)(k+2)$. We show that a simple code construction exists in all cases. We describe codes and their encoding and decoding, both quite simple.

In Sect. 6 we considered $q = (k+1)(k+2)$. If $k+1$ is not a prime power, then we have found a code construction and again describe the codes, their encoding and decoding. This is the main result in this paper. For $k+1$ a prime power, we conjecture that no codes exist.

References

1. Elarief, N., Bose, B.: Optimal, systematic, q-ary codes correcting all asymmetric and symmetric errors of limited magnitude. IEEE Trans. Inf. Theory **56**, 979–983 (2010)
2. Hardy, G.H., Wright, E.M.: An Introduction to the Theory of Numbers, 4th edn. Oxford University Press, London (1962)
3. Jiang, A., Mateescu, R., Schwartz, M., Bruck, J.: Rank modulation for flash memories. IEEE Trans. Inf. Theory **55**, 2659–2673 (2009)
4. Kløve, T., Elarief, N., Bose, B.: Systematic, single limited magnitude error correcting codes for flash memories. IEEE Trans. Inf. Theory **57**, 4477–4487 (2011)
5. Kløve, T., Luo, J., Yari, S.: Codes correcting single errors of limited magnitude. IEEE Trans. Inf. Theory **58**, 2206–2219 (2012)
6. Schwartz, M.: Quasi-cross lattice tilings with applications to flash memory. IEEE Trans. Inf. Theory **58**, 2397–2405 (2012)
7. Yari, S., Kløve, T., Bose, B.: Some linear codes correcting single errors of limited magnitude for flash memories. IEEE Trans. Inf. Theory **59**, 7278–7287 (2013)

Boolean Functions

Bent and Semi-bent Functions via Linear Translators

Neşe Koçak[1,2], Sihem Mesnager[3,4,5], and Ferruh Özbudak[2,6(✉)]

[1] ASELSAN Inc., Ankara, Turkey
[2] Institute of Applied Mathematics, Middle East Technical University,
Ankara, Turkey
nesekocak@aselsan.com.tr
[3] Department of Mathematics, University of Paris VIII, Paris, France
[4] University of Paris XIII, LAGA, UMR 7539, CNRS, Paris, France
[5] Telecom ParisTech, Paris, France
smesnager@univ-paris8.fr
[6] Department of Mathematics, Middle East Technical University, Ankara, Turkey
ozbudak@metu.edu.tr

Abstract. The paper is dealing with two important subclasses of plateaued functions: bent and semi-bent functions. In the first part of the paper, we construct mainly bent and semi-bent functions in the Maiorana-McFarland class using Boolean functions having linear structures (linear translators) systematically. Although most of these results are rather direct applications of some recent results, using linear structures (linear translators) allows us to have certain flexibilities to control extra properties of these plateaued functions. In the second part of the paper, using the results of the first part and exploiting these flexibilities, we modify many secondary constructions. Therefore, we obtain new secondary constructions of bent and semi-bent functions not belonging to the Maiorana-McFarland class. Instead of using bent (semi-bent) functions as ingredients, our secondary constructions use only Boolean (vectorial Boolean) functions with linear structures (linear translators) which are very easy to choose. Moreover, all of them are very explicit and we also determine the duals of the bent functions in our constructions. We show how these linear structures should be chosen in order to satisfy the corresponding conditions coming from using derivatives and quadratic/cubic functions in our secondary constructions.

Keywords: Boolean functions · Bent functions · Semi-bent functions · Walsh-hadamard transform · Linear structures · Linear translators and derivatives

1 Introduction

The classes of bent and semi-bent functions are special subclasses of the so-called plateaued functions [25]. They are studied in cryptography because, besides

© Springer International Publishing Switzerland 2015
J. Groth (Ed.): IMACC 2015, LNCS 9496, pp. 205–224, 2015.
DOI: 10.1007/978-3-319-27239-9_13

having low Walsh-Hadamard transform magnitude which provides protection against fast correlation attacks and linear cryptanalysis, they can also possess other desirable properties. For their relations to coding theory and applications in cryptography bent and semi-bent functions have attracted a lot of research.

Bent functions are nice combinatorial objects. They are maximally nonlinear Boolean functions with an even number of variables. They were defined by Rothaus [23] in 1976 but already studied by Dillon [15] since 1974. Open problems on binary bent functions can be found in [6]. A very recent survey on bent functions can be found in [7]. A book devoted especially to bent functions and containing a complete survey on bent functions (including its variations and generalizations) is [22]. The term semi-bent function was introduced by Chee, Lee and Kim at Asiacrypt' 94 [14]. These functions had been previously investigated under the name of 3-valued almost optimal Boolean functions. A survey containing open problems on semi-bent functions can be found in [20]. Despite the amount of research in the theory of bent and semi-bent functions, the classification of those functions is still elusive, therefore, not only their characterization, but also their construction are challenging problems. Several constructions of explicit bent and semi-bent functions have been proposed in the literature but investigation of such kind of functions is still needed.

The concept of a linear translator exists of p-ary function (see for instance [16]) but it was introduced in cryptography, mainly for Boolean functions (see for instance [10]). Functions with linear structures are considered as weak for some cryptographic applications. For instance, a recent attack on hash functions proposed in [1] exploits a similar weakness of the involved mappings. All Boolean functions using a linear translator have been characterized by Lai [17]. Further, Charpin and Kyureghyan have done the characterization for the functions in univariate variables from $\mathbb{F}_{p^n}$ to $\mathbb{F}_p$ of the form $Tr_{\mathbb{F}_{p^n}/\mathbb{F}_p}(F(x))$, where $F(x)$ is a function over $\mathbb{F}_{p^n}$ and $Tr_{\mathbb{F}_{p^n}/\mathbb{F}_p}$ denotes the trace function from $\mathbb{F}_{p^n}$ to $\mathbb{F}_p$. The result of Lai in [17] has been formulated recently by Charpin and Sarkar [13].

For a Boolean map, linear structures or linear translators are not desirable and are generally considered as a defect. In this paper, we show that one can recycle such Boolean functions to get Boolean functions with optimal or very high nonlinearity. More precisely, we show that one can obtain primary constructions of bent and semi-bent functions from Boolean maps having linear structures or linear translator in Sects. 3, 4 and 5. All the primary constructions proposed in the paper belong to the well-known class of Maiorana-McFarland. However, an important feature of the bent functions presented in this paper is that their dual functions can be explicitly computed. Next, we focus on secondary constructions presented in [3] and in [4] (see also [19]). Note that several primary constructions have been derived in [19] and in [21] from the Carlet's result ([4], Theorem 3) which has been completed in ([19], Theorem 4). We show how to obtain new secondary constructions by reusing bent functions presented in the paper. Our new secondary constructions are very explicit and they use Boolean functions (vectorial Boolean functions) with certain linear structures (linear translators) as ingredients instead of bent or semi-bent functions. The conditions on such linear structures (linear translators) in our secondary constructions are easily satisfied. Finally, we show that one can construct bent functions from bent functions of Sects. 3 and 4 by adding a quadratic or cubic function appropriately chosen.

This paper is organized as follows: We provide a short background in Sect. 2. We present explicit constructions of bent and semi-bent functions in Maiorana-McFarland type in Sects. 3, 4 and 5. We give various secondary constructions in Sects. 6, 7 and 8.

2   Notation and Preliminaries

For any set E, $E^\star = E \setminus \{0\}$ and $\#E$ will denote the cardinality of E. A Boolean function on the finite field $\mathbb{F}_{2^n}$ of order 2^n is a mapping from $\mathbb{F}_{2^n}$ to the prime field $\mathbb{F}_2$. Recall that for any positive integers k, and r dividing k, the trace function from $\mathbb{F}_{2^k}$ to $\mathbb{F}_{2^r}$, denoted by Tr_r^k, is the mapping defined for every $x \in \mathbb{F}_{2^k}$ as $Tr_r^k(x) := x + x^{2^r} + x^{2^{2r}} + \cdots + x^{2^{k-r}}$. For a Boolean function f on $\mathbb{F}_{2^n}$, the Walsh-Hadamard transform of f is the discrete Fourier transform of the sign function $\chi_f := (-1)^f$ of f, whose value at $\omega \in \mathbb{F}_{2^n}$ is defined as $\widehat{\chi_f}(\omega) = \sum_{x \in \mathbb{F}_{2^n}} (-1)^{f(x) + Tr_1^n(\omega x)}$.

Definition 1. *Let n be an even integer. A Boolean function f on $\mathbb{F}_{2^n}$ is said to be bent if its Walsh transform satisfies $\widehat{\chi_f}(a) = \pm 2^{\frac{n}{2}}$ for all $a \in \mathbb{F}_{2^n}$.*

Bent functions come in pairs. For a bent function f on $\mathbb{F}_{2^n}$, we define its *dual function* $\tilde{f}$ as a Boolean function on $\mathbb{F}_{2^n}$ satisfying the equation : $(-1)^{\tilde{f}(x)} 2^{\frac{n}{2}} = \widehat{\chi_f}(x)$ for all $x \in \mathbb{F}_{2^n}$.

The dual $\tilde{f}$ of a bent function is also bent.

Definition 2. *Let n be an even integer. A Boolean function f on $\mathbb{F}_{2^n}$ is said to be semi-bent if its Walsh transform satisfies $\widehat{\chi_f}(a) \in \{0, \pm 2^{\frac{n+2}{2}}\}$ for all $a \in \mathbb{F}_{2^n}$.*

Definition 3. [25] *A Boolean function f on $\mathbb{F}_{2^n}$ is said to be k-plateaued if its Walsh transform satisfies $\widehat{\chi_f}(a) \in \{0, \pm 2^{\frac{n+k}{2}}\}$ for all $a \in \mathbb{F}_{2^n}$ and for some fixed k, $0 \le k \le n$.*

When n is even, bent functions correspond to 0-*plateaued* functions and semi-bent functions correspond to 2-*plateaued* functions.

We refer to [5] for further background and important notions like algebraic representation, trace representation and bivariate representation of Boolean functions. We will mainly use bivariate representation of bent and semi-bent functions in this paper.

Next we recall definitions of linear translator and linear structure.

Definition 4. *Let $n = rk$, $1 \le k \le n$. Let f be a function from $\mathbb{F}_{2^n}$ to $\mathbb{F}_{2^k}$, $\gamma \in \mathbb{F}_{2^n}^*$ and b be a constant of $\mathbb{F}_{2^k}$. Then γ is a b-linear translator of f if $f(x) + f(x + u\gamma) = ub$ for all $x \in \mathbb{F}_{2^n}$ and $u \in \mathbb{F}_{2^k}$. If $f(x) + f(x + \gamma) = b$ for all $x \in \mathbb{F}_{2^n}$, then γ is called a b-linear structure of f.*

The notion of b-*linear translator* is well known in the literature (see for example [16]). The notion of b-*linear structure is usually given for functions* $f : \mathbb{F}_{2^n} \to \mathbb{F}_2$, *that is* $k = 1$ *(see for example* [9]*)*.

Remark 1. Note that being b-linear translator is stronger than being b-linear structure if $k > 1$ and they are the same if $k = 1$. For example, let $f : \mathbb{F}_{2^4} \to \mathbb{F}_{2^2}$ be a function defined as $f(x) = Tr_2^4(x^2 + \gamma x)$ where $\gamma \in \mathbb{F}_{2^4} \backslash \mathbb{F}_{2^2}$. Then, γ is a 0-linear structure of f but it is not a 0-linear translator of f as $f(x + u\gamma) \neq f(x)$ for $u \in \mathbb{F}_{2^2} \backslash \mathbb{F}_2$.

The notions of linear structures, linear translators and derivatives are related.

Definition 5. *Let* $F : \mathbb{F}_{2^n} \to \mathbb{F}_{2^m}$. *For* $a \in \mathbb{F}_{2^n}$, *the function* $D_a F$ *given by* $D_a F(x) = F(x) + F(x + a), \forall x \in \mathbb{F}_{2^n}$ *is called the derivative of* F *in the direction of* a.

Note that $D_\gamma f(x) = b$ for each $x \in \mathbb{F}_{2^n}$ if and only if γ is a b-linear structure of f. Similarly, $D_{u\gamma} f(x) = ub$ for each $x \in \mathbb{F}_{2^n}$ and each $u \in \mathbb{F}_{2^k}$ if and only if γ is a b-linear translator of f.

In the literature, usually, Boolean functions are denoted by small letters (like f) and vector Boolean functions are denoted by capital letters (like F). Nevertheless, for the sake of simplicity of notation, we also denote Boolean functions (vector Boolean functions) by capital letters (small letters) when it seems more appropriate in this paper.

3   Constructions of Bent and Semi-bent Boolean Functions from the Class of Maiorana-McFarland Using One Linear Structure

A function $H : \mathbb{F}_{2^m} \times \mathbb{F}_{2^m} \to \mathbb{F}_2$ is said to be in the class of Maiorana-McFarland if it can be written in bivariate form as

$$H(x, y) = Tr_1^m (x\phi(y)) + h(y) \tag{3.1}$$

where ϕ is a map from $\mathbb{F}_{2^m}$ to $\mathbb{F}_{2^m}$ and h is a Boolean function on $\mathbb{F}_{2^m}$. It is well-known that we can choose ϕ so that H is bent or H is semi-bent. Indeed, it is well-known that bent functions of the form (3.1) come from one-to-one maps while 2-to-1 maps lead to semi-bent functions.

Proposition 1. ([5,15,18]) *Let* H *be defined by (3.1). Then,*

1. H *is bent if and only if* ϕ *is a permutation and its dual function is* $\tilde{H}(x, y) = Tr_1^m (y\phi^{-1}(x)) + h(\phi^{-1}(x))$.
2. H *is semi-bent if* ϕ *is 2-to-1.*

As a first illustration of Proposition 1, let us consider a first class of maps from $\mathbb{F}_{2^m}$ to itself: $\phi : y \mapsto y + \gamma f(y)$ where γ is a linear structure of f. This class has the property that it only contains one-to-one maps or 2-to-1 maps. Therefore, by Proposition 1, one can obtain the following infinite families of bent and semi-bent functions.

Proposition 2. *Let f and h be two Boolean functions over $\mathbb{F}_{2^m}$. Let H be the Boolean function defined on $\mathbb{F}_{2^m} \times \mathbb{F}_{2^m}$ by*

$$H(x,y) = Tr_1^m(xy + \gamma x f(y)) + h(y), \gamma \in \mathbb{F}_{2^m}.$$

H is bent (resp. semi-bent) if and only if γ is a 0-linear (resp. 1-linear) structure of f. Furthermore, if H is bent, then its dual is

$$\tilde{H}(x,y) = Tr_1^m\big(yx + \gamma y f(x))\big) + h(x + \gamma f(x)).$$

Proof. Properties of $\phi : y \mapsto y + \gamma f(y)$ are well-known and firstly developed in [8,9] (see also [11,16]). Bijectivity is given by Theorem 2 of [8]. For the 2-to-1 property, see Theorems 3, 6 in [9]. The proof is then immediately obtained. Also, note that since ϕ is an involution (see also [11,12,16]), we have $\tilde{H}(x,y) = Tr_1^m\big(y\phi(x))\big) + h(\phi(x))$.

In order to show that the hypotheses of Proposition 2 hold in certain cases, we give the following examples which are direct applications of Theorems 3, 4 in [8].

Example 1. Let $\gamma \in \mathbb{F}_{2^m}^\star$ and $\beta \in \mathbb{F}_{2^m}$ such that $Tr_1^m(\beta\gamma) = 0$ (resp. $Tr_1^m(\beta\gamma) = 1$). Let $H : \mathbb{F}_{2^m} \to \mathbb{F}_{2^m}$ be an arbitrary mapping and h be any Boolean function on $\mathbb{F}_{2^m}$. Then the function g defined over $\mathbb{F}_{2^m} \times \mathbb{F}_{2^m}$ by

$$g(x,y) = Tr_1^m(xy + \gamma x Tr_1^m(H(y^2 + \gamma y) + \beta y)) + h(y)$$

is bent (resp. semi-bent).

Example 2. Let $0 \leq i \leq m - 1$, $i \notin \{0, \frac{m}{2}\}$ and $\delta, \gamma \in \mathbb{F}_{2^m}$ such that $\delta^{2^i - 1} = \gamma^{1 - 2^{2i}}$. Let h be any Boolean function on $\mathbb{F}_{2^m}$ and g be the Boolean function defined on $\mathbb{F}_{2^m} \times \mathbb{F}_{2^m}$ by

$$g(x,y) = Tr_1^m(xy + \gamma x Tr_1^m(\delta y^{2^i + 1})) + h(y).$$

If $Tr_1^m(\delta\gamma^{2^i + 1}) = 0$ (resp. $Tr_1^m(\delta\gamma^{2^i + 1}) = 1$) then g is bent (resp. semi-bent).

Observe that if we compose ϕ at left by a linearized permutation polynomial L, any output has the same number of preimages under ϕ than under $L \circ \phi$. Hence, one can slightly generalize Proposition 2 as follows.

Proposition 3. *Let f and h be two Boolean functions over $\mathbb{F}_{2^m}$ and $\gamma \in \mathbb{F}_{2^m}$. Let L be a linearized permutation polynomial of $\mathbb{F}_{2^m}$. The Boolean function H defined by*

$$H(x,y) = Tr_1^m\left(xL(y) + L(\gamma)x f(y)\right) + h(y)$$

is bent (resp. semi-bent) if and only if γ is a 0-linear (resp. 1-linear) structure of f. Moreover, if H is bent then its dual function $\tilde{H}$ is given by

$$\tilde{H}(x,y) = Tr_1^m(yL^{-1}(x) + \gamma y f(L^{-1}(x)) + h(L^{-1}(x) + \gamma f(L^{-1}(x))).$$

4   Constructions of Bent and Semi-bent Boolean Functions from the Class of Maiorana-McFarland Using Two Linear Structures

In this section we consider the functions H of the form (3.1):

$$H(x, y) = Tr_1^m \left(x\phi(y) \right) + h(y) \text{ with } \phi(y) = \pi_1 \left(\pi_2(y) + \gamma f(\pi_2(y)) + \delta g(\pi_2(y)) \right) \tag{4.1}$$

where f, g and h are Boolean functions over $\mathbb{F}_{2^m}$, $\gamma, \delta \in \mathbb{F}_{2^m}^\star$, $\gamma \neq \delta$ and π_1, π_2 are permutations of $\mathbb{F}_{2^m}$ (not necessarily linear). The class (4.1) contains the functions involved in Proposition 1 and in Proposition 3 (which corresponds to the case where $f = g$). In the line of Sect. 3, we study the cases where γ and δ are linear structures of the Boolean functions involved in ϕ. Then one can exhibit conditions of bentness or semi-bentness as those of Propositions 1 and 3 that we present in the following two propositions. We indicate that, despite their similarities with Propositions 1 and 3, we obtain bent functions that do not fall in the scope of Propositions 1 and 3.

Proposition 4. *Let H be defined by Eq. (4.1). Then H is bent if one of the following conditions holds:*

  (i) *γ is a 0-linear structure of f, δ is a 0-linear structure of f and g,*
 (ii) *γ is a 0-linear structure of f, δ is a 1-linear structure of f and $\delta + \gamma$ is a 0-linear structure of g,*
(iii) *δ is a 0-linear structure of g, γ is a 0-linear structure of f and g,*
 (iv) *δ is a 0-linear structure of g, γ is a 1-linear structure of g and $\delta + \gamma$ is a 0-linear structure of f,*
  (v) *δ is a 1-linear structure of f, γ is a 1-linear structure of f and g,*
 (vi) *γ is a 1-linear structure of g, δ is a 1-linear structure of f and g.*

Moreover, if H is bent then its dual is $\tilde{H}(x, y) = Tr_1^m \left(y\phi^{-1}(x) \right) + h(\phi^{-1}(x))$ where $\phi^{-1} = \pi_2^{-1} \circ \rho^{-1} \circ \pi_1^{-1}$ and ρ^{-1} is given explicitly in the Appendix as Proposition 8. In particular, choosing $\pi_1(x) = L(x)$ as a linearized permutation polynomial and π_2 as the identity, we get that

$$H(x, y) = Tr_1^m \left(xL(y) + L(\gamma)xf(y) + L(\delta)xg(y) \right) + h(y) \tag{4.2}$$

is bent in the conditions above and $\tilde{H}(x, y) = Tr_1^m \left(y\rho^{-1}(L^{-1}(x)) \right) + h(\rho^{-1}(L^{-1}(x)))$.

Proof. We give the proof for only case (i) since the proofs for the other cases are very similar. It suffices to show that $\rho : y \mapsto y + \gamma f(y) + \delta g(y)$ is a permutation. Suppose that $\rho(y) = \rho(z)$, i.e.,

$$y + \gamma f(y) + \delta g(y) = z + \gamma f(z) + \delta g(z). \tag{4.3}$$

Taking f of both sides we obtain $f\big(y + \gamma f(y) + \delta g(y)\big) = f\big(z + \gamma f(z) + \delta g(z)\big)$. Since γ and δ are 0-linear structures of f, we have

$$f(y) = f(z). \tag{4.4}$$

Combining Eqs. (4.3) and (4.4), we get $y + \delta g(y) = z + \delta g(z)$. Taking g of both sides we obtain $g(y + \delta g(y)) = g(z + \delta g(z))$. Since δ is a 0-linear structure of g, we conclude

$$g(y) = g(z). \tag{4.5}$$

Combining Eqs. (4.3), (4.4) and (4.5), we reach that $y = z$. For the dual function, ρ^{-1} is written explicitly in the Appendix as Proposition 8 and the proof for ρ^{-1} for case (i) is given.

Remark 2. The converse of Proposition 4 is not always true. For example, for $f(x) = Tr_1^3(x^3 + \alpha^5 x)$, $g(x) = Tr_1^3(\alpha x^3 + \alpha^5 x)$, $\gamma = \alpha$ and $\delta = \alpha^3$ where α is a primitive element of $\mathbb{F}_{2^3}$, ϕ is a permutation but none of the conditions given in Proposition 4 is satisfied.

The following result shows in which cases ϕ is 2-to-1 and hence H is semi-bent.

Proposition 5. *Let H be defined by (4.1). Then H is semi-bent if one of the following conditions holds:*

(i) γ, δ are 1-linear structures of f and γ is a 0-linear structure of g,
(ii) δ is a 1-linear structure of f and γ, δ are 0-linear structures of g,
(iii) γ, δ are 0-linear structures of f and δ is a 1-linear structure of g,
(iv) δ is a 0-linear structure of f and γ, δ are 1-linear structures of g,
(v) γ is a 0-linear structure of f, δ is a 1-linear structure of f and $\gamma + \delta$ is a 1-linear structure of g,
(vi) γ is a 1-linear structure of g, δ is a 0-linear structure of g and $\gamma + \delta$ is a 1-linear structure of f.

In particular, choosing $\pi_1(x) = L(x)$ as a linearized permutation polynomial and π_2 as the identity, we get that $H(x,y) = Tr_1^m\big(xL(y) + L(\gamma)xf(y) + L(\delta)xg(y)\big) + h(y)$ is semi-bent in the conditions above.

Proof. We give the proof for case (i) only since the proofs for other cases are similar. Now, we need to show that $\rho(y) : y \mapsto y + \gamma f(y) + \delta g(y)$ is 2-to-1. Let $\rho(y) = a$ for some $a \in \mathbb{F}_{2^m}$. Then, $y \in \{a, a + \gamma, a + \delta, a + \gamma + \delta\}$. As γ is a 1-linear structure of f and 0-linear structure of g, we have $\rho(a) = \rho(a + \gamma)$ and $\rho(a + \delta) = \rho(a + \gamma + \delta)$. Moreover, $\rho(a + \delta) = a + \delta + \gamma f(a + \delta) + \delta g(a + \delta) = a + \delta + \gamma + \gamma f(a) + \delta g(a + \delta)$ where we use that δ is a 1-linear structure of f. We observe that $\rho(a) = a + \gamma f(a) + \delta g(a) \neq \rho(a + \delta)$. Indeed, otherwise if the equality holds, then $\gamma + \delta + \delta\big(g(a) + g(a + \delta)\big) = 0$. This is a contradiction as $\gamma \neq \delta$ and $\gamma \neq 0$. This implies that $\rho^{-1}(a) = \{a, a + \gamma\}$ or $\rho^{-1}(a) = \{a + \delta, a + \gamma + \delta\}$ which shows that ρ is 2-to-1.

Remark 3. The converse of Proposition 5 is not always true. For example, for $f(x) = Tr_1^3(\alpha^4 x^3 + \alpha^4 x)$, $g(x) = Tr_1^3(\alpha x^3 + \alpha^2 x)$, $\gamma = \alpha$ and $\delta = \alpha^3$ where α is a primitive element of $\mathbb{F}_{2^3}$, ϕ is 2-to-1 but none of the conditions given in Proposition 5 is satisfied.

5   Constructions of Bent and k-Plateaued Functions Using Linear Translators

In the preceding sections, we have shown that one can construct bent and semi-bent functions from Boolean functions having linear structures, that is, having constant derivatives. An extension of these constructions is to consider Boolean maps taking its values in a subfield of the ambient field instead of Boolean functions in (3.1). In that case, the natural notion replacing linear structures is the notion of linear translators. We still adopt the approach of the preceding sections and aim to construct bent functions in the class of Maiorana-McFarland. To this end, one can apply results on permutations constructed from Boolean maps having linear translators presented in [16] and obtain the following infinite families of bent and plateaued functions.

Proposition 6. *Let m be a positive integer and k be a divisor of m. Let f be a function from $\mathbb{F}_{2^m}$ to $\mathbb{F}_{2^k}$ and h be a Boolean function on $\mathbb{F}_{2^m}$. Let H be the function defined on $\mathbb{F}_{2^m} \times \mathbb{F}_{2^m}$ by*

$$H(x,y) = Tr_1^m(xy + \gamma x f(y)) + h(y), \quad \gamma \in \mathbb{F}_{2^m}^\star.$$

(i) If γ is a c-linear translator of f where $c \in \mathbb{F}_{2^m}$ and $c \neq 1$, then H is bent and its dual function is given as

$$\tilde{H}(x,y) = Tr_1^m\left(y\left(x + \gamma \frac{f(x)}{1+c}\right)\right) + h\left(x + \gamma \frac{f(x)}{1+c}\right).$$

Moreover, $H(x,y) = Tr_1^m(xL(y) + L(\gamma)xf(y)) + h(y)$ where L is an $\mathbb{F}_{2^k}$-linearized permutation polynomial, is also bent under these conditions and its dual is

$$\tilde{H}(x,y) = Tr_1^m\left(y\left(L^{-1}(x) + \gamma \frac{f(L^{-1}(x))}{1+c}\right)\right) + h\left(L^{-1}(x) + \gamma \frac{f(L^{-1}(x))}{1+c}\right).$$

(ii) If γ is a 1-linear translator of f and $h = 0$ then H is k-plateaued with Walsh transform values

$$\widehat{\chi_H}(a,b) = \begin{cases} \pm 2^{m+k} & \text{if } Tr_k^m(b\gamma) = 0, \\ 0 & \text{otherwise.} \end{cases}$$

Note that Proposition 6 generalizes partially Proposition 2 (extending the condition 0-linear structure to c-linear translator with $c \neq 1$). Furthermore, one can derive from Propositions 4 and 5 similar statements if $f : \mathbb{F}_{2^m} \rightarrow \mathbb{F}_{2^k}$ instead of being a Boolean function. Indeed, it suffices to change the 0-linear structures (resp. 1-linear structures) with 0-linear translators. (resp. 1-linear translators).

6 Bent Functions Not Belonging to the Class of Maiorana-McFarland Using Linear Translators

In the following we are now interested in investigating constructions of bent functions that do not necessarily belong to the class of Maiorana- McFarland contrary to the preceding sections. To this end, we are particularly interested in the secondary construction of the form $f(x) = \phi_1(x)\phi_2(x) + \phi_1(x)\phi_3(x) + \phi_2(x)\phi_3(x)$ presented in [4] and next completed in [19]. More precisely, it is proven in [4] that if ϕ_1, ϕ_2 and ϕ_3 are bent, then if $\psi := \phi_1 + \phi_2 + \phi_3$ is bent and if $\tilde{\psi} = \tilde{\phi}_1 + \tilde{\phi}_2 + \tilde{\phi}_3$, then f is bent, and $\tilde{f} = \tilde{\phi}_1\tilde{\phi}_2 + \tilde{\phi}_1\tilde{\phi}_3 + \tilde{\phi}_2\tilde{\phi}_3$. Next, it is proven in [19] that the converse is also true: if ϕ_1, ϕ_2, ϕ_3 and ψ are bent, then f is bent if and only if $\tilde{\psi} + \tilde{\phi}_1 + \tilde{\phi}_2 + \tilde{\phi}_3 = 0$ (where $\psi := \phi_1 + \phi_2 + \phi_3$). In this section, we show that one can reuse Boolean functions of the shape presented in the preceding sections in the construction of [19,21].

Firstly, one can derive easily bent functions f, whose dual functions are very simple, by choosing functions H_i in the class of Maiorana-McFarland such that the permutation involving in each H_i is built in terms of an involution and a linear translator. More explicitly, each H_i is a Boolean function over $\mathbb{F}_{2^m}$ defined by $H_i(y) = Tr_1^m\Big(L(y) + L(\gamma_i)h(g(y))\Big)$ where L is a $\mathbb{F}_{2^k}$-linear involution on $\mathbb{F}_{2^m}$ (k being a divisor of m); carefully chosen according to the hypothesis of [12, Corollary 2], g is a function from $\mathbb{F}_{2^m}$ to $\mathbb{F}_{2^k}$, h is a mapping from $\mathbb{F}_{2^k}$ to itself, and γ_1, γ_2 and γ_3 are three pairwise distinct elements of $\mathbb{F}_{2^m}^\star$ which are 0-linear translators of g such that $\gamma_1 + \gamma_2 + \gamma_3 \neq 0$. Bent functions f are therefore obtained from a direct application of [19, Theorem 4], [12, Corollary 2].

Secondly, we extend a result from [21] by considering two linear structures instead of one. This result uses linear structures as in the first case of Proposition 4. Similarly, for the other five cases we can construct bent functions and their duals. Due to space limitations these results are presented in the Appendix as Propositions 9, 10, 11, 12 and 13.

Proposition 7. *Let f and g be functions from $\mathbb{F}_{2^m}$ to $\mathbb{F}_2$. For $i \in \{1, 2, 3\}$ set $\phi_i(y) := y + \gamma_i f(y) + \delta_i g(y)$ where*

(i) δ_1, δ_2, δ_3 *are elements of $\mathbb{F}_{2^m}^\star$ which are 0-linear structures of f and g;*
(ii) γ_1, γ_2 *and γ_3 are elements of $\mathbb{F}_{2^m}^\star$ which are 0-linear structures of f;*
(iii) $\gamma_1 + \gamma_2$ *and $\gamma_1 + \gamma_3$ are 0-linear structures of g.*

Then the function h defined on $\mathbb{F}_{2^m} \times \mathbb{F}_{2^m}$ by $h(x, y) = Tr_1^m\Big(x\phi_1(y)\Big)Tr_1^m$ $\Big(x\phi_2(y)\Big) + Tr_1^m\Big(x\phi_1(y)\Big)Tr_1^m\Big(x\phi_3(y)\Big) + Tr_1^m\Big(x\phi_2(y)\Big)Tr_1^m\Big(x\phi_3(y)\Big)$ is bent and the dual of h is given by $\tilde{h}(x, y) = Tr_1^m\Big(y\phi_1^{-1}(x)\Big)Tr_1^m\Big(y\phi_2^{-1}(x)\Big) + Tr_1^m\Big(y\phi_1^{-1}(x)\Big)Tr_1^m\Big(y\phi_3^{-1}(x)\Big) + Tr_1^m\Big(y\phi_2^{-1}(x)\Big)Tr_1^m\Big(y\phi_3^{-1}(x)\Big)$ where $\phi_i^{-1}(x) = x + \gamma_i f(x) + \delta_i \left[g(x)(1 + f(x)) + g(x + \gamma_i)f(x)\right].$

Proof. Let $\psi_i(x,y) = Tr_1^m\big(x\phi_i(y)\big)$. Then by Proposition 4, ψ_i is bent for $i = 1, 2, 3$. Let $\gamma = \gamma_1 + \gamma_2 + \gamma_3$ and $\delta = \delta_1 + \delta_2 + \delta_3$. Then, $\psi(x,y) = Tr_1^m\big(x(y + \gamma f(y) + \delta g(y))\big)$ is bent since γ is a 0-linear structure of f and δ is a 0-linear structure of f and g. Now, it remains to show that $\tilde{\psi} = \tilde{\psi}_1 + \tilde{\psi}_2 + \tilde{\psi}_3$. $\tilde{\psi} = Tr_1^m\big(x\phi^{-1}(y)\big)$ and $\phi^{-1}(x)$ is given in Proposition 8 in the Appendix.

Note that $\tilde{\psi} = \tilde{\psi}_1 + \tilde{\psi}_2 + \tilde{\psi}_3$ if and only if $g(x + \gamma_1) = g(x + \gamma_2) = g(x + \gamma_3) = g(x + \gamma_1 + \gamma_2 + \gamma_3)$ which means $\gamma_1 + \gamma_2$ and $\gamma_1 + \gamma_3$ are 0-linear structures of g.

7 A Secondary Construction of Bent and Semi-bent Functions Using Derivatives and Linear Translators

In this section, we consider a new kind of secondary construction. That construction has been proposed by Carlet and Yucas [3] and is presented below.

Theorem 1. *Let f and g be two bent functions over $\mathbb{F}_{2^n}$. Assume that there exists $a \in \mathbb{F}_{2^n}$ such that $D_a f = D_a g$. Then the function $h : \mathbb{F}_{2^n} \to \mathbb{F}_2$ defined by $h(x) = f(x) + D_a f(x)\big(f(x) + g(x)\big)$ is bent and its dual is $\tilde{h}(x) = \tilde{f}(x) + Tr_1^n(ax)(\tilde{f}(x) + \tilde{g}(x))$.*

In the line of Theorem 1 and of the preceding sections, we shall derive from Theorem 1 new secondary constructions of bent and semi-bent functions in Theorems 2 and 3. To this end, we will use the following lemma.

Lemma 1. *Let $b \in \mathbb{F}_{2^m}$ and $\mathcal{W} \subseteq \mathbb{F}_{2^m}$ be an $m - 1$ dimensional linear subspace with $b \notin \mathcal{W}$. Let $\mu : \mathbb{F}_{2^m} \to \mathbb{F}_2$ be a Boolean function such that b is a 0-linear structure of μ. Choose arbitrary functions $h_1 : \mathbb{F}_{2^m} \to \mathbb{F}_2$ and $u : \mathcal{W} \to \mathbb{F}_2$ and define the Boolean function $h_2 : \mathbb{F}_{2^m} \to \mathbb{F}_2$ by $h_2(w) = u(w) + h_1(w)$ and $h_2(w+b) = u(w) + h_1(w+b) + \mu(w)$ for $w \in \mathcal{W}$. Then $D_b h_1(y) + D_b h_2(y) = \mu(y)$ for all $y \in \mathbb{F}_{2^m}$.*

Proof. We observe that $h_2(w + b) + h_2(w) = h_1(w + b) + h_1(w) + \mu(w)$ for all $w \in \mathcal{W}$ by definition. Using the fact that b is a 0-linear structure of μ we complete the proof.

Note that Lemma 1 gives a construction of a Boolean function $h_2 : \mathbb{F}_{2^m} \to \mathbb{F}_2$ with the property $D_b h_1(y) + D_b(h_2(y) = \mu(y)$ for all $y \in \mathbb{F}_{2^m}$ for given $b \in \mathbb{F}_{2^m}$, $h_1 : \mathbb{F}_{2^m} \to \mathbb{F}_2$ and μ having b with 0-linear structure. The construction uses $m - 1$ free variables in the form of the function $u : \mathcal{W} \to \mathbb{F}_2$.

Using Lemma 1, Theorem 1 and results from Sect. 5, we present below a new secondary construction of bent functions.

Theorem 2. *Let $1 \leq k < m$ be integers with $k \mid m$. Let f, g be functions from $\mathbb{F}_{2^m}$ to $\mathbb{F}_{2^k}$. Assume that $\gamma, \delta \in \mathbb{F}_{2^m}^{\star}$ are 0-linear translators of f and g,* *respectively. Further assume that $b \in \mathbb{F}_{2^m}$ is a 0-linear structure of f and g. Let $a \in \mathbb{F}_{2^m}$ be an arbitrary element. For arbitrary function $h_1 : \mathbb{F}_{2^m} \to \mathbb{F}_2$ construct $h_2 : \mathbb{F}_{2^m} \to \mathbb{F}_2$ satisfying $D_b h_1(y) = D_b h_2(y) + Tr_1^m(a(\gamma f(y+b) + \delta g(y+b)))$ for all $y \in \mathbb{F}_{2^m}$ using Lemma 1. Set $F(x,y) := Tr_1^m(xy + \gamma x f(y)) + h_1(y)$ and $G(x,y) := Tr_1^m(xy + \delta x g(y)) + h_2(y)$. The function defined by*

$$H(x,y) = F(x,y) + D_{a,b}F(x,y)\big(F(x,y) + G(x,y)\big)$$

is bent and its dual is

$$\begin{aligned}
\tilde{H}(x,y) = {} & Tr_1^m\big(yx + \gamma y f(x)\big) + h_1(x + \gamma f(x)) \\
& + Tr_1^m(ax + by)\big[Tr_1^m\big(y(\gamma f(x) + \delta g(x))\big) + h_1(x + \gamma f(x)) \\
& + h_2(x + \delta g(x))\big].
\end{aligned}$$

Proof. F and G are bent by Proposition 6. Using the fact that b is a 0-linear structure of f and g we get that $D_{a,b}F(x,y) = Tr_1^m\big(xb + a(y + b + \gamma f(y + b))\big) + D_b h_1(y)$ and $D_{a,b}G(x,y) = Tr_1^m\big(xb + a(y + b + \delta g(y + b))\big) + D_b h_2(y)$. Hence $D_{a,b}F(x,y) = D_{a,b}G(x,y)$ and the proof follows from Theorem 1 and Proposition 6.

Using [24, Theorem 16] instead of Theorem 1 we obtain the following secondary construction of semi-bent functions.

Theorem 3. *Under notation and assumptions of Theorem 2 we construct $h_2 : \mathbb{F}_{2^m} \to \mathbb{F}_2$ satisfying $D_b h_1(y) = D_b h_2(y) + Tr_1^m(a(\gamma f(y+b) + \delta g(y+b))) + 1$ (instead of $D_b h_1(y) = D_b h_2(y) + Tr_1^m(a(\gamma f(y+b) + \delta g(y+b)))$) for all $y \in \mathbb{F}_{2^m}$. Set F and G in the same way. Then the function defined by $H(x,y) = F(x,y) + G(x,y) + D_{a,b}F(x,y) + D_{a,b}FG(x,y)$ is semi-bent.*

Note that Theorem 3 gives a secondary construction of semi-bent functions of high degree by choosing the arbitrary function $h_1 : \mathbb{F}_{2^m} \to \mathbb{F}_2$ of large degree. Moreover it gives a different construction than the one given in [20, Sect. 4.2.5] and hence it is an answer to Problem 4 of [20].

8   A Secondary Construction of Bent Functions Using Certain Quadratic and Cubic Functions Together with Linear Structures

In this section we consider Boolean functions that are the sum of a bent function of Sects. 3 or 4 and a quadratic or cubic function. We show that one can choose appropriately the quadratic and cubic function so that those Boolean functions are bent again. Furthermore, the dual functions of those bent functions can be explicitly computed as in the preceding sections. The main results are Theorems 4, 5, 6 and 7.

Theorem 4 is based on [2, Lemma 1]. We note that the bent functions of Theorem 4 is different from the two classes of plateaued functions in Sect. 6 of

[2]. First of all we obtain bent functions while two classes of functions in Sect. 6 of [2] produce only plateaued functions.

Theorem 6 is a further generalization of Theorem 4 using cubic functions instead of quadratic functions.

Lemma 2. [2] *Let $w_1, w_2, u \in \mathbb{F}_{2^m}$ with $\{w_1, w_2\}$ linearly independent over $\mathbb{F}_{2^m}$. We have*

$$\sum_{x \in \mathbb{F}_{2^m}} (-1)^{Tr_1^m(w_1 x) Tr_1^m(w_2 x) + Tr_1^m(ux)}$$

$$= \begin{cases} 0 & \text{if } u \notin \langle w_1, w_2 \rangle = \{0, w_1, w_2, w_1 + w_2\}, \\ 2^{m-1} & \text{if } u \in \{0, w_1, w_2\}, \\ -2^{m-1} & \text{if } u = w_1 + w_2. \end{cases}$$

In Lemma 2, for any given $\mathbb{F}_2$-linearly independent set, the Boolean function on $\mathbb{F}_{2^m}$ given by $x \mapsto Tr_1^m(w_1 x) Tr_1^m(w_2 x)$ is a quadratic function.

Theorem 4. *Let $w_1, w_2, \gamma \in \mathbb{F}_{2^m}$ with $\{w_1, w_2\}$ linearly independent over $\mathbb{F}_2$. Assume that $f, h : \mathbb{F}_{2^m} \to \mathbb{F}_2$ are Boolean functions such that w_1 and w_2 are 0-linear structures of f and h. Moreover, we assume that γ is a 0-linear structure of f. Then the Boolean function F defined on $\mathbb{F}_{2^m} \times \mathbb{F}_{2^m}$ by*

$$F(x, y) = Tr_1^m(xw_1) Tr_1^m(xw_2) + Tr_1^m(xy + \gamma x f(y)) + h(y) \qquad (8.1)$$

is bent and its dual function is

$$\tilde{F}(x, y) = Tr_1^m(yw_1) Tr_1^m(yw_2) + Tr_1^m(yx + \gamma y f(x)) + h(x + \gamma f(x)).$$

Moreover, $F(x, y) = Tr_1^m(xw_1) Tr_1^m(xw_2) + Tr_1^m(xL(y) + L(\gamma) x f(y)) + h(y)$ where L is a linearized permutation polynomial of $\mathbb{F}_{2^m}$ is also bent under the same conditions and its dual function is

$$\tilde{F}(x, y) = Tr_1^m(yw_1) Tr_1^m(yw_2) + Tr_1^m(yL^{-1}(x) + \gamma y f(L^{-1}(x)))$$
$$+ h(L^{-1}(x) + \gamma f(L^{-1}(x))).$$

Proof. One has for every $(a, b) \in \mathbb{F}_{2^m} \times \mathbb{F}_{2^m}$,

$$\widehat{\chi_F}(a, b) = \sum_{y \in \mathbb{F}_{2^m}} (-1)^{h(y) + Tr_1^m(by)} \sum_{x \in \mathbb{F}_{2^m}} (-1)^{Tr_1^m(xw_1) Tr_1^m(xw_2) + Tr_1^m(xy + \gamma x f(y) + ax)}$$

Let $\phi(y) = y + \gamma f(y)$ and $\mathcal{S} = \sum_{x \in \mathbb{F}_{2^m}} (-1)^{Tr_1^m(xw_1) Tr_1^m(xw_2) + Tr_1^m(x(\phi(y) + a))}$. Then by Lemma 2, we have

$$\mathcal{S} = \begin{cases} 0 & \text{if } \phi(y) + a \notin \{0, w_1, w_2, w_1 + w_2\}, \\ 2^{m-1} & \text{if } \phi(y) + a \in \{0, w_1, w_2\}, \\ -2^{m-1} & \text{if } \phi(y) + a = w_1 + w_2. \end{cases}$$

Now, $f(a) = f(a + w_1) = f(a + w_2) = f(a + w_1 + w_2)$ since w_1 and w_2 are 0-linear structures of f. We have two cases, namely $f(a) = 0$ and $f(a) = 1$. Here, only the proof for the case $f(a) = 0$ is given since the proof for the other case is very similar.

Assume $f(a) = 0$. Then $\phi(y) + a \in \{0, w_1, w_2\}$ when $y \in \mathcal{A} = \{a, a + w_1, a + w_2\}$ and $\phi(y) + a = w_1 + w_2$ when $y = a + w_1 + w_2$. Hence,

$$\widehat{\chi_F}(a,b) = 2^{m-1}\left[\sum_{y \in \mathcal{A}}(-1)^{h(y)+Tr_1^m(by)} - (-1)^{h(a+w_1+w_2)+Tr_1^m(b(a+w_1+w_2))}\right].$$

Since w_1 and w_2 are 0-linear structures of h, we obtain

$$\widehat{\chi_F}(a,b) = 2^{m-1}\left[(-1)^{h(a)+Tr_1^m(ba)}\right]\mathcal{S}_1$$

where

$$\mathcal{S}_1 = \left[1 + (-1)^{Tr_1^m(bw_1)} + (-1)^{Tr_1^m(bw_2)} - (-1)^{Tr_1^m(b(w_1+w_2))}\right]. \qquad (8.2)$$

Note that

$$\mathcal{S}_1 = \begin{cases} 2 & \text{if } Tr_1^m(bw_1)Tr_1^m(bw_2) = 0, \\ -2 & \text{if } Tr_1^m(bw_1)Tr_1^m(bw_2) = 1. \end{cases}$$

Combining these we obtain that F is bent and its dual $\tilde{F}$ satisfies that

$$\tilde{F}(x,y) = Tr_1^m(yw_1)Tr_1^m(yw_2) + Tr_1^m\big(yx + y\gamma f(x)\big) + h(x + \gamma f(x)).$$

Remark 4. In Theorem 4, for given $\mathbb{F}_2$-linearly independent subset $\{w_1, w_2\}$, the Boolean function on $\mathbb{F}_{2^m} \times \mathbb{F}_{2^m}$ given by $(x, y) \mapsto Tr_1^m(xw_1)Tr_1^m(xw_2)$ is a quadratic function, which is used as the first summand in the definition of $F(x, y)$ in Eq. (8.1). In the proof of Theorem 4, we apply Lemma 2 for this quadratic function. Note that if $\gamma \neq 0$ and $1 + deg(f)$, $deg(h)$ and 2 are distinct, then the degree of $F(x, y)$ is $max\{1 + deg(f), deg(h), 2\}$, which may be much larger than 2.

In the following we present a straightforward generalization of Theorem 4.

Theorem 5. *Let $w_1, w_2, \gamma, \delta \in \mathbb{F}_{2^m}$ with $\{w_1, w_2\}$ linearly independent over $\mathbb{F}_2$. Assume that $f, g, h : \mathbb{F}_{2^m} \to \mathbb{F}_2$ are Boolean functions such that w_1 and w_2 are 0-linear structures of f, g and h. Moreover, we assume that γ is a 0-linear structure of f and δ is a 0-linear structure of f and g. Then the Boolean function F defined on $\mathbb{F}_{2^m} \times \mathbb{F}_{2^m}$ by*

$$F(x,y) = Tr_1^m(xw_1)Tr_1^m(xw_2) + Tr_1^m\big(x(L(y) + L(\gamma)f(y) + L(\delta)g(y))\big) + h(y)$$

is bent and its dual function is

$$\tilde{F}(x,y) = Tr_1^m(yw_1)Tr_1^m(yw_2) + Tr_1^m\big(y\rho^{-1}(x)\big) + h(\rho^{-1}(x)) \text{ where}$$
$$\rho^{-1}(x) = L^{-1}(x) + \gamma f(L^{-1}(x)) + \delta\left[g(L^{-1}(x))\big(1 + f(L^{-1}(x))\big) + g(L^{-1}(x)\right.$$
$$\left. + \gamma)f(L^{-1}(x))\right].$$

We now give the analogue of Lemma 2 which improves Lemma 1 of [2].

Lemma 3. *Let $w_1, w_2, w_3, u \in \mathbb{F}_{2^m}$ with $\{w_1, w_2, w_3\}$ linearly independent over $\mathbb{F}_{2^m}$. We have*

$$\sum_{x \in \mathbb{F}_{2^m}} (-1)^{Tr_1^m(w_1 x) Tr_1^m(w_2 x) Tr_1^m(w_3 x) + Tr_1^m(ux)}$$

$$= \begin{cases} 0 & if \ u \notin \langle w_1, w_2, w_3 \rangle, \\ 3.2^{m-2} & if \ u = 0, \\ 2^{m-2} & if \ u \in \{w_1, w_2, w_3, w_1 + w_2 + w_3\}, \\ -2^{m-2} & if \ u \in \{w_1 + w_2, w_1 + w_3, w_2 + w_3\}. \end{cases}$$

Proof. Let $\mathcal{T}$ denotes the sum in the statement of the lemma. Let $\mathcal{T}_1$ and $\mathcal{T}_2$ be the sums as

$$\mathcal{T}_1 = \sum_{x \in \mathbb{F}_{2^m} | Tr_1^m(w_1 x) = 0} (-1)^{Tr_1^m(ux)}$$

and

$$\mathcal{T}_2 = \sum_{x \in \mathbb{F}_{2^m} | Tr_1^m(w_1 x) = 1} (-1)^{Tr_1^m(w_2 x) Tr_1^m(w_3 x) + Tr_1^m(ux)}.$$

We have that $\mathcal{T} = \mathcal{T}_1 + \mathcal{T}_2$. It is clear that

$$\mathcal{T}_1 = \begin{cases} 0 & if \ u \notin \langle 0, w_1 \rangle = \{0, w_1\}, \\ 2^{m-1} & if \ u \in \{0, w_1\}. \end{cases}$$

Using Lemma 2 we obtain that

$$\mathcal{T}_2 = \begin{cases} 0 & if \ u \notin \langle w_1, w_2, w_3 \rangle, \\ 2^{m-2} & if \ u \in \{0, w_1, w_2, w_3, w_1 + w_2 + w_3\}, \\ -2^{m-2} & if \ u \in \{w_1 + w_2, w_1 + w_3, w_2 + w_3\}. \end{cases}$$

Combining $\mathcal{T}_1$ and $\mathcal{T}_2$ we complete the proof.

Remark 5. This remark is analogous to Remark 4. In Theorem 6, for given $\mathbb{F}_2$-linearly independent subset $\{w_1, w_2, w_3\}$, the Boolean function on $\mathbb{F}_{2^m} \times \mathbb{F}_{2^m}$ given by

$$(x, y) \mapsto Tr_1^m(xw_1) Tr_1^m(xw_2) Tr_1^m(xw_3)$$

is a cubic function, which is used as the first summand in the definition of $F(x, y)$ in Eq. (8.3). In the proof of Theorem 6, we apply Lemma 3 for this cubic function. As in Remark 4, the degree of $F(x, y)$ is $max\{1 + deg(f), deg(h), 3\}$ under suitable conditions, which may be much larger than 3.

Theorem 6. *Let f and h be two Boolean functions on $\mathbb{F}_{2^m}$. Let $w_1, w_2, w_3 \in \mathbb{F}_{2^m}$ be linearly independent and $\gamma \in \mathbb{F}_{2^m}$. Assume that γ is a 0-linear structure of f, and w_1, w_2, w_3 are 0-linear structures of f and h. Then, the function F defined on $\mathbb{F}_{2^m} \times \mathbb{F}_{2^m}$ by*

$$F(x, y) = Tr_1^m(xw_1) Tr_1^m(xw_2) Tr_1^m(xw_3) + Tr_1^m\big(x(L(y) + L(\gamma)f(y))\big) + h(y) \tag{8.3}$$

is bent and its dual is

$$\tilde{F}(x, y) = Tr_1^m(yw_1) Tr_1^m(yw_2) Tr_1^m(yw_3) + Tr_1^m\big(y(L^{-1}(x) + \gamma f(L^{-1}(x)))\big)$$
$$+ h(L^{-1}(x) + \gamma f(L^{-1}(x))).$$

Proof. Let $\phi(y) = y + \gamma f(y)$. For every $(a, b) \in \mathbb{F}_{2^m} \times \mathbb{F}_{2^m}$,

$$\widehat{\chi_F}(a, b) = \sum_{y \in \mathbb{F}_{2^m}} (-1)^{h(y) + Tr_1^m(by)}$$

$$\sum_{x \in \mathbb{F}_{2^m}} (-1)^{Tr_1^m(w_1 x) Tr_1^m(w_2 x) Tr_1^m(w_3 x) + Tr_1^m\left(x(\phi(y) + a)\right)}.$$

For the case $f(a) = 0$,

- $\phi(y) + a = 0$ when $y = a$,
- $\phi(y) + a \in \{w_1, w_2, w_3, w_1 + w_2 + w_3\}$ when
  $y \in \mathcal{A}_1 = \{a + w_1, a + w_2, a + w_3, a + w_1 + w_2 + w_3\}$
- $\phi(y) + a \in \{w_1 + w_2, w_1 + w_3, w_2 + w_3\}$
  when $y \in \mathcal{A}_2 = \{a + w_1 + w_2, a + w_1 + w_3, a + w_2 + w_3\}$.

Then, following the steps in proof of Theorem 4 and using Lemma 3, we get

$$\widehat{\chi_F}(a, b) = 3.2^{m-2}(-1)^{Tr_1^m(ba) + h(a)} + 2^{m-2} \sum_{y \in \mathcal{A}_1} (-1)^{Tr_1^m(by) + h(y)}$$

$$-2^{m-2} \sum_{y \in \mathcal{A}_2} (-1)^{Tr_1^m(by) + h(y)}$$

$$= 2^{m-2} \left[(-1)^{Tr_1^m(ba) + h(a)} \right] \mathcal{S}$$

where

$$\mathcal{S} = [3 + \mathcal{S}_1 + \mathcal{S}_2], \tag{8.4}$$

$\mathcal{S}_1 = (-1)^{Tr_1^m(bw_1)} + (-1)^{Tr_1^m(bw_2)} + (-1)^{Tr_1^m(bw_3)} + (-1)^{Tr_1^m(b(w_1 + w_2 + w_3))}$ and
$\mathcal{S}_2 = (-1)^{Tr_1^m(b(w_1 + w_2))} + (-1)^{Tr_1^m(b(w_1 + w_3))} + (-1)^{Tr_1^m(b(w_2 + w_3))}$.
Let $(-1)^{Tr_1^m(bw_i)} = c_i$ where $c_i \in \mathbb{F}_2$, for $i = 1, 2, 3$. Then, $3 + \mathcal{S}_1 + \mathcal{S}_2 = \pm 4$ and
hence $\widehat{\chi_F}(a, b) = \pm 2^m$.

The proof for the case $f(a) = 1$ is very similar.

As in Theorem 5, in the following we get a modification of Theorem 6 using two linear structures instead of one linear structure.

Theorem 7. *Let f, g and h be Boolean functions on $\mathbb{F}_{2^m}$. Let $w_1, w_2, w_3 \in \mathbb{F}_{2^m}$ be linearly independent and $\gamma, \delta \in \mathbb{F}_{2^m}$, $\gamma \neq \delta$. Assume that γ is a 0-linear structure of f, δ is a 0-linear structure of f and g. Moreover, assume that w_1, w_2, w_3 are 0-linear structures of f, g and h. Then, the function F defined on $\mathbb{F}_{2^m} \times \mathbb{F}_{2^m}$ by*

$$F(x, y) = Tr_1^m(xw_1) Tr_1^m(xw_2) Tr_1^m(xw_3) + Tr_1^m\left(x(L(y) + L(\gamma)f(y)\right.$$
$$\left. + L(\delta)g(y))\right) + h(y)$$

is bent and its dual is

$$\tilde{F}(x, y) = Tr_1^m(yw_1) Tr_1^m(yw_2) Tr_1^m(yw_3) + Tr_1^m\left(y\rho^{-1}(x)\right) + h(\rho^{-1}(x)) \text{ where}$$
$$\rho^{-1}(x) = L^{-1}(x) + \gamma f(L^{-1}(x)) + \delta \left[g(L^{-1}(x))\left(1 + f(L^{-1}(x))\right) + g(L^{-1}(x)) \right.$$
$$\left. + \gamma)f(L^{-1}(x)) \right].$$

Acknowledgment. The authors would like to thank the anonymous reviewers and the program committee for the detailed and constructive comments which improved the paper a lot.

A   Appendix

The following proposition is related to Proposition 4 in Sect. 4.

Proposition 8. *Let H be defined by* Eq. (4.1), *γ and δ be defined as in Proposition 4. Then the dual of H is $\tilde{H}(x,y) = Tr_1^m \left(y\phi^{-1}(x) \right) + h(\phi^{-1}(x))$ where $\phi^{-1} = \pi_2^{-1} \circ \rho^{-1} \circ \pi_1^{-1}$ and $\rho^{-1}(x)$ is given as follows.*

(i) If γ is a 0-linear structure of f, δ is a 0-linear structure of f and g, then

$$\rho^{-1}(x) = x + \gamma f(x) + \delta \left[g(x)(1 + f(x)) + g(x + \gamma)f(x) \right].$$

(ii) If γ is a 0-linear structure of f, δ is a 1-linear structure of f and $\delta + \gamma$ is a 0-linear structure of g, then

$$\rho^{-1}(x) = x + \gamma \left[g(x) + f(x)\left(1 + g(x) + g(x + \gamma)\right) \right] + \delta \left[g(x)(1 + f(x)) \right.$$
$$\left. + g(x + \gamma)f(x) \right].$$

(iii) If δ is a 0-linear structure of g, γ is a 0-linear structure of f and g, then

$$\rho^{-1}(x) = x + \gamma \left[f(x)(1 + g(x)) + f(x + \delta)g(x) \right] + \delta g(x).$$

(vi) If δ is a 0-linear structure of g, γ is a 1-linear structure of g and $\delta + \gamma$ is a 0-linear structure of f, then

$$\rho^{-1}(x) = x + \gamma \left[f(x)(1 + g(x)) + f(x + \delta)g(x) \right] + \delta \left[f(x)(1 + g(x)) \right.$$
$$\left. + \left(1 + f(x + \delta)\right)g(x) \right].$$

(v) If δ is a 1-linear structure of f or δ is a 0-linear structure of g, then

$$\rho^{-1}(x) = x + \gamma \left[f(x)(1 + g(x + \delta)) + \left(1 + f(x)\right)g(x) \right] + \delta f(x).$$

(vi) If γ is a 1-linear structure of g, δ is a 1-linear structure of f and g, then

$$\rho^{-1}(x) = x + \gamma g(x) + \delta \left[f(x)(1 + g(x)) + f(x + \gamma)g(x) \right].$$

Proof. We give only the proof for the case (i). Assume that γ is a 0-linear structure of f, δ is a 0-linear structure of f and g, then we claim that

$$\rho^{-1}(x) = \begin{cases} x & \text{if } f(x) = 0 \text{ and } g(x) = 0 \\ x + \delta & \text{if } f(x) = 0 \text{ and } g(x) = 1 \\ x + \gamma & \text{if } f(x) = 1 \text{ and } g(x + \gamma) = 0 \\ x + \gamma + \delta & \text{if } f(x) = 1 \text{ and } g(x + \gamma) = 1 \end{cases} \tag{A.1}$$

Let $\rho(y) = a$. Then,

$$y + \gamma f(y) + \delta g(y) = a \tag{A.2}$$

Taking f of both sides gives $f(y + \gamma f(y) + \delta g(y)) = f(a)$. Since γ and δ are 0-linear structures of f, we get

$$f(y) = f(a). \tag{A.3}$$

Note that, $(f(a), g(a)) \in \{(0,0), (0,1), (1,0), (1,1)\}$. These four cases correspond to the cases in Eq. (A.1). We prove only the first case in Eq. (A.1) and the proofs of other cases are similar. Hence, we assume that $(f(a), g(a)) = (0,0)$. Then, by Eq. (A.3), $f(y) = 0$ and by Eq. (A.2), $y + \delta g(y) = a$. Taking g of both sides and using that δ is a 0-linear structure of g, we obtain that $g(y + \delta g(y)) = g(y) = g(a)$. As $g(a) = 0$ by our assumption, we get $g(y) = 0$ and putting $f(y) = g(y) = 0$ in Eq. (A.2) we conclude that $y = a$.

Finally, the Eq. (A.1) can be written in the form

$$\rho^{-1}(x) = x + \gamma f(x) + \delta \left[g(x)(1 + f(x)) + g(x + \gamma)f(x) \right].$$

The following five propositions are related to Proposition 7 in Sect. 6.

Proposition 9. *Let f and g be functions from $\mathbb{F}_{2^m}$ to $\mathbb{F}_2$. For $i \in \{1, 2, 3\}$ set $\phi_i(y) := y + \gamma_i f(y) + \delta_i g(y)$ where*

(i) γ_1, γ_2, γ_3 are elements of $\mathbb{F}_{2^m}^\star$ which are 0-linear structures of f;
(ii) δ_1, δ_2 and δ_3 are elements of $\mathbb{F}_{2^m}^\star$ which are 1-linear structures of f;
(iii) $\gamma_1 + \delta_1$, $\gamma_2 + \delta_2$, $\gamma_3 + \delta_3$ are 0-linear structures of g;
(vi) $\gamma_1 + \gamma_2$ and $\gamma_1 + \gamma_3$ are 0-linear structures of g.

Then the function h defined on $\mathbb{F}_{2^m} \times \mathbb{F}_{2^m}$ by

$$h(x, y) = Tr_1^m \left(x\phi_1(y) \right) Tr_1^m \left(x\phi_2(y) \right) + Tr_1^m \left(x\phi_1(y) \right) Tr_1^m \left(x\phi_3(y) \right)$$
$$+ Tr_1^m \left(x\phi_2(y) \right) Tr_1^m \left(x\phi_3(y) \right)$$

is bent and the dual of h is given by

$$\tilde{h}(x, y) = Tr_1^m \left(y\phi_1^{-1}(x) \right) Tr_1^m \left(y\phi_2^{-1}(x) \right) + Tr_1^m \left(y\phi_1^{-1}(x) \right) Tr_1^m \left(y\phi_3^{-1}(x) \right)$$
$$+ Tr_1^m \left(y\phi_2^{-1}(x) \right) Tr_1^m \left(y\phi_3^{-1}(x) \right)$$

where

$$\phi_i^{-1}(x) = x + \gamma \left[g(x) + f(x)(1 + g(x) + g(x + \gamma)) \right]$$
$$+ \delta \left[g(x)(1 + f(x)) + g(x + \gamma)f(x) \right].$$

Proposition 10. *Let f and g be functions from $\mathbb{F}_{2^m}$ to $\mathbb{F}_2$. For $i \in \{1, 2, 3\}$ set $\phi_i(y) := y + \gamma_i f(y) + \delta_i g(y)$ where*

(i) γ_1, γ_2, γ_3 are elements of $\mathbb{F}_{2^m}^\star$ which are 0-linear structures of f and g;
(ii) δ_1, δ_2 and δ_3 are elements of $\mathbb{F}_{2^m}^\star$ which are 0-linear structures of g;
(iii) $\delta_1 + \delta_2$ and $\delta_1 + \delta_3$ are 0-linear structures of f.

Then the function h defined on $\mathbb{F}_{2^m} \times \mathbb{F}_{2^m}$ by

$$h(x,y) = Tr_1^m\Big(x\phi_1(y)\Big)Tr_1^m\Big(x\phi_2(y)\Big) + Tr_1^m\Big(x\phi_1(y)\Big)Tr_1^m\Big(x\phi_3(y)\Big)$$
$$+ Tr_1^m\Big(x\phi_2(y)\Big)Tr_1^m\Big(x\phi_3(y)\Big)$$

is bent and the dual of h is given by

$$\tilde{h}(x,y) = Tr_1^m\Big(y\phi_1^{-1}(x)\Big)Tr_1^m\Big(y\phi_2^{-1}(x)\Big) + Tr_1^m\Big(y\phi_1^{-1}(x)\Big)Tr_1^m\Big(y\phi_3^{-1}(x)\Big)$$
$$+ Tr_1^m\Big(y\phi_2^{-1}(x)\Big)Tr_1^m\Big(y\phi_3^{-1}(x)\Big)$$

where $\phi_i^{-1}(x) = x + \gamma\left[f(x)(1 + g(x)) + f(x + \delta)g(x)\right] + \delta g(x)$.

Proposition 11. Let f and g be functions from $\mathbb{F}_{2^m}$ to $\mathbb{F}_2$. For $i \in \{1,2,3\}$ set $\phi_i(y) := y + \gamma_i f(y) + \delta_i g(y)$ where

(i) γ_1, γ_2, γ_3 are elements of $\mathbb{F}_{2^m}^\star$ which are 1-linear structures of g;
(ii) δ_1, δ_2 and δ_3 are elements of $\mathbb{F}_{2^m}^\star$ which are 0-linear structures of g;
(iii) $\gamma_1 + \delta_1$, $\gamma_2 + \delta_2$, $\gamma_3 + \delta_3$ are 0-linear structures of f;
(vi) $\delta_1 + \delta_2$ and $\delta_1 + \delta_3$ are 0-linear structures of f.

Then the function h defined on $\mathbb{F}_{2^m} \times \mathbb{F}_{2^m}$ by

$$h(x,y) = Tr_1^m\Big(x\phi_1(y)\Big)Tr_1^m\Big(x\phi_2(y)\Big) + Tr_1^m\Big(x\phi_1(y)\Big)Tr_1^m\Big(x\phi_3(y)\Big)$$
$$+ Tr_1^m\Big(x\phi_2(y)\Big)Tr_1^m\Big(x\phi_3(y)\Big)$$

is bent and the dual of h is given by

$$\tilde{h}(x,y) = Tr_1^m\Big(y\phi_1^{-1}(x)\Big)Tr_1^m\Big(y\phi_2^{-1}(x)\Big) + Tr_1^m\Big(y\phi_1^{-1}(x)\Big)Tr_1^m\Big(y\phi_3^{-1}(x)\Big)$$
$$+ Tr_1^m\Big(y\phi_2^{-1}(x)\Big)Tr_1^m\Big(y\phi_3^{-1}(x)\Big)$$

where

$$\phi_i^{-1}(x) = x + \gamma\left[f(x)(1 + g(x)) + f(x + \delta)g(x)\right]$$
$$+ \delta\left[f(x)(1 + g(x)) + (1 + f(x + \delta))g(x)\right].$$

Proposition 12. Let f and g be functions from $\mathbb{F}_{2^m}$ to $\mathbb{F}_2$. For $i \in \{1,2,3\}$ set $\phi_i(y) := y + \gamma_i f(y) + \delta_i g(y)$ where

(i) γ_1, γ_2, γ_3 are elements of $\mathbb{F}_{2^m}^\star$ which are 1-linear structures of f and g;
(ii) δ_1, δ_2 and δ_3 are elements of $\mathbb{F}_{2^m}^\star$ which are 1-linear structures of f;
(iii) $\delta_1 + \delta_2$ and $\delta_1 + \delta_3$ are 0-linear structures of g.

Then the function h defined on $\mathbb{F}_{2^m} \times \mathbb{F}_{2^m}$ by

$$h(x,y) = Tr_1^m\Big(x\phi_1(y)\Big)Tr_1^m\Big(x\phi_2(y)\Big) + Tr_1^m\Big(x\phi_1(y)\Big)Tr_1^m\Big(x\phi_3(y)\Big)$$
$$+ Tr_1^m\Big(x\phi_2(y)\Big)Tr_1^m\Big(x\phi_3(y)\Big)$$

is bent and the dual of h is given by

$$\tilde{h}(x,y) = Tr_1^m\Big(y\phi_1^{-1}(x)\Big)Tr_1^m\Big(y\phi_2^{-1}(x)\Big) + Tr_1^m\Big(y\phi_1^{-1}(x)\Big)Tr_1^m\Big(y\phi_3^{-1}(x)\Big)$$
$$+ Tr_1^m\Big(y\phi_2^{-1}(x)\Big)Tr_1^m\Big(y\phi_3^{-1}(x)\Big)$$

where $\phi_i^{-1}(x) = x + \gamma\left[f(x)(1 + g(x + \delta)) + \big(1 + f(x)\big)g(x)\right] + \delta f(x)$.

Proposition 13. *Let f and g be functions from $\mathbb{F}_{2^m}$ to $\mathbb{F}_2$. For $i \in \{1, 2, 3\}$ set $\phi_i(y) := y + \gamma_i f(y) + \delta_i g(y)$ where*

(i) γ_1, γ_2, γ_3 are elements of $\mathbb{F}_{2^m}^{\star}$ which are 1-linear structures of g;
(ii) δ_1, δ_2 and δ_3 are elements of $\mathbb{F}_{2^m}^{\star}$ which are 1-linear structures of f and g;
(iii) $\gamma_1 + \gamma_2$ and $\gamma_1 + \gamma_3$ are 0-linear structures of f.

Then the function h defined on $\mathbb{F}_{2^m} \times \mathbb{F}_{2^m}$ by

$$h(x,y) = Tr_1^m\Big(x\phi_1(y)\Big)Tr_1^m\Big(x\phi_2(y)\Big) + Tr_1^m\Big(x\phi_1(y)\Big)Tr_1^m\Big(x\phi_3(y)\Big)$$
$$+ Tr_1^m\Big(x\phi_2(y)\Big)Tr_1^m\Big(x\phi_3(y)\Big)$$

is bent and the dual of h is given by

$$\tilde{h}(x,y) = Tr_1^m\Big(y\phi_1^{-1}(x)\Big)Tr_1^m\Big(y\phi_2^{-1}(x)\Big) + Tr_1^m\Big(y\phi_1^{-1}(x)\Big)Tr_1^m\Big(y\phi_3^{-1}(x)\Big)$$
$$+ Tr_1^m\Big(y\phi_2^{-1}(x)\Big)Tr_1^m\Big(y\phi_3^{-1}(x)\Big)$$

where $\phi_i^{-1}(x) = x + \gamma g(x) + \delta\left[f(x)(1 + g(x)) + f(x + \gamma)g(x)\right]$.

References

1. Canteaut, A., Naya-Plasencia, M.: Structural weakness of mappings with a low differential uniformity. In: Conference on Finite Fields and Applications (2009)
2. Carlet, C., Prouff, E.: On plateaued functions and their constructions. In: Johansson, T. (ed.) FSE 2003. LNCS, vol. 2887, pp. 54–73. Springer, Heidelberg (2003)
3. Carlet, C., Yucas, J.L.: Piecewise constructions of bent and almost optimal boolean functions. Des. Codes Crypt. **37**, 449–464 (2005)
4. Carlet, C.: On bent and highly nonlinear balanced/resilient functions and their algebraic immunities. In: Fossorier, M.P.C., Imai, H., Lin, S., Poli, A. (eds.) AAECC 2006. LNCS, vol. 3857, pp. 1–28. Springer, Heidelberg (2006)

5. Carlet, C.: Boolean functions for cryptography and error correcting codes. In: Crama, Y., Hammer, P.L. (eds.) Boolean Models and Methods in Mathematics, Computer Science, and Engineering, pp. 257–397. Cambridge University Press, Cambridge (2010)

6. Carlet, C.: Open problems in mathematics and computational science. In: Kaya Koç, Ç. (ed.) Open Problems on Binary Bent Functions. Springer, Switzerland (2014)

7. Carlet, C., Mesnager, S.: Four decades of research on bent functions. J. Des. Codes Crypt. (to appear)

8. Charpin, P., Kyureghyan, G.M.: On a class of permutation polynomials over $\mathbb{F}_{2^n}$. In: Golomb, S.W., Parker, M.G., Pott, A., Winterhof, A. (eds.) SETA 2008. LNCS, vol. 5203, pp. 368–376. Springer, Heidelberg (2008)

9. Charpin, P., Kyureghyan, G.M.: When does $G(x) + \gamma\,Tr(H(x))$ permute $\mathbb{F}_2$? Finite Fields Appl. **15**(5), 615–632 (2009)

10. Charpin, P., Kyureghyan, G.M.: Monomial functions with linear structure and permutation polynomials. In: Finite Fields: Theory and Applications - Fq9 - Contemporary Mathematics, vol. 518, pp. 99–111. AMS (2010)

11. Charpin, P., Kyureghyan, G.M., Suder, V.: Sparse permutations with low differential uniformity. Finite Fields Appl. **28**, 214–243 (2014)

12. Charpin, P., Mesnager, S., Sarkar, S.: On involutions of finite fields. In: Proceedings of 2015 IEEE International Symposium on Information Theory, ISIT (2015)

13. Charpin, P., Sarkar, S.: Polynomials with linear structure and Maiorana-McFarland construction. IEEE Trans. Inf. Theory **57**(6), 3796–3804 (2011)

14. Chee, S., Lee, S., Kim, K.: Semi-bent functions. In: Safavi-Naini, R., Pieprzyk, J.P. (eds.) ASIACRYPT 1994. LNCS, vol. 917, pp. 107–118. Springer, Heidelberg (1995)

15. Dillon, J.: Elementary hadamard difference sets. Ph.D. Dissertation, University of Maryland, College Park (1974)

16. Kyureghyan, G.: Constructing permutations of finite fields via linear translators. J. Comb. Theory Ser. A **118**(3), 1052–1061 (2011)

17. Lai, X.: Additive and linear structures of cryptographic functions. In: Preneel, B. (ed.) FSE 1994. LNCS, vol. 1008, pp. 75–85. Springer, Heidelberg (1995)

18. Mesnager, S.: Semi-bent functions from oval polynomials. In: Stam, M. (ed.) IMACC 2013. LNCS, vol. 8308, pp. 1–15. Springer, Heidelberg (2013)

19. Mesnager, S.: Several new infinite families of bent functions and their duals. IEEE Trans. Inf. Theory. **60**(7), 4397–4407 (2014)

20. Mesnager, S.: Open problems in mathematics and computational science. In: Kaya Koç, Ç. (ed.) On Semi-bent Functions and Related Plateaued Functions Over the Galois Field F_{2^n}. Springer, Switzerland (2014)

21. Mesnager, S.: Further constructions of infinite families of bent functions from new permutations and their duals. J. Crypt. Commun. (CCDS). Springer (to appear)

22. Mesnager, S.: Bent functions: fundamentals and results. Springer (2015, to appear)

23. Rothaus, O.-S.: On bent functions. J. Combin. Theory Ser. A. **20**, 300–305 (1976)

24. Sun, G., Wu, C.: Construction of semi-bent boolean functions in even number of variables. Chin. J. Electron. **18**(2), 231–237 (2009)

25. Zheng, Y., Zhang, X.-M.: Plateaued functions. In: Varadharajan, V., Mu, Y. (eds.) ICICS 1999. LNCS, vol. 1726, pp. 284–300. Springer, Heidelberg (1999)

Comparison of Cube Attacks Over Different Vector Spaces

Richard Winter[1]($\boxtimes$), Ana Salagean[1], and Raphael C.-W. Phan[2]

[1] Department of Computer Science, Loughborough University, Loughborough, UK
{R.Winter,A.M.Salagean}@lboro.ac.uk
[2] Faculty of Engineering, Multimedia University, Cyberjaya, Malaysia
raphael@mmu.edu.my

Abstract. We generalise the cube attack of Dinur and Shamir (and the similar AIDA attack of Vielhaber) to a more general higher order differentiation attack, by summing over an arbitrary subspace of the space of initialisation vectors. The Moebius transform can be used for efficiently examining all the subspaces of a big space, similar to the method used by Fouque and Vannet for the usual cube attack.

Secondly we propose replacing the Generalised Linearity Test proposed by Dinur and Shamir with a test based on higher order differentiation/Moebius transform. We show that the proposed test provides all the information provided by the Generalised Linearity Test, at the same computational cost. In addition, for functions that do not pass the linearity test it also provides, at no extra cost, an estimate of the degree of the function. This is useful for guiding the heuristics for the cube/AIDA attacks.

Finally we implement our ideas and test them on the stream cipher Trivium.

Keywords: Cube/AIDA attack · Trivium · Linearity testing · Moebius transform · Higher order differentiation

1 Introduction

The cube attack introduced by Dinur and Shamir [3] and the similar AIDA attack introduced by Vielhaber [11] have received much attention over the last few years. They can be viewed as higher order differential attacks (see [5,9]). The idea of higher order differentials was introduced in cryptography by Lai [10] and was used in many different attacks, most of them being statistical attacks, whereas the cube/AIDA attacks are primarily algebraic.

Several techniques were proposed in order to make the cube attack more efficient. Of particular interest to the present work are the Moebius transform used by Fouque and Vannet [7] and the Generalised Linearity Test introduced by Dinur and Shamir [4].

We propose generalising the cube attack by using higher order differentiation in its general form. In other words, rather than summing over a "cube", we sum

© Springer International Publishing Switzerland 2015
J. Groth (Ed.): IMACC 2015, LNCS 9496, pp. 225–238, 2015.
DOI: 10.1007/978-3-319-27239-9_14

over an arbitrary subspace of the space of public (tweakable) variables. The usual cube attack becomes then the particular case where the subspace is generated by vectors from the canonical basis. The Moebius transform can again be used to make computations more efficient, by reusing values to compute the summations over many subspaces at once.

Secondly we propose an alternative to the Generalised Linearity Test proposed by Dinur and Shamir [4]. Given a set of t linearly independent keys, our linearity test computes the higher order derivatives of order $2, 3, \ldots, t$ with respect to any subset of keys and then checks whether all the results are zero. If a function fails this linearity test, the lowest order of a non-zero derivative gives a lower bound for the degree of the function, so we obtain extra information at no extra cost. We show that the set of functions that pass the General Linearity Test in [4] is exactly the same as the set of functions that pass our proposed test. The extra information about the degree is useful for guiding the heuristics in the cube attack, as it gives us information as to whether we are close or not to obtaining a linear function. Also, in some implementations of the cube attack, quadratic equations are used if insufficient linear equations are found.

We implemented our ideas and tested the implementation on the stream cipher Trivium [1], which is a popular candidate for testing cube attacks. We looked at between 640 and 703 initialisation rounds. We tested several spaces of initialisation vectors of dimension 28 (including the space corresponding to a usual cube attack), and all their subspaces, using the Moebius transform. We estimated the degrees of the results using our proposed linearity test.

We found one particular vector space which, compared to the usual cube attack, produces significantly more linear equations, but at a slightly higher dimension of subspaces. However for most vector spaces the results are significantly worse than the usual cube attack, which leads us to believe that the success of the usual cube attack on Trivium is not only due to the relatively low degree of the polynomial, but also to the fact that the monomials of that polynomial are not uniformly distributed, as would be expected if it was a random polynomial. In other words, the polynomials in Trivium are "aligned" with the canonical basis rather than being in a generic position. We suggest therefore that preceding Trivium by a (secret) linear change of coordinates on the initialisation vectors would improve its resistance to cube attacks. We did some preliminary experimental testing of this idea, but a full exploration would be a topic of future work.

2   Preliminaries

2.1   Cube Attack

The Cube attack was originally proposed by Dinur and Shamir in [3] and is closely related to the AIDA attack introduced by Vielhaber in [11].

Let $f : \mathbb{F}_2^n \to \mathbb{F}_2$ be a Boolean function in variables $x_1 \ldots x_n$. Any Boolean function can be written in Algebraic Normal Form, i.e. as a polynomial function of degree at most one in each variable. Choosing a subset of indices $I = \{i_1 \ldots i_k\} \subseteq \{1, 2, \ldots, n\}$, the "cube" C_I is defined by choosing the 2^k possible

0/1 combinations for the variables with indices in I, with the other variables left undetermined. Summing over all vectors in C_I we obtain a function

$$f_I = \sum_{v \in C_I} f(\mathbf{v}).$$

which depends only on the variables which are not in I.

Factoring out the term $t_I = x_{i_1} \cdots x_{i_k}$, we can write f as

$$f(x_1, \ldots, x_n) = t_I f_{S(I)} + r(x_1, \ldots, x_n).$$

where $f_{S(I)}$ is a polynomial that shares no common variables with t_I, whereas r is a polynomial in which each term misses at least one variable in t_I. The main results on which the cube attack is based are:

Theorem 1 *([3, Theorem 1]). For any polynomial f and subset of variables I, $f_I \equiv f_{S(I)} \pmod 2$.*

Corollary 1. *If $\deg(f) = d$ and I contains $d - 1$ elements, then f_I has degree at most one.*

When mounting an actual attack, we have two types of variables, the secret variables $x_1, \ldots, x_n$ and the public, or "tweakable" variables $v_1, \ldots, v_m$, which the attacker can control. The cipher consists of a "black box" function $g : \mathbb{F}_2^n \times \mathbb{F}_2^m \to \mathbb{F}_2$. The attacker chooses a set I of indices of the public variables, sets the other public variables to constant values (usually zero) and computes g_I, which will now only depend on the secret variables. In the preprocessing phase, the attacker studies the cipher, so they can evaluate g_I for any chosen values of the secret variables. It is hoped that, for suitable choices of I (particularly the ones of cardinality approaching $\deg(g) - 1$, assuming $\deg(g) \le m$), g_I is linear (but not constant) in the secret variables. Linearity tests are discussed in the next section.

If the preprocessing phase found a large number of sets I for which g_I is linear and non-constant (ideally n linearly independent g_I), one can then use this information in the online phase. Now the secret variables are unknown, but the attacker can still control the public variables. The attacker computes g_I for the values of I identified in the preprocessing phase, and then they can determine the secret variables by solving a system of linear equations.

2.2   Generalised Linearity Test

Consider a function $f : \mathbb{F}_2^n \to \mathbb{F}_2$. We want to decide whether f is an affine function, i.e. it is a polynomial of degree one or less. We assume n is large and evaluations of f are costly, so we cannot evaluate f for all its inputs. We are looking therefore for a probabilistic test.

In the original cube attack paper [3], Dinur and Shamir used the BLR test, i.e. the textbook definition of linearity: test whether $f(\mathbf{a}) + f(\mathbf{b}) = f(\mathbf{a} + \mathbf{b}) + f(\mathbf{0})$. If f fails this test, then it is not an affine function. If it passes the test for "sufficiently many" pairs $\mathbf{a}, \mathbf{b}$, then we conclude that f is probably affine.

Since the test above needs 3 evaluations of f for each test (assuming we store and reuse $f(\mathbf{0})$), in [4, Sect. 4], Dinur and Shamir proposed the following Generalised Linearity Test, which has the advantage that it reuses many evaluations of f so it is overall much more computationally efficient.

Consider a set $\{\mathbf{b}_1, \ldots, \mathbf{b}_t\} \subseteq \mathbb{F}_2^n$ of linearly independent elements. The Generalised Linearity Test consists of the following set of $2^t - t - 1$ equations:

$$\left\{ f\left(\sum_{i=0}^{t} c_i \mathbf{b}_i\right) + \sum_{i=0}^{t} c_i f(\mathbf{b}_i) + ((\mathrm{w}(\mathbf{c}) - 1)) \bmod 2) f(\mathbf{0}) = 0 \,\middle|\, \mathbf{c} \in \mathbb{F}_2^n, \mathrm{w}(c) \geq 2 \right\}$$
(1)

where w( ) denotes the Hamming weight. Again, if there are equations which are not satisfied by f, then f is not affine, otherwise we conclude that f is probably affine (assuming t is "large enough"). Note that here we need 2^t evaluations of f for $2^t - t - 1$ tests, so an amortised cost of just over one evaluation per test, compared to 3 evaluations for the previous test.

Remark 1. In the original description of this test in [4] there is a mistake, in that the term $(\mathrm{w}(\mathbf{c}) - 1) \bmod 2$ is missing. This would make the test incorrect whenever the weight is odd, so affine functions with non-zero constant term would wrongly fail the test.

2.3   Moebius Transform

Let $f : \mathbb{F}_2^n \to \mathbb{F}_2$. The Moebius transform of f is a function $f^M : \mathbb{F}_2^n \to \mathbb{F}_2$ defined as $f^M(\mathbf{y}) = \sum_{\mathbf{x} \preceq \mathbf{y}} f(\mathbf{x})$ where $\mathbf{x} = (x_1, \ldots, x_n)$ and the partial order relation $\preceq$ is defined as $(x_1, \ldots, x_n) \preceq (y_1, \ldots, y_n)$ iff $x_i \leq y_i$ for all $i = 1, \ldots, n$. It is well known that the Moebius transform has the property that for any $\mathbf{a} \in \mathbb{F}_2^n$, $f^M(\mathbf{a})$ equals the coefficient of the term $x_1^{a_1} \ldots x_n^{a_n}$ in f. Further details about the Moebius transform can be found, for example, in [8]. An efficient algorithm which, given the truth table of f computes the truth table of f^M in-place in $n2^{n-1}$ operations is also given in [8].

In connection with the cube attack, note that when choosing a set of variable indices $I = \{i_1, \ldots, i_k\}$, if we define $\mathbf{a}$ as having ones in the positions in I and zeroes elsewhere, we have $f_I(\mathbf{0}) = f^M(\mathbf{a})$. Hence the algorithm for computing the Moebius transform can also be used for efficiently computing f_J for all the subsets J of a large set I.

The idea of making the cube attack more efficient by reusing computations for cubes which are all subcubes of a very large cube was sketched by Dinur and Shamir [4]. Fouque and Vannet [6] fully developed this powerful technique via Moebius transforms, thus obtaining results for Trivium for a larger number of initialisation rounds than previous cube attacks.

2.4   Higher Order Differentiation

The notion of higher order derivative (or higher order differentiation) was introduced in the cryptographic context by Lai [10].

Definition 1. *Let $f : \mathbb{F}_2^n \to \mathbb{F}_2$ be a function in n variables $x_1, \ldots, x_n$. Let $\mathbf{a} = (a_1, \ldots, a_n) \in \mathbb{F}_2^n \setminus \{\mathbf{0}\}$. The differentiation operator (or finite difference operator) along a vector $\mathbf{a}$ associates to each function f the function $\Delta_\mathbf{a} f$ (the derivative of f) defined as*

$$\Delta_\mathbf{a} f(x_1, \ldots, x_n) = f(x_1 + a_1, \ldots, x_n + a_n) + f(x_1, \ldots, x_n).$$

Denoting $\mathbf{x} = (x_1, \ldots, x_n)$ we can also write $\Delta_\mathbf{a} f(\mathbf{x}) = f(\mathbf{x} + \mathbf{a}) + f(\mathbf{x})$.

Higher order differentiation (higher order derivative) refers to repeated application of this operator and will be denoted as:

$$\Delta_{\mathbf{a_1}, \ldots, \mathbf{a_k}}^{(k)} f = \Delta_{\mathbf{a_1}} \Delta_{\mathbf{a_2}} \cdots \Delta_{\mathbf{a_k}} f$$

where $\mathbf{a_1}, \ldots, \mathbf{a_k} \in \mathbb{F}_2^n \setminus \{\mathbf{0}\}$ are linearly independent. An explicit expression for computing higher order derivatives follows directly from the definition:

$$\Delta_{\mathbf{a_1}, \ldots, \mathbf{a_k}}^{(k)} f = \sum_{(c_1, \ldots, c_k) \in \{0,1\}^k} f(\mathbf{x} + c_1 \mathbf{a_1} + \ldots + c_k \mathbf{a_k}) \qquad (2)$$

Differentiation decreases the degree of polynomials:

Theorem 2 *[10]. Let $f : \mathbb{F}_2^n \to \mathbb{F}_2$ and $\mathbf{a} \in \mathbb{F}_2^n \setminus \{\mathbf{0}\}$. Then $\deg(\Delta_\mathbf{a} f) \le \deg(f) - 1$.*

The main construction of the cube attack can be reformulated in terms of higher order differentiation, see [5,9]. Namely for a set of indices $I = \{i_1, \ldots, i_k\}$

$$f_I = \Delta_{\mathbf{e}_{i_1}, \ldots, \mathbf{e}_{i_k}}^{(k)} f$$

where $\mathbf{e}_i$ are the vectors of the canonical basis, i.e. they have a one in position i and zeroes elsewhere.

3   General Differentiation Attack

As in Subsect. 2.1 we assume the cipher consists of a "black box" function $g(\mathbf{x}, \mathbf{v})$ with $g : \mathbb{F}_2^n \times \mathbb{F}_2^m \to \mathbb{F}_2$ and $\mathbf{x}$ denoting secret variables and $\mathbf{v}$ denoting public variables. We generalise the cube/AIDA attacks by choosing an arbitrary subspace $V \subseteq \mathbb{F}_2^m$ of the space of public variables and defining a function g_V as

$$g_V(\mathbf{x}) = \sum_{\mathbf{v} \in V} g(\mathbf{x}, \mathbf{v}).$$

Denote by k the dimension of V and let $\{\mathbf{v}_1, \ldots, \mathbf{v}_k\}$ be a basis for V. Using Eq. (2) we can give an equivalent formula for g_V using higher order differentiation, namely $g_V(\mathbf{x}) = (\Delta^{(k)}_{\mathbf{v}_1, \ldots, \mathbf{v}_k} g)(\mathbf{x}, \mathbf{0})$. Note that the usual cube attack becomes a particular case of this attack, for $V = \langle \mathbf{e}_{i_1}, \ldots, \mathbf{e}_{i_k} \rangle$, where $\mathbf{e}_i$ are the vectors of the canonical basis, and $I = \{i_1, \ldots, i_k\}$ are the positions chosen for the usual cube attack.

Using Theorem 2 we have that $\deg(g_V) \leq \deg(g) - \dim(V)$. Hence, as in the cube attack (see Corollary 1), if the dimension of V is $k = \deg(g) - 1$ we are guaranteed that g_V is linear or constant.

We can therefore search for spaces V such that the resulting g_V is a linear function in the secret variables. Linearity tests can detect whether the result is linear, like in the usual cube attack. This search space is a superset of the search space of the usual cube attack.

Moebius transform can be used here again for improved efficiency. Namely we start with a large vector space $V = \langle \mathbf{v}_1, \ldots, \mathbf{v}_k \rangle$ For any fixed value $\mathbf{x}$ of the secret variables we compute the truth table of the function $h(y_1, \ldots, y_k) = g(\mathbf{x}, y_1 \mathbf{v}_1 + \ldots + y_k \mathbf{v}_k)$. We then apply the Moebius transform to h. For any subspace $V' = \langle \mathbf{v}_{i_1}, \ldots, \mathbf{v}_{i_j} \rangle$ of V let $\mathbf{a}$ be the vector with ones in exactly the positions $i_1, \ldots, i_j$ and zeroes elsewhere. We have $h^M(\mathbf{a}) = \sum_{(c_1, \ldots, c_j) \in \{0,1\}^j} g(\mathbf{x}, c_1 \mathbf{v}_{i_1} + \ldots + c_j \mathbf{v}_{i_j}) = \sum_{\mathbf{v}' \in V'} g(\mathbf{x}, \mathbf{v}') = g_{V'}(\mathbf{x}, \mathbf{0})$. Hence, again, the Moebius transform h^M computes simultaneously all the $g_{V'}(\mathbf{x}, \mathbf{0})$ for all subspaces V' of V.

Remark 2. In [3] Dinur and Shamir also consider the possibility of setting some of the non-cube public variables to 1 rather than zero. That is not the same as our approach, as the set to sum over in that case is no longer a vector space. We can include that generalisation in our approach as follows. Let $\mathbf{c}$ be a fixed vector of public variables and V a vector space. Instead of computing f_V as before, we can compute instead the sum

$$\sum_{\mathbf{v} \in V} g(\mathbf{x}, \mathbf{c} + \mathbf{v})$$

which, using Eq. (2), can be proved to equal $(\Delta^{(k)}_{\mathbf{v}_1, \ldots, \mathbf{v}_k} g)(\mathbf{x}, \mathbf{c})$. The attack can work equally well in this scenario.

4 Proposed Linearity Test

We propose an alternative to the Generalised Linearity Test presented in Subsect. 2.2. Again, let $f : \mathbb{F}_2^n \to \mathbb{F}_2$ and let $\{\mathbf{b}_1, \ldots, \mathbf{b}_t\} \subseteq \mathbb{F}_2^n$ be a set of t linearly independent elements. (The question of how to choose a suitable value for t is, as we shall see, exactly the same as in [3], and a further discussion of this choice is beyond the scope of this paper.) For each d with $1 \leq d \leq t$ consider the following set of equations:

$$L_d = \left\{ \sum_{\mathbf{u} \preceq \mathbf{c}} f\left(\sum_{i=0}^{t} u_i \mathbf{b}_i\right) = 0 \,\Big|\, \mathbf{c} \in \mathbb{F}_2^t, \mathrm{w}(\mathbf{c}) = d \right\} \tag{3}$$

where $\mathbf{u} \preceq \mathbf{c}$ means $u_i \le c_i$ for all $i = 1, \ldots, t$, as defined in Sect. 2.3. Each L_d has $\binom{t}{d}$ equations. Each equation can alternatively be written using higher order derivatives:

$$L_d = \left\{ \Delta^{(d)}_{\mathbf{b}_{i_1}, \ldots, \mathbf{b}_{i_d}} f(\mathbf{0}) = 0 | \{i_1, \ldots, i_d\} \subseteq \{1, \ldots, t\} \right\}.$$

Note that the equations in L_2 correspond to the usual (BLR) linearity tests. The equations in L_3 have been proposed for testing whether the function is quadratic in [4, Sect. 4]. The following result is quite straightforward but we prove it for completeness.

Proposition 1. *If f has degree d, then it satisfies all the equations in the sets $L_{d+1}, \ldots, L_t$.*

Proof. By Theorem 2, $\deg(\Delta^{(j)}_{\mathbf{b}_{i_1}, \ldots, \mathbf{b}_{i_j}} f) \le \deg(f) - j = d - j$. Hence for all $j > d$ we have $\Delta^{(j)}_{\mathbf{b}_{i_1}, \ldots, \mathbf{b}_{i_j}} f \equiv 0$, so the equations L_j are satisfied for all $j > d$.

One can give an alternative proof of this result using the Moebius transform.

Based on the result above we propose an alternative to the Generalised Linearity Test. Namely we test whether a function f has degree one or less by testing whether it satisfies the equations in the sets $L_2, \ldots, L_t$. An advantage of this test is that if a function fails the test (i.e. has degree more than one), we can get, at no additional cost, an indication of its degree. Namely, if d is the highest number for which some equations in L_d are not satisfied, then we know that $\deg(f) \ge d$. We will estimate the degree of f as being d.

The proposed test contains the same number of equations (namely $2^t - t - 1$) and needs the same number of evaluations of f (namely 2^t) as the Generalised Linearity Test. If the function f is affine (has degree at most one), then it passes both types of test, so there are no false negatives. If f has degree 2 or more it might still pass one of the types of tests (i.e. we can have false positives). We can ask ourselves whether some functions can pass our proposed test but fail the Generalised Linearity Test or vice-versa. We show that this is not possible, in other words the tests are equivalent. More precisely:

Proposition 2. *A function f satisfies the Generalised Linearity Test (1) iff it satisfies the sets of equations $L_2, L_3, \ldots L_t$.*

Proof. We rename $y_{\{i_1, \ldots, i_j\}} = f(\mathbf{b}_{i_1} + \ldots + \mathbf{b}_{i_j})$. Both the Generalised Linearity Test and the set of equations $L_2, L_3, \ldots L_t$ can be rewritten as homogeneous systems of $2^t - t - 1$ linear equations in the 2^t unknowns y_J for all $J \subseteq \{1, \ldots, t\}$. Both sets of equations are in triangular form, so both solution spaces have dimension $t + 1$. To prove that the two solution spaces are equal, it suffices therefore to prove one inclusion. We show that a solution of the first set of equations is also a solution for the second. The first system gives immediately the solution

$$y_{\{i_1, \ldots, i_j\}} = \sum_{\ell=1}^{j} y_{\{i_\ell\}} + ((j-1) \bmod 2) y_\emptyset. \tag{4}$$

Consider now an equation from the second set corresponding to a vector $\mathbf{c}$ of weight $d \geq 2$. Let $I = \{i_1, \ldots, i_d\}$ be the positions of the non-zero entries of $\mathbf{c}$. The equation becomes $\sum_{J \subseteq I} y_J = 0$. Substituting the solution (4) of the first system of equations in this equation we obtain an equation that only contains the variables $y_{\{i_1\}}, \ldots, y_{\{i_d\}}$ and $y_\emptyset$. We count how many times each of these variables appears: $y_{\{i_1\}}$ will appear a number of times equal to the number of subsets J of I that contain $y_{\{i_1\}}$; this is half of the subsets, i.e. 2^{d-1} times. The variable $y_\emptyset$ appears once for each subset $J \subseteq I$ of even cardinality, i.e. $\sum_{\ell=0}^{\lfloor s/2 \rfloor} \binom{s}{2\ell} = 2^{d-1}$ times. Hence the left hand side of the equation becomes $2^{d-1}(y_{\{i_1\}} + \ldots + y_{\{i_1\}} + y_\emptyset)$. Since $d \geq 2$ and we are in $\mathbb{F}_2$, we have $2^{d-1} = 0$ so the equation is satisfied.

Finally note that the Moebius transform can again be used for efficiency. Namely putting $h(y_1, \ldots, y_t) = f(y_1 \mathbf{b_1} + \ldots + y_t \mathbf{b_t})$ and computing the Moebius transform h^M of h, the set of equations L_d are precisely the equations $h^M(y_1, \ldots, y_t) = 0$ for all $(y_1, \ldots, y_t)$ of weight d. Moreover we obtain automatically the sets of equations L_1 and L_0. If f satisfies L_1 then f is a constant, which is another test needed in the cube attack.

5   Implementation

We implemented our ideas and tested them on the stream cipher Trivium [1]. The public variables are in this case the initialisation vector and the secret variables are the key.

To test the cube attack over different vector spaces, as described in Sect. 3 we generated a large vector space V of initialisation vectors of dimension 28 giving us $\binom{28}{k}$ subspaces of each dimension $k = 0, \ldots, 28$. Linearity testing is performed as described in Sect. 4 using a basis of 6 linearly independent keys, meaning we evaluate at 2^6 keys for a total of $2^6 - 6 - 1$ evaluations. This will allow us to detect results that are constant, of estimated degree 1 to 5 or degree 6 or more.

We utilised a 64-bit parallelised implementation of Trivium in order to analyse 64 rounds simultaneously. The first bit of output represents round 640, with the 64th bit of output representing round 703. This allows us to compare our results with the results presented by Dinur and Shamir in their original cube paper [3] which found cubes of size 12 between 672 and 680 rounds. We also used parallelisation to implement the preprocessing phase of the cube attack. We utilised a multicore machine so that each core receives one of the 2^6 keys and runs Trivium for the given key and all the 2^{28} initialisation vectors in the large vector space. Each core then applies the Moebius transform on the data it computed. This significantly improved the efficiency of the preprocessing phase.

The vector space V is specified by a basis of 28 vectors, and it will be helpful to think of them as the rows of a 28×80 matrix A. The implementation can run the standard cube attack, by choosing 28 variable indices $i_1, \ldots, i_{28}$ and setting the entries of A so that columns $i_1, \ldots, i_{28}$ form a diagonal matrix and the remaining columns are all zero. When running the attack on an arbitrary vector space, we again choose 28 variable indices $i_1, \ldots, i_{28}$, set the columns $i_1, \ldots, i_{28}$ to form a diagonal matrix, but specify two probabilities p and q which define whether the entries in the remaining $80 - 28 = 52$ columns set to 0 or 1.

We define q as the probability that a column in matrix A will follow the probability p or be set to all zeroes. The probability p defines the probability that an entry will be set to 1 in matrix A. This means that when p is set to 0, q becomes irrelevant and when q is set to 0, p becomes irrelevant. Setting either p or q to 0 is the equivalent of running the standard cube attack.

We run the attack where $q = 1$ and $p = 0, 0.03, 0.5, 0.97, 1$, meaning that all columns in the basis which don't correspond to any of the $i_1, \ldots, i_{28}$ indices are chosen according to the probability p (for $p = 1$, all the remaining 52 columns are set to all ones in the basis). A further test is run where $q = 0.0625$ and $p = 0.5$ which generates a fairly sparse matrix A as the probability of a column of variables being chosen using probability p is low therefore most variables are set to 0. We kept the choice of $i_1, \ldots, i_{28}$ the same in all cases.

6   Discussion

Figures 1 and 2 show how many subspaces of each dimension (as a percentage of all subspaces of that dimension) were found to return constant results where $q = 1$ and $p = 0$ or $p = 1$.

Multiple lines show the results over different numbers of rounds, from 641 to 703. Predictably there are fewer constant results found at smaller dimensions as the number of rounds increases, indicating that the degree of the underlying polynomial is increasing.

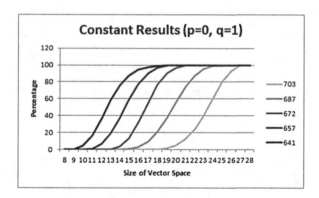

Fig. 1. Percentage of Constant Vector Spaces where $p = 0, q = 1$ for selected rounds

When comparing the two figures, it is clear that constant results are found at smaller dimensions when $p = 0$ compared to $p = 1$. There were no constant results found for 703 rounds when $p = 1$ whereas constant results were found in

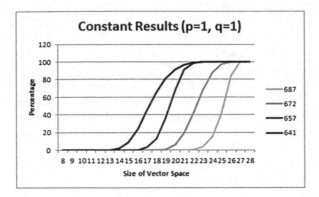

Fig. 2. Percentage of Constant Vector Spaces where $p = 1, q = 1$ for selected rounds

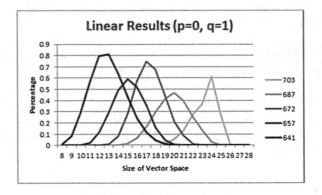

Fig. 3. Percentage of Linear Vector Spaces where $p = 0, q = 1$ for selected rounds

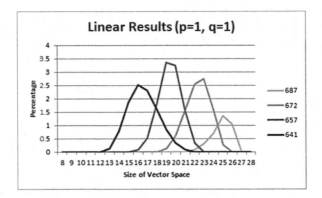

Fig. 4. Percentage of Linear Vector Spaces where $p = 1, q = 1$ for selected rounds

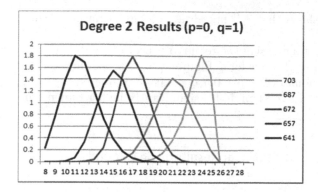

Fig. 5. Percentage of Degree 2 Vector Spaces where $p = 0, q = 1$ for selected rounds

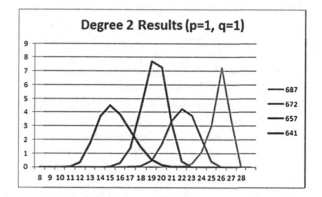

Fig. 6. Percentage of Degree 2 Vector Spaces where $p = 1, q = 1$ for selected rounds

cubes as small as 19 when $p = 0$. This shows that changing the vector space can have a negative effect on an attacker's ability to find linear results.

This result is confirmed by Figs. 3 and 4, which are similar to Figs. 1 and 2 but show the percentages of subspaces that produce linear (rather than constant) results. For each of the rounds presented, the peak dimension where linear results are most frequent is slightly larger (by 3 to 5 units) when $p = 1$ (Fig. 4) compared to where $p = 0$ (Fig. 3).

There are however some benefits to changing the vector space, as shown in Fig. 7. When analysed on the same scale, we can see that a higher percentage of subspaces produce linear results when we increase the search space by changing the vector space of the cube attack. Across all rounds, the test where $p = 1$ consistently showed a 3 to 4 times higher percentage of linear results being found as compared to where $p = 0$ (Fig. 5).

Furthermore, the percentage of subspaces found at dimension 14 where $p = 1$ in Fig. 7 is equivalent to the percentage of cubes found at cube size 12 and 13 where $p = 0$. While again reinforcing the result that the required dimension does increase, this shows that it is not always of significant detriment to the attacker (Fig. 6).

The trend of increasing the required vector space dimension continues when we test using a small value for q. Figure 7 shows a large increase in the percentage of linear cubes found when $p = 0.5, q = 0.0625$ compared to $p = 1, q = 1$ although these linear results are found at a significantly higher dimension.

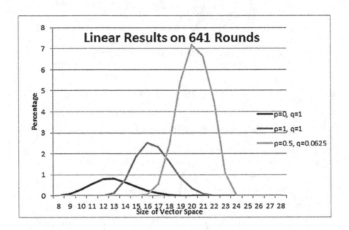

Fig. 7. Percentage of Linear Vector Spaces where $q = 1$ when $p = 0$ and $p = 1$, and $q = 0.0625$ when $p = 0.5$

When the value of p is set to a value other than 0 or 1 while $q = 1$, the attack becomes significantly less effective. We tested with both very dense set of basis vectors ($p = 0.97$) and very sparse ($p = 0.03$) as well as uniform ($p = 0.5$) and the results were nearly identical in all cases. Table 1 shows that in these cases there were no constant, linear or degree 2 results found, as well as an insignificant number of degree 3 results using 641 rounds. The similar behaviour between all values of p in this range when $q = 1$ could be due to the fact that although we controlled the density of the basis vectors, the rest of the vectors in the space will have quite high density irrespective of the density of the basis vectors due to the high value of q. This is in contrast to the result presented in Fig. 7 which showed a significant number of linear results when $p = 0.5$ and q is small.

Table 1. Percentage of Vector Spaces of degree 0 to 6 when $0 < p < 1$ and $q = 1$

Degree	0	1	2	3	4	5	6
Percentage	0	0	0	$<0.001\,\%$	$0.78\,\%$	$49.21\,\%$	$50\,\%$

The fact that for Trivium the cube attack over arbitrary vector spaces V performs in general worse than the usual cube attack when q is large is, on one hand, disappointing from an attacker's point of view. However on the other hand it offers insights into the properties of Trivium. Namely it shows that cube attacks work for Trivium not only because the degree increases relatively slowly through the rounds, but also because for the degree that is achieved, the distribution of the monomials is not uniform, as would be expected for a random function. This phenomenon has been observed in other contexts, for example the density of terms of each degree is estimated by Fouque and Vannet [6]. Another manifestation of this phenomenon is that out of all the linear equations that were found by different cube attacks on Trivium reported in the literature, the vast majority contain only one or two secret key variables, instead of around 40 variables as would be expected.

An alternative way to look at this would be to create a new, enhanced "black box" for Trivium, which contains a multiplication of the vector of public IV variables by an arbitrary fixed invertible matrix. The result of the multiplication is fed into the usual Trivium black box function. It is expected that the usual cube attack on this enhanced black box would have a much reduced chance of success, as it would be, in effect, equivalent to running our generalised cube attack on the usual Trivium black box. Obviously, the matrix needs to be secret, as otherwise the attacker could undo its effect. Preliminary tests on small cubes seem to confirm this idea, but a full investigation will be the subject of future work.

Acknowledgements. The authors would like to thank the referees for useful comments. One of the referees brought to our attention the recent paper [2] that we had not been aware of, and in which higher order derivatives with respect to an arbitrary vector space (as explored in Sect. 3) were used for statistical attacks on the NORX cipher.

References

1. De Cannière, C.: TRIVIUM: a stream cipher construction inspired by block cipher design principles. In: Katsikas, S.K., López, J., Backes, M., Gritzalis, S., Preneel, B. (eds.) ISC 2006. LNCS, vol. 4176, pp. 171–186. Springer, Heidelberg (2006)
2. Das, S., Maitra, S., Meier, W.: Higher order differential analysis of NORX. IACR Cryptology ePrint Arch. **2015**, 186 (2015)
3. Dinur, I., Shamir, A.: Cube attacks on tweakable black box polynomials. In: Joux, A. (ed.) EUROCRYPT 2009. LNCS, vol. 5479, pp. 278–299. Springer, Heidelberg (2009)
4. Dinur, I., Shamir, A.: Applying cube attacks to stream ciphers in realistic scenarios. Crypt. Commun. **4**(3–4), 217–232 (2012)
5. Duan, M., Lai, X.: Higher order differential cryptanalysis framework and its applications. In: International Conference on Information Science and Technology (ICIST), pp. 291–297 (2011)
6. Fouque, P.-A., Vannet, T.: Improving key recovery to 784 and 799 rounds of trivium using optimized cube attacks. In: Moriai, S. (ed.) FSE 2013. LNCS, vol. 8424, pp. 502–517. Springer, Heidelberg (2014)

7. Fouque, P.-A., Vannet, T.: Improving key recovery to 784 and 799 rounds of trivium using optimized cube attacks. IACR Cryptology ePrint Arch. **2015**, 312 (2015)
8. Joux, A.: Algorithmic Cryptanalysis, 1st edn. Chapman and Hall/CRC, Boca Raton (2009)
9. Knellwolf, S., Meier, W.: High order differential attacks on stream ciphers. Crypt. Commun. **4**(3–4), 203–215 (2012)
10. Lai, X.: Higher order derivatives and differential cryptanalysis. In: Blahut, R.E., Costello Jr., D.J., Maurer, U., Mittelholzer, T. (eds.) Communications and Cryptography. The Springer International Series in Engineering and Computer Science, vol. 276, pp. 227–233. Springer, US (1994)
11. Vielhaber, M.: Breaking ONE.FIVIUM by AIDA an algebraic IV differential attack. Cryptology ePrint Archive, Report 2007/413 (2007). http://eprint.iacr.org/

On the Diffusion Property of Iterated Functions

Jian Liu[1,2], Sihem Mesnager[3]([✉]), and Lusheng Chen[4]

[1] School of Computer Software, Tianjin University,
Tianjin 300072, People's Republic of China
jianliu.nk@gmail.com
[2] CNRS, UMR 7539 LAGA, Paris, France
[3] Department of Mathematics, University of Paris VIII, University of Paris XIII,
CNRS, UMR 7539 LAGA and Telecom ParisTech, Paris, France
smesnager@univ-paris8.fr
[4] School of Mathematical Sciences, Nankai University,
Tianjin 300071, People's Republic of China
lschen@nankai.edu.cn

Abstract. For vectorial Boolean functions, the behavior of iteration has consequence in the diffusion property of the system. We present a study on the diffusion property of iterated vectorial Boolean functions. The measure that will be of main interest here is the notion of the degree of completeness, which has been suggested by the NESSIE project. We provide the first (to the best of our knowledge) two constructions of (n, n)-functions having perfect diffusion property and optimal algebraic degree. We also obtain the complete enumeration results for the constructed functions.

Keywords: Boolean functions · Degree of completeness · Perfect diffusion property · Algebraic degree · Balancedness

1 Introduction

Vectorial Boolean functions have been extensively studied for their applications in cryptography, coding theory, combinatorial design etc., see [4] for a survey. Let $\mathbb{F}_2^n$ denote the n-dimensional vector space over the finite field $\mathbb{F}_2$ with two elements. Vectorial Boolean functions are functions from the vector space $\mathbb{F}_2^n$ to the vector space $\mathbb{F}_2^m$, for given positive integers n and m. These functions are called (n, m)-functions and include the single-output Boolean functions (which correspond to the case $m = 1$).

In 1949, Shannon [16] used the term diffusion to denote the quantitative spreading of information. For an (n, m)-function, the diffusion property describes the influence of input bits on the output bits. The exact meaning of diffusion relates strongly to the methods of cryptanalysis. An intuitive measure related

This work is supported by the National Key Basic Research Program of China under Grant 2013CB834204.

© Springer International Publishing Switzerland 2015
J. Groth (Ed.): IMACC 2015, LNCS 9496, pp. 239–253, 2015.
DOI: 10.1007/978-3-319-27239-9_15

to diffusion is of considerable importance for vectorial Boolean functions: the *degree of completeness*, denoted by D_c, which is given by the comments from the NESSIE project [15]. In [10], Kam and Davida introduced the concept of *complete functions*, which means that every output bit depends on every input bit (also see [7,9]). For a vectorial Boolean function, the degree of completeness quantifies the rate of input bits that the output bits depend on. A complete function possesses optimal degree of completeness, i.e., $D_c = 1$. In this paper, we use the notion of the degree of completeness to indicate the diffusion property of a vectorial Boolean function. In this sense, by perfect diffusion property we mean iterated vectorial Boolean functions which possess optimal degree of completeness. There are other indicators of diffusion property which shall not be discussed here. For instance, the *branch number of diffusion layers* (see [6]) relates closely to differential cryptanalysis [2] and linear cryptanalysis [12] on block ciphers. The *diffusion factor* (see [6]) quantifies the average number of changed output bits when a single input bit is changed. The degree of completeness can be seen as some kind of weakened version of diffusion factor.

The investigation on the degree of completeness of iterated (n, n)-functions helps in general the understanding of the evolution of diffusion property of cryptographic systems. Some methods of cryptanalysis on cryptosystems are based on the idea of identifying the relation between a particular output bit with the input bits. If every output bit depends on only a few of the input bits, there may exist some potential attacks, such as algebraic attacks [1,5], since one may convert the cipher-text into a system of polynomial equations and solve it directly. For example, in a product cryptosystem [16] such as block cipher and hash function, the degrees of completeness of iterated round functions (seen as vectorial Boolean functions) have consequence in the diffusion property of the whole system. A round function is preferable to have perfect diffusion property for providing complete diffusion, see the general model in Sect. 3.2. In the context of stream ciphers, the model of augmented functions should be considered (see [8]), where an update function L is iterated to generate keystreams by composing an output function. If the degrees of completeness of the iterated update functions $L^{(i)}$, $i = 1, 2, \ldots$, are very low, then the algebraic attack is expected to be efficient. Other potential applications of functions with perfect diffusion property could be found.

Though the degree of completeness has been observed from a cryptographic point of view, it seems that as a mathematical object, vectorial Boolean function with good diffusion property has rarely been studied in the literature. In this paper, we mainly study the diffusion property of vectorial Boolean functions. A function is called to have *perfect diffusion property* (see Definition 2) if the degree of completeness always attains 1 (under the affine permutations) after the function has been iterated some number of times. We provide two constructions of vectorial Boolean functions which have perfect diffusion property, and prove that the iterated functions always have optimal algebraic degree. To the best of our knowledge, this is the first time when such constructions are proposed. We first construct a class of rotation symmetric (n, n)-functions (see Definition 3) with perfect diffusion property. These functions are generalizations of rotation symmetric Boolean functions, which have practical advantages that the evaluations are efficient and the representations are short. In our second construction, a class of almost balanced (n, n)-functions with perfect diffusion property is given. Moreover, complete enumeration results for the constructed functions are obtained, which show that there are many (n, n)-functions with perfect diffusion property.

This paper is organized as follows. Formal definitions and necessary preliminaries are introduced in Sect. 2. In Sect. 3, two constructions of vectorial Boolean functions with perfect diffusion property are proposed for the first time, and the complete enumeration results for these functions are presented. To avoid being too theoretical, we give an explicit example to show that it is possible to construct recursive round functions to provide complete diffusion. We summarize this paper in the last section.

2   Background and Preliminaries

In this paper, additions and multiple sums calculated modulo 2 will be denoted by $\oplus$ and $\bigoplus_i$ respectively, additions and multiple sums calculated in characteristic 0 or in the additions of elements of the finite field $\mathbb{F}_{2^n}$ will be denoted by $+$ and $\sum_i$ respectively. The functions from the vector space $\mathbb{F}_2^n$ to $\mathbb{F}_2$ are called n-variable Boolean functions, and the set of all the n-variable Boolean functions is denoted by $\mathcal{B}_n$. For $f \in \mathcal{B}_n$, the Hamming weight (in brief, weight) of f is $\mathrm{wt}(f) = |\{x \in \mathbb{F}_2^n \mid f(x) = 1\}|$, and the $(0,1)$-sequence defined by $(f(\mathbf{v}_0), f(\mathbf{v}_1), \ldots, f(\mathbf{v}_{2^n-1}))$ is called the truth table of f, where $\mathbf{v}_0 = (0, \ldots, 0, 0)$, $\mathbf{v}_1 = (0, \ldots, 0, 1), \ldots, \mathbf{v}_{2^n-1} = (1, \ldots, 1, 1)$ are ordered by lexicographical order. An n-variable Boolean function f can be uniquely represented in the algebraic normal form (in brief, ANF) that

$$f(x) = \bigoplus_{v \in \mathbb{F}_2^n} c_v x_1^{v_1} x_2^{v_2} \cdots x_n^{v_n},$$

where $x = (x_1, \ldots, x_n) \in \mathbb{F}_2^n$, $v = (v_1, \ldots, v_n) \in \mathbb{F}_2^n$, $c_v \in \mathbb{F}_2$. Let $\mathrm{wt}(v)$ denote the Hamming weight (or weight) of a vector v, that is the number of its nonzero coordinates, then $\deg(f) = \max_{v \in \mathbb{F}_2^n}\{\mathrm{wt}(v) \mid c_v \neq 0\}$ is called the algebraic degree of f.

Proposition 1. [3]   Let $f \in \mathcal{B}_n$ and $f(x) = \bigoplus_{v \in \mathbb{F}_2^n} c_v x_1^{v_1} x_2^{v_2} \cdots x_n^{v_n}$. Then, $c_v = \bigoplus_{x \in \mathbb{F}_2^n, x \preceq v} f(x)$, where $x \preceq v$ means that $x = (x_1, \ldots, x_n)$ is covered by $v = (v_1, \ldots, v_n)$, i.e., for any $i = 1, \ldots, n$, $x_i = 1$ implies $v_i = 1$. In particular, $\deg(f) = n$ if and only if $\mathrm{wt}(f)$ is odd.

An affine permutation L on $\mathbb{F}_2^n$ is defined as $L(x) = xM \oplus a$, where M is a nonsingular $n \times n$ matrix over $\mathbb{F}_2$ and $a \in \mathbb{F}_2^n$. Moreover, if $a = \mathbf{0}$, then L is called a linear permutation.

Proposition 2. [3] *The algebraic degree of an n-variable Boolean function f is affine invariant, i.e., for every affine permutation L, we have $\deg(f \circ L) = \deg(f)$.*

For $i = 1, \ldots, n$, denote by e_i the vector in $\mathbb{F}_2^n$ whose ith component equals 1, and 0 elsewhere. The *degree of completeness* of an n-variable Boolean function f is defined as

$$D_c(f) = 1 - \frac{|\{i \mid a_i = 0, 1 \leqslant i \leqslant n\}|}{n}, \tag{1}$$

where $a_i = |\{x \in \mathbb{F}_2^n \mid f(x) \oplus f(x \oplus e_i) = 1\}|$, $i = 1, \ldots, n$. Equivalently, let

$$\mathcal{V}(f) = \{i \mid \exists x \in \mathbb{F}_2^n \text{ such that } f(x) \oplus f(x \oplus e_i) = 1, 1 \leqslant i \leqslant n\}, \tag{2}$$

be the set of indices of the variables appearing in the ANF of f, then $D_c(f) = |\mathcal{V}(f)|/n$.

Let n and m be two positive integers. The functions from $\mathbb{F}_2^n$ to $\mathbb{F}_2^m$ are called (n, m)-*functions* (or *vectorial Boolean functions*). Such a function F is given by $F = (f_1, \ldots, f_m)$, where the Boolean functions $f_1, \ldots, f_m$ are called the *coordinate functions* of F. An (n, m)-function is called *balanced* if for any $b \in \mathbb{F}_2^m$, the size of the pre-image set $|F^{-1}(b)| = 2^{n-m}$. The *derivative* of F at direction a is defined as

$$\triangle_a F(x) = F(x) \oplus F(x \oplus a), \quad a \in \mathbb{F}_2^n \backslash \{\mathbf{0}\}.$$

The *algebraic degree* of F, denoted by $\mathrm{Deg}(F)$, is defined as

$$\mathrm{Deg}(F) = \max_{1 \leqslant i \leqslant m} \deg(f_i).$$

The *degree of completeness* of F is defined as

$$D_c(F) = \frac{1}{m}(D_c(f_1) + \cdots + D_c(f_m)). \tag{3}$$

Since the degree of completeness of an n-variable Boolean function can also be described as $D_c(f) = |\mathcal{V}(f)|/n$, where $\mathcal{V}(f)$ is defined as in (2), then for an (n, m)-function $F = (f_1, \ldots, f_m)$, we have $D_c(F) = (|\mathcal{V}(f_1)| + \cdots + |\mathcal{V}(f_m)|)/nm$. Also, the following equivalent definition is easy to obtain, which is originally given by the NESSIE project [15].

Definition 1. [15] *For an (n, m)-function $F = (f_1, \ldots, f_m)$, the* degree of completeness *is defined as*

$$D_c(F) = 1 - \frac{|\{(i, j) \mid a_{ij} = 0, 1 \leqslant i \leqslant n, 1 \leqslant j \leqslant m\}|}{mn},$$

where $a_{ij} = |\{x \in \mathbb{F}_2^n \mid f_j(x) \oplus f_j(x \oplus e_i) = 1\}|$, $i = 1, \ldots, n$, $j = 1, \ldots, m$.

For an (n, m)-function F, it is obvious that $0 \leqslant D_c(F) \leqslant 1$, and F is called *complete* if $D_c(F) = 1$ (see [10]), which provides the highest possible level of diffusion. Note that $D_c(F)$ defined in (3) takes the mean value of all the $D_c(f_i)$'s with $i = 1, \ldots, m$, while the following two meaningful measures are also intuitive,

$$D_c^{max}(F) = \max_{1 \leqslant i \leqslant m} \{D_c(f_i)\}, \quad D_c^{min}(F) = \min_{1 \leqslant i \leqslant m} \{D_c(f_i)\}.$$

Clearly, $D_c^{min}(F) \leqslant D_c(F) \leqslant D_c^{max}(F)$, and $D_c^{min}(F) = 1$ if and only if $D_c(F) = 1$. Hence, D_c^{min} is the strongest measure of completeness for vectorial Boolean functions. In this paper, we mainly discuss the measure D_c suggested by the NESSIE project [15].

For an n-variable Boolean function f, since for any $b \in \mathbb{F}_2^n$,

$$a_i = |\{x \in \mathbb{F}_2^n \mid f(x) \oplus f(x \oplus e_i) = 1\}| = |\{x \in \mathbb{F}_2^n \mid f(x \oplus b) \oplus f(x \oplus b \oplus e_i) = 1\}|,$$

where $i = 1, \ldots, n$, then from (1), we have $D_c(f(x)) = D_c(f(x \oplus b))$. In general, the degree of completeness is not invariant under composition on the right by linear permutations. For example, let $f(x_1, \ldots, x_n) = x_1 \oplus \cdots \oplus x_n \in \mathcal{B}_n$, and $L(x_1, \ldots, x_n) = (x_1 \oplus \cdots \oplus x_n, x_2, \ldots, x_n)$ which is a linear permutation on $\mathbb{F}_2^n$, then $f \circ L(x_1, \ldots, x_n) = x_1$, and thus $D_c(f) = 1 > D_c(f \circ L) = 1/n$. For a positive integer r, let $F^{(r)} = \overbrace{F \circ \cdots \circ F}^{r}$ denote the rth iterated function of F.

Definition 2. *An (n, m)-function F is called* non-degenerate *if for every linear permutation L on $\mathbb{F}_2^n$, $D_c(F \circ L) = 1$. Moreover, F is said to have* perfect diffusion *property if $m = n$ and for any positive integer k, $F^{(k)}$ is non-degenerate.*

Theorem 1. *For an (n, m)-function $F = (f_1, \ldots, f_m)$, if for all $i = 1, \ldots, m$, $\deg(f_i) = n$, then F is non-degenerate.*

Proof. According to Proposition 2, one gets that for every linear permutation L on $\mathbb{F}_2^n$ and every $1 \leqslant i \leqslant m$, $\deg(f_i \circ L) = \deg(f_i) = n$, thus $D_c(f_i \circ L) = 1$. From (3), we have that $D_c(F \circ L) = (D_c(f_1 \circ L) + \cdots + D_c(f_m \circ L))/m = 1$. Hence, F is non-degenerate. $\qquad\square$

Remark 1. When we choose a basis $\{\alpha_1, \ldots, \alpha_n\}$ of $\mathbb{F}_{2^n}$ over $\mathbb{F}_2$, then the vector space $\mathbb{F}_2^n$ can be endowed with the structure of finite field $\mathbb{F}_{2^n}$ by an isomorphism $\pi : x = (x_1, \ldots, x_n) \in \mathbb{F}_2^n \to x_1\alpha_1 + \cdots + x_n\alpha_n \in \mathbb{F}_{2^n}$. For a Boolean function f on $\mathbb{F}_{2^n}$, we identify $D_c(f)$ with $D_c(f \circ \pi)$. Since for any $i = 1, \ldots, n$,

$$\begin{aligned}
a_i &= |\{x \in \mathbb{F}_2^n \mid f \circ \pi(x) + f \circ \pi(x \oplus e_i) = 1\}| \\
&= |\{x \in \mathbb{F}_2^n \mid f(\pi(x)) + f(\pi(x) + \pi(e_i)) = 1\}| \\
&= |\{x \in \mathbb{F}_{2^n} \mid f(x) + f(x + \alpha_i) = 1\}|,
\end{aligned}$$

then from (1), one gets that $D_c(f) = D_c(f \circ \pi) = 1$ if and only if for any $i = 1, \ldots, n$, the derivative $\Delta_{\alpha_i} f(x)$ is not a zero function (i.e., there exists $x \in \mathbb{F}_{2^n}$ such that $\Delta_{\alpha_i} f(x) = 1$). Note that L is a linear permutation on $\mathbb{F}_2^n$ if and only if $\pi \circ L \circ \pi^{-1}$ is an additive automorphism on $\mathbb{F}_{2^n}$. Hence, from Definition 2, a Boolean function f is non-degenerate if for any $i = 1, \ldots, n$ and any additive automorphism L of $\mathbb{F}_{2^n}$, $\Delta_{\alpha_i} f \circ L(x)$ is not a zero function.

Remark 2. The trace function from $\mathbb{F}_{2^n}$ to $\mathbb{F}_2$ is defined as

$$\mathrm{Tr}_1^n(x) = x + x^2 + x^{2^2} + \cdots + x^{2^{n-1}},$$

where $x \in \mathbb{F}_{2^n}$. Given a basis $\{\alpha_1, \ldots, \alpha_n\}$ of $\mathbb{F}_{2^n}$ over $\mathbb{F}_2$, a function F from $\mathbb{F}_{2^n}$ to itself can be written as $F(x) = f_1(x)\alpha_1 + \cdots + f_n(x)\alpha_n$, where $f_i(x) = \mathrm{Tr}_1^n(\beta_i F(x))$, $i = 1, \ldots, n$, are the n-variable coordinate Boolean functions of F, and $\{\beta_1, \ldots, \beta_n\}$ is the *dual basis* of $\{\alpha_1, \ldots, \alpha_n\}$ satisfying

$$\mathrm{Tr}_1^n(\alpha_i \beta_j) = \begin{cases} 0 \text{ for } i \neq j, \\ 1 \text{ for } i = j. \end{cases}$$

It is known that for any basis of $\mathbb{F}_{2^n}$ over $\mathbb{F}_2$, there exists a dual basis (see [11, Chap. 2]). From Definition 2, we know that an (n, n)-function F has perfect diffusion property if and only if for every k, every coordinate function of $F^{(k)}$ is non-degenerate, which is equivalent to saying that, for any $j \in \{1, \ldots, n\}$, $f_j^{(k)}(x) = \mathrm{Tr}_1^n\left(\beta_j F^{(k)}(x)\right)$ is non-degenerate. From Remark 1, we have that $f_j^{(k)}(x)$ is non-degenerate if and only if for any $i \in \{1, \ldots, n\}$ and any additive automorphism L of $\mathbb{F}_{2^n}$,

$$
\begin{aligned}
\Delta_{\alpha_i} f_j^{(k)} \circ L(x) &= f_j^{(k)} \circ L(x) + f_j^{(k)} \circ L(x + \alpha_i) \\
&= \mathrm{Tr}_1^n\left(\beta_j F^{(k)} \circ L(x)\right) + \mathrm{Tr}_1^n\left(\beta_j F^{(k)} \circ L(x + \alpha_i)\right) \\
&= \mathrm{Tr}_1^n\left(\beta_j \Delta_{\alpha_i} F^{(k)} \circ L(x)\right)
\end{aligned}
$$

is not a zero function. As a consequence, from Definition 2, an (n, n)-function F is said to have perfect diffusion property if for any positive integer k, any $i, j \in \{1, \ldots, n\}$, and any additive automorphism L of $\mathbb{F}_{2^n}$, $\mathrm{Tr}_1^n\left(\beta_j \Delta_{\alpha_i} F^{(k)} \circ L(x)\right)$ is not a zero function.

3   Constructions of Vectorial Boolean Functions with Perfect Diffusion Property

In this section, we construct two classes of (n, n)-functions which have perfect diffusion property. Moreover, the enumeration results for the constructed functions are obtained.

3.1   Rotation Symmetric (n, n)-Functions with Perfect Diffusion Property

Let $(x_1, x_2, \ldots, x_n) \in \mathbb{F}_2^n$. For $1 \leqslant k \leqslant n - 1$, define

$$\rho_n^k(x_1, x_2, \ldots, x_n) = (x_{k+1}, \ldots, x_n, x_1, \ldots, x_k),$$

and $\rho_n^0(x_1, x_2, \ldots, x_n) = (x_1, x_2, \ldots, x_n)$. Inspired by the concept of rotation symmetric Boolean functions used in fast hashing algorithms [14], we present the following definition of rotation symmetric (n, n)-functions.

Definition 3. *Let f be an n-variable Boolean function. An (n, n)-function F is called* rotation symmetric *(in brief, RS) if it has the form*

$$F(x) = \left(f(x), f \circ \rho_n^1(x), f \circ \rho_n^2(x), \dots, f \circ \rho_n^{n-1}(x)\right). \qquad (4)$$

Let $f \in \mathcal{B}_n$ and $F = (f, f \circ \rho_n^1, \dots, f \circ \rho_n^{n-1})$. For any $x \in \mathbb{F}_2^n$ and any integer $l \geqslant 1$,

$$
\begin{aligned}
F \circ \rho_n^l(x) &= \left(f \circ \rho_n^l(x), f \circ \rho_n^{l+1}(x), \dots, f \circ \rho_n^{l-1}(x)\right) \\
&= \rho_n^l\left(f(x), f \circ \rho_n^1(x), \dots, f \circ \rho_n^{n-1}(x)\right) = \rho_n^l \circ F(x). \qquad (5)
\end{aligned}
$$

An (n, n)-function F satisfying Eq. (5) is called *shift-invariant* in [6]. Recall that an n-variable rotation symmetric Boolean function f is defined as $f \circ \rho_n^1(x) = f(x)$ for any $x \in \mathbb{F}_2^n$ (see [14]). By Eq. (5), the following equivalent definition of RS (n, n)-functions is easy to obtain.

Proposition 3. *An (n, n)-function F is RS if and only if for any $x \in \mathbb{F}_2^n$,*

$$F \circ \rho_n^1(x) = \rho_n^1 \circ F(x).$$

Proposition 4. *If F is an RS (n, n)-function, then for any integer $k \geqslant 1$, $F^{(k)}$ is an RS (n, n)-function.*

Proof. We prove it by induction on k. The result is already true for $k = 1$. Suppose that $F^{(k)}$ is RS for $k = s$, where $s \geqslant 1$, then $F^{(s)}$ has the form

$$F^{(s)}(x) = \left(f(x), f \circ \rho_n^1(x), \dots, f \circ \rho_n^{n-1}(x)\right),$$

which implies that

$$
\begin{aligned}
F^{(s+1)}(x) = F^{(s)}(F(x)) &= \left(f\left(F(x)\right), f \circ \rho_n^1\left(F(x)\right), \dots, f \circ \rho_n^{n-1}\left(F(x)\right)\right) \\
&= \left(f \circ F(x), f \circ F \circ \rho_n^1(x), \dots, f \circ F \circ \rho_n^{n-1}(x)\right), \qquad (6)
\end{aligned}
$$

where Eq. (6) follows from Eq. (5) since F is RS. Hence, for $k = s+1$, $F^{(k)}$ is an RS (n, n)-function. By the mathematical induction, we get that for $k \geqslant 1$, $F^{(k)}$ is RS. $\qquad \square$

Remark 3. From Propositions 3 and 4, one can see that rotation symmetric (n, n)-functions possess many desirable properties like (i) the algebraic representations are short; (ii) the evaluation of the functions is efficient (since a circular shift of the input bits leads to the corresponding shift of the output bits); (iii) the iterated functions are still rotation symmetric.

Under the action of ρ_n^k, $0 \leqslant k \leqslant n - 1$, the *orbit* generated by the vector $x = (x_1, x_2, \dots, x_n)$ is defined as

$$\mathcal{O}_n(x) = \left\{\rho_n^k(x_1, x_2, \dots, x_n) \mid 0 \leqslant k \leqslant n - 1\right\}. \qquad (7)$$

It is easy to check that the cardinality of an orbit generated by $x = (x_1, \dots, x_n)$ is a factor of n. In fact, let $|\mathcal{O}_n(x)| = t$, and suppose that $n = p \cdot t + r$ for $p, r \in \mathbb{Z}$ and

$0 < r < t$. Then $\rho_n^t(x) = x$, which implies that $\rho_n^{p \cdot t}(x) = x$, thus $\rho_n^{n-p \cdot t}(x) = x$, i.e., $\rho_n^r(x) = x$, a contradiction to the fact that $|\mathcal{O}_n(x)| = t > r$. Clearly, all the orbits generate a partition of $\mathbb{F}_2^n$. Every orbit can be represented by its lexicographically first element, called the *representative element*. It is proved that (see e.g. [6, Appendix A.1]) the number of distinct orbits is $\Psi_n = \frac{1}{n} \sum_{k|n} \phi(k) 2^{n/k}$, where $\phi(k)$ is the Euler's *phi*-function. Let $\{\Lambda_1^{(n)}, \Lambda_2^{(n)}, \ldots, \Lambda_{\Psi_n}^{(n)}\}$ denote the set of all the representative elements in lexicographical order, where $\Lambda_1^{(n)} = \mathbf{0}$ and $\Lambda_{\Psi_n}^{(n)} = \mathbf{1}$, and we use $\{\Lambda_1, \Lambda_2, \ldots, \Lambda_{\Psi_n}\}$ for short if there is no risk of confusion.

For $f \in \mathcal{B}_n$ and $1 \leqslant i \leqslant \Psi_n$, let $f|_{\mathcal{O}_n(\Lambda_i)}$ denote the restriction of f to $\mathcal{O}_n(\Lambda_i)$, i.e., for $x \in \mathcal{O}_n(\Lambda_i)$, $f|_{\mathcal{O}_n(\Lambda_i)}(x) = f(x)$. Then, we have the following theorem.

Theorem 2. *For any n-variable Boolean function f satisfying the following conditions:*

(i) For $i = 1, 2, \ldots, \Psi_n - 1$, wt $\left(f|_{\mathcal{O}_n(\Lambda_i)}\right) = t_i \cdot \text{wt}(\Lambda_i)/n$, where $t_i = |\mathcal{O}_n(\Lambda_i)|$;
(ii) $f(\mathbf{1}) = 0$,

the RS (n, n)-function $F(x) = \left(f(x), f \circ \rho_n^1(x), \ldots, f \circ \rho_n^{n-1}(x)\right)$ has perfect diffusion property, and for every $k \geqslant 1$, $\text{Deg}\left(F^{(k)}\right) = n$.

Proof. For $i = 1, 2, \ldots, \Psi_n - 1$, let $F|_{\mathcal{O}_n(\Lambda_i)}$ denote the $t_i \times n$ matrix over $\mathbb{F}_2$ that

$$F|_{\mathcal{O}_n(\Lambda_i)} = \begin{pmatrix} F(\Lambda_i) \\ F(\rho_n^1(\Lambda_i)) \\ \vdots \\ F(\rho_n^{t_i-1}(\Lambda_i)) \end{pmatrix}_{t_i \times n} = \begin{pmatrix} F(\Lambda_i) \\ \rho_n^1(F(\Lambda_i)) \\ \vdots \\ \rho_n^{t_i-1}(F(\Lambda_i)) \end{pmatrix}_{t_i \times n},$$

where $t_i = |\mathcal{O}_n(\Lambda_i)|$ and the last equality is from Eq. (5). It is obvious that the number of 1's in every column (as well as every row) of $F|_{\mathcal{O}_n(\Lambda_i)}$ is the same. Since wt $\left(f|_{\mathcal{O}_n(\Lambda_i)}\right) = t_i \cdot \text{wt}(\Lambda_i)/n$, then every row of $F|_{\mathcal{O}_n(\Lambda_i)}$ has the same weight that

$$\frac{\text{wt}\left(f|_{\mathcal{O}_n(\Lambda_i)}\right) \cdot n}{t_i} = \frac{t_i \cdot \text{wt}(\Lambda_i) \cdot n}{n \cdot t_i} = \text{wt}(\Lambda_i).$$

Hence, for $x = \rho_n^l(\Lambda_i)$, where $i = 1, 2, \ldots, \Psi_n - 1$ and $l = 0, \ldots, |\mathcal{O}_n(\Lambda_i)| - 1$,

$$\text{wt}(F(x)) = \text{wt}\left(F\left(\rho_n^l(\Lambda_i)\right)\right) = \text{wt}(\Lambda_i) = \text{wt}\left(\rho_n^l(\Lambda_i)\right) = \text{wt}(x),$$

i.e., for every $x \in \mathbb{F}_2^n \backslash \{\mathbf{1}\}$, $\text{wt}(F(x)) = \text{wt}(x)$. Therefore, for every $k \geqslant 1$ and every $x \in \mathbb{F}_2^n \backslash \{\mathbf{1}\}$, we have

$$\text{wt}\left(F^{(k)}(x)\right) = \text{wt}(x). \tag{8}$$

Thanks to Proposition 4, we know that $F^{(k)}$ is still an RS (n, n)-function, thus we can write $F^{(k)}$ as $\left(f^{(k)}, f^{(k)} \circ \rho_n^1, \ldots, f^{(k)} \circ \rho_n^{n-1}\right)$, where $f^{(k)} \in \mathcal{B}_n$. From

Eq. (8), we have that for $i = 1, 2, \ldots, \Psi_n - 1$, every column of the matrix

$$
F^{(k)}\big|_{\mathcal{O}_n(\Lambda_i)} = \begin{pmatrix} F^{(k)}(\Lambda_i) \\ F^{(k)}(\rho_n^1(\Lambda_i)) \\ \vdots \\ F^{(k)}(\rho_n^{t_i-1}(\Lambda_i)) \end{pmatrix}_{t_i \times n} = \begin{pmatrix} F^{(k)}(\Lambda_i) \\ \rho_n^1\left(F^{(k)}(\Lambda_i)\right) \\ \vdots \\ \rho_n^{t_i-1}\left(F^{(k)}(\Lambda_i)\right) \end{pmatrix}_{t_i \times n}
$$

has weight $t_i \cdot \mathrm{wt}(\Lambda_i)/n$. Since $\bigcup_{i=1}^{\Psi_n} \mathcal{O}_n(\Lambda_i) = \mathbb{F}_2^n$, then it is easy to prove that

$$
\sum_{i=1}^{\Psi_n} \frac{t_i \cdot \mathrm{wt}(\Lambda_i)}{n} = 2^{n-1}.
$$

Condition (ii) implies $\mathrm{wt}\left(F^{(k)}(\mathbf{1})\right) = 0$, then we have

$$
\mathrm{wt}\left(f^{(k)}\right) = \sum_{i=1}^{\Psi_n - 1} \frac{t_i \cdot \mathrm{wt}(\Lambda_i)}{n} = 2^{n-1} - 1.
$$

Hence, according to Proposition 1, $\deg\left(f^{(k)}\right) = n$, which leads to $\mathrm{Deg}\left(F^{(k)}\right) = n$. Note that for $l = 0, \ldots, n-1$, ρ_n^l is an affine permutation on $\mathbb{F}_2^n$, then from Proposition 2, we have $\deg\left(f^{(k)} \circ \rho_n^l\right) = \deg(f^{(k)}) = n$. Thus, Theorem 1 implies that $F^{(k)}$ is non-degenerate. Therefore, $F(x)$ has perfect diffusion property. $\quad\square$

In Theorem 3, we will calculate the number of all the functions constructed in Theorem 2. Before that, we introduce the following lemma which is given by Maximov [13, Lemma 1].

Lemma 1. *[13] For $\mathbb{F}_2^n$, the number of orbits with t elements of weight w is*

$$
\eta_{n,t,w} = \begin{cases} \frac{1}{t} \sum_{k|t, q_k|w} \mu(t/k) \cdot \binom{n/q_k}{w/q_k}, & \text{for } t, w = 1, \ldots, n, \text{ where } q_k = \frac{n}{\gcd(n,k)}, \\ 1, & \text{for } t = 1, w = 0, \\ 0, & \text{otherwise,} \end{cases}
$$

$$(9)$$

where $\mu(\cdot)$ is the Möbius function, i.e., for integer $t \geqslant 1$, $\mu(t) = 1$, if $t = 1$; $\mu(t) = (-1)^m$, if $t = p_1 p_2 \cdots p_m$, where $p_1, \ldots, p_m$ are distinct primes; $\mu(t) = 0$, for all other cases.

Theorem 3. *The number of distinct RS (n, n)-functions constructed in Theorem 2 is*

$$
\mathcal{N}_n = \prod_{w=1}^{n-1} \prod_{t=1}^{n} \left(\frac{t}{t \cdot w}{n}\right)^{\eta_{n,t,w}},
$$

$$(10)$$

where $\eta_{n,t,w} = \frac{1}{t} \sum_{k|t, q_k|w} \mu(t/k) \cdot \binom{n/q_k}{w/q_k}$, $q_k = \frac{n}{\gcd(n,k)}$, and $\mu(\cdot)$ is the Möbius function.

Proof. In Theorem 2, for any $i = 1, 2, \ldots, \Psi_n - 1$, wt $\left(f|_{\mathcal{O}_n(\Lambda_i)}\right) = t_i \cdot \text{wt}(\Lambda_i)/n$, which implies that one can construct $\binom{t_i}{t_i \cdot \text{wt}(\Lambda_i)/n}$ distinct $f|_{\mathcal{O}_n(\Lambda_i)}$'s. Moreover, Lemma 1 claims that the number of orbits with $t = t_i$ elements of weight $w = \text{wt}(\Lambda_i)$ is $\eta_{n,t,w}$. Then, one can get that the number of distinct RS (n, n)-functions constructed in Theorem 2 is $\mathcal{N}_n$ in Eq. (10). □

Example 1. For $\mathbb{F}_2^6$, all the orbits and the values of $\eta_{6,t,w}$ in Eq. (9) are listed in Table 1, where t and w are respectively the number and the weight of elements in an orbit.

Then, from Theorem 3, we have

$$\mathcal{N}_6 = \prod_{w=1}^{5} \prod_{t=1}^{6} \left(\frac{t}{\frac{t \cdot w}{6}} \right)^{\eta_{6,t,w}} = 2.6244 \times 10^{11} \approx 2^{37.9},$$

while the number of all the $(6, 6)$-functions is $2^{2^6 \cdot 6} = 2^{384}$.

Table 1. All the orbits of $\mathbb{F}_2^6$ and the values of $\eta_{6,t,w}$

	t	w		t	w
$\mathcal{O}_6(000000)$	1	0	$\mathcal{O}_6(010011)$	6	3
$\mathcal{O}_6(000001)$	6	1	$\mathcal{O}_6(010101)$	2	3
$\mathcal{O}_6(000011)$	6	2	$\mathcal{O}_6(001111)$	6	4
$\mathcal{O}_6(000101)$	6	2	$\mathcal{O}_6(010111)$	6	4
$\mathcal{O}_6(001001)$	3	2	$\mathcal{O}_6(011011)$	3	4
$\mathcal{O}_6(000111)$	6	3	$\mathcal{O}_6(011111)$	6	5
$\mathcal{O}_6(001011)$	6	3	$\mathcal{O}_6(111111)$	1	6

$\eta_{6,t,w}$	t 1 2 3 4 5 6
w	
0	1 0 0 0 0 0
1	0 0 0 0 0 1
2	0 0 1 0 0 2
3	0 1 0 0 0 3
4	0 0 1 0 0 2
5	0 0 0 0 0 1
6	1 0 0 0 0 0

Example 2. In Example 1, we have shown all the orbits $\{\mathcal{O}_6(\Lambda_i), i = 1, \ldots, 14\}$ of $\mathbb{F}_2^6$. Let f be a 6-variable Boolean function defined as

$$
\begin{aligned}
f|_{\mathcal{O}_6(000000)} &= (0), & f|_{\mathcal{O}_6(010011)} &= (1, 0, 0, 1, 0, 1), \\
f|_{\mathcal{O}_6(000001)} &= (0, 0, 0, 1, 0, 0), & f|_{\mathcal{O}_6(010101)} &= (0, 1), \\
f|_{\mathcal{O}_6(000011)} &= (0, 0, 1, 1, 0, 0), & f|_{\mathcal{O}_6(001111)} &= (1, 1, 0, 1, 1, 0), \\
f|_{\mathcal{O}_6(000101)} &= (0, 1, 0, 1, 0, 0), & f|_{\mathcal{O}_6(010111)} &= (0, 1, 1, 1, 0, 1), \\
f|_{\mathcal{O}_6(001001)} &= (1, 0, 0), & f|_{\mathcal{O}_6(011011)} &= (1, 0, 1), \\
f|_{\mathcal{O}_6(000111)} &= (0, 1, 0, 0, 1, 1), & f|_{\mathcal{O}_6(011111)} &= (1, 1, 1, 1, 0, 1), \\
f|_{\mathcal{O}_6(001011)} &= (1, 0, 0, 1, 0, 1), & f|_{\mathcal{O}_6(111111)} &= (0),
\end{aligned}
$$

where the binary vectors on the right-hand side denote the truth tables of the restriction functions $f|_{\mathcal{O}_6(\Lambda_i)}$, $i = 1, \ldots, 14$, i.e.,

$$f|_{\mathcal{O}_6(\Lambda_i)} = \left(f(\Lambda_i), f \circ \rho_6^1(\Lambda_i), \ldots, f \circ \rho_6^{t_i-1}(\Lambda_i)\right),$$

where $t_i = |\mathcal{O}_6(\Lambda_i)|$. Since the function f satisfies the conditions in Theorem 2, then the RS $(6, 6)$-function

$$F = \left(f, f \circ \rho_6^1, \ldots, f \circ \rho_6^5\right)$$

has perfect diffusion property. Due to Proposition 4, the iterated function $F^{(k)}$ is RS for $k \geqslant 2$. Let $f^{(k)} = f\left(F^{(k-1)}\right)$, then $F^{(k)} = \left(f^{(k)}, f^{(k)} \circ \rho_6^1, \ldots, f^{(k)} \circ \rho_6^5\right)$. By Proposition 1, the ANFs of the following Boolean functions can be obtained from the truth tables of f, $f^{(2)}$, $f^{(3)}$, $f^{(4)}$ respectively.

$f(x_1, \ldots, x_6)$

$= x_1x_2x_3x_4x_5x_6 \oplus x_1x_2x_3x_4x_5 \oplus x_2x_3x_4x_5x_6 \oplus x_1x_2x_3x_4 \oplus x_1x_2x_3x_5$

$\oplus\; x_1x_2x_4x_5 \oplus x_1x_2x_4x_6 \oplus x_1x_2x_5x_6 \oplus x_1x_3x_4x_6 \oplus x_2x_3x_4x_5 \oplus x_2x_3x_4x_6$

$\oplus\; x_2x_4x_5x_6 \oplus x_3x_4x_5x_6 \oplus x_1x_2x_5 \oplus x_1x_2x_6 \oplus x_1x_3x_4 \oplus x_1x_3x_6 \oplus x_1x_5x_6$

$\oplus\; x_3x_4x_5 \oplus x_3x_4x_6 \oplus x_4x_5x_6 \oplus x_4,$

$f^{(2)}(x_1, \ldots, x_6)$

$= x_1x_2x_3x_4x_5x_6 \oplus x_1x_2x_3x_5x_6 \oplus x_1x_2x_4x_5x_6 \oplus x_1x_2x_3x_5 \oplus x_1x_2x_3x_6$

$\oplus\; x_1x_2x_4x_5 \oplus x_1x_3x_4x_6 \oplus x_2x_4x_5x_6 \oplus x_3x_4x_5x_6 \oplus x_1x_2x_3$

$\oplus\; x_1x_3x_4 \oplus x_2x_4x_5 \oplus x_3x_4x_5 \oplus x_1,$

$f^{(3)}(x_1, \ldots, x_6)$

$= x_1x_2x_3x_4x_5x_6 \oplus x_1x_2x_3x_4x_5 \oplus x_1x_2x_3x_4x_6 \oplus x_1x_3x_4x_5x_6 \oplus x_2x_3x_4x_5x_6$

$\oplus\; x_1x_2x_3x_5 \oplus x_1x_2x_4x_5 \oplus x_1x_2x_4x_6 \oplus x_1x_2x_5x_6 \oplus x_1x_3x_4x_5 \oplus x_1x_3x_4x_6$

$\oplus\; x_2x_3x_4x_5 \oplus x_2x_4x_5x_6 \oplus x_1x_2x_5 \oplus x_1x_2x_6 \oplus x_1x_3x_4 \oplus x_1x_3x_6 \oplus x_1x_5x_6$

$\oplus\; x_2x_3x_4 \oplus x_3x_4x_6 \oplus x_4x_5x_6 \oplus x_4,$

$f^{(4)}(x_1, \ldots, x_6) = f^{(2)}(x_1, \ldots, x_6).$

The ANFs of the functions directly show that the RS $(6, 6)$-function F has perfect diffusion property, and for every $k \geqslant 1$, $\mathrm{Deg}\left(F^{(k)}\right) = 6$.

3.2   Almost Balanced (n, n)-Functions with Perfect Diffusion Property

We have presented a construction of RS (n, n)-functions with perfect diffusion property. These functions are of interest from a practical point of view as their representations are short and the evaluations are efficient. In this part, we propose a large set of almost balanced (n, n)-functions with perfect diffusion property. Here we call an (n, m)-function F *almost balanced*, if for every $b \in \mathbb{F}_{2^m}$, $|F^{-1}(b) - 2^{n-m}|$ takes a small value. For a finite set E with cardinality $|E| = N$, the set of all the permutations on E forms a symmetric group $\mathcal{S}_N$ whose group operation is the function composition.

Note that for $n \geqslant 2$, there is no balanced (n, n)-function (i.e., permutation on $\mathbb{F}_2^n$) with perfect diffusion property. In fact, let F be a permutation on $\mathbb{F}_2^n$, then since all the permutations on $\mathbb{F}_2^n$ form a finite symmetric group, there must exist some $i \geqslant 1$ such that $F^{(i)} = \mathrm{id}$, where id denotes the identity permutation. Hence, we have $F^{(i)}(x) = x$ for every $x \in \mathbb{F}_2^n$, which implies $\mathrm{D}_c\left(F^{(i)}\right) = 1/n < 1$. Thus, F cannot have perfect diffusion property. Therefore, finding almost balanced (n, n)-functions with perfect diffusion property is attractive.

Theorem 4. *For any σ that belongs to the symmetric group on the set $\mathbb{F}_2^n \setminus \{\mathbf{0}, \mathbf{1}\}$, the almost balanced (n, n)-function*

$$F(x) = \begin{cases} \mathbf{0}, & x = \mathbf{0} \text{ or } \mathbf{1}, \\ \sigma(x), & \text{otherwise}, \end{cases} \tag{11}$$

has perfect diffusion property, and for every $k \geqslant 1$, $\mathrm{Deg}\left(F^{(k)}\right) = n$.

Proof. From Eq. (11), one gets that for any $k \geqslant 1$,

$$F^{(k)}(x) = \begin{cases} \mathbf{0}, & x = \mathbf{0} \text{ or } \mathbf{1}, \\ \sigma^{(k)}(x), & \text{otherwise}. \end{cases}$$

Since $\sigma^{(k)}$ is a permutation on $\mathbb{F}_2^n \setminus \{\mathbf{0}, \mathbf{1}\}$, then it is easy to see that every coordinate function of $F^{(k)}$ has weight $2^{n-1} - 1$, which implies from Proposition 1 and Theorem 1 that $\mathrm{Deg}\left(F^{(k)}\right) = n$ and $F^{(k)}$ is non-degenerate. Therefore, F has perfect diffusion property. □

The following enumeration result is obvious.

Theorem 5. *The number of distinct almost balanced (n, n)-functions constructed in Theorem 4 is $\mathcal{P}_n = (2^n - 2)!$.*

Example 3. The number of almost balanced $(6, 6)$-functions with perfect diffusion property constructed in Theorem 4 is $\mathcal{P}_6 = (2^6 - 2)! \approx 2^{284}$, compared with the enumeration result in Example 1 that the number of RS $(6, 6)$-functions with perfect diffusion property constructed in Theorem 2 is $\mathcal{N}_6 \approx 2^{37.9}$.

Remark 4. Denote by $\mathscr{F}_n$, $\mathscr{G}_n$ the sets of all the (n, n)-functions constructed in Theorems 2 and 4 respectively. Then, it is easy to check that $\mathscr{F}_2 = \mathscr{G}_2$, $\mathscr{F}_3 \subseteq \mathscr{G}_3$, and for $n \geqslant 4$, $\mathscr{F}_n \bigcap \mathscr{G}_n \neq \emptyset$ but neither $\mathscr{F}_n \subseteq \mathscr{G}_n$ nor $\mathscr{G}_n \subseteq \mathscr{F}_n$.

As an application in product cryptosystems, we consider the following model.

Model. *Let G be an (n, n)-function, K_i, $i = 0, 1, \ldots$, be vectors in $\mathbb{F}_2^n$. Then, in a product cryptosystem, the ith round function F_i is*

$$F_i(x) = \begin{cases} G(x \oplus K_0), & \text{if } i = 1, \\ G(F_{i-1}(x) \oplus K_{i-1}), & \text{if } i \geqslant 2. \end{cases} \tag{12}$$

Suppose that $K_0 = K_1 = \cdots = K$, and we define $F(x) = G(x \oplus K)$. Then, by (12), we have for $i \geqslant 1$, $F_i(x) = F^{(i)}(x)$. The function F is preferable to have perfect diffusion property, which leads to $D_c(F_i) = 1$ for each $i \geqslant 1$. If the K_i's are not identical, then the case is more complicated. In the following, we use a simple example to illustrate that by using (n, n)-functions in (11), one can get $D_c(F_i) = 1$ for i odd. The example given here is suggestive if not very practical.

Example 4. In the above model, let

$$G(x) = \begin{cases} \mathbf{0}, & x = \mathbf{0} \text{ or } \mathbf{1}, \\ \sigma(x), & otherwise, \end{cases}$$

be an almost balanced function in (11), where σ is a permutation on $E = \mathbb{F}_2^n \backslash \{\mathbf{0}, \mathbf{1}\}$ satisfying $\{\mathbf{0}, \mathbf{1}\} \bigcup U(\sigma)$ is a $\mathbb{F}_2$-subspace of $\mathbb{F}_2^n$, where $U(\sigma) = \{x \in E \mid \sigma(x) = x\}$ is the set of fixed points of σ. Let K_{i-1}, F_i, $i \geqslant 1$, be defined in (12). We now prove that if $U(\sigma) \neq \emptyset$ and for $i \geqslant 1$, $K_i \in U(\sigma) \backslash A_i$, where $A_1 = \emptyset$ and

$$A_i = \left\{ \bigoplus_{j=1}^{k} K_{i-j}, \bigoplus_{j=1}^{k} K_{i-j} \oplus \mathbf{1} \;\middle|\; k = 1, \ldots, i-1 \right\}, \quad i \geqslant 2,$$

then $\mathrm{Deg}(F_i) = n$ and $D_c(F_i) = 1$ for all odd i. One can easily check that for $i \geqslant 2$, A_i is a set, i.e., all the elements in A_i are distinct.

For $i \geqslant 1$, let $f_l^{(i)}$ be the lth coordinate function of F_i, where $1 \leqslant l \leqslant n$. It is clear that $\mathrm{wt}\left(f_l^{(1)}\right) = 2^{n-1} - 1$ which is odd, then from the proof of Theorem 4, we have $\mathrm{Deg}(F_1) = n$ and $D_c(F_1) = 1$. Moreover, there exist exactly two $x \in \mathbb{F}_2^n$ such that $F_1(x) = \mathbf{0}$, and for each $y \in E \backslash A_1$, there exists exactly one $x \in \mathbb{F}_2^n$ such that $F_1(x) = y$. By calculating iteratively, one can get that for every $i \geqslant 2$ and every $k = 1, \ldots, i-1$, there exist exactly two $x \in \mathbb{F}_2^n$ such that $F_i(x) = \bigoplus_{j=1}^{k} K_{i-j}$ or $\mathbf{0}$, and for each $y \in E \backslash A_i$, there exists exactly one $x \in \mathbb{F}_2^n$ such that $F_i(x) = y$. Since for $i \geqslant 1$, $K_i \notin A_i$, then $K_i \oplus \mathbf{1} \in E \backslash A_i$, thus there exists exactly one $x \in \mathbb{F}_2^n$, denoted by $x_{i,0}$, such that $F_i(x_{i,0}) = K_i \oplus \mathbf{1}$, which implies $G(F_i(x_{i,0}) \oplus K_i) = \mathbf{0}$. Recall that $\{\mathbf{0}, \mathbf{1}\} \bigcup U(\sigma)$ is a $\mathbb{F}_2$-subspace of $\mathbb{F}_2^n$, then for $i \geqslant 1$, since $K_i \in U(\sigma)$, we have $A_i \subseteq U(\sigma)$. From the above discussion, we obtain that for $i \geqslant 1$, the multiset $\{* \; G(F_i(x) \oplus K_i) \mid x \in \mathbb{F}_2^n \backslash \{x_{i,0}\} \; *\}$ is equal to the multiset $\{* \; F_i(x) \oplus K_i \mid x \in \mathbb{F}_2^n \backslash \{x_{i,0}\} \; *\}$. Hence, for $i \geqslant 1$, denote by $g_l^{(i)}$ the lth coordinate function of $F_i(x) \oplus K_i$, where $1 \leqslant l \leqslant n$, then we have

$$\mathrm{wt}\left(f_l^{(i+1)}\right) = \mathrm{wt}\left(g_l^{(i)}\right) - 1.$$

It is obvious that $\mathrm{wt}\left(f_l^{(i)}\right)$ and $\mathrm{wt}\left(g_l^{(i)}\right)$ have the same parity, which leads to that $\mathrm{wt}\left(f_l^{(i)}\right)$ and $\mathrm{wt}\left(f_l^{(i+1)}\right)$ have different parities. Therefore, if $i \geqslant 1$ is odd, then $\mathrm{wt}\left(f_l^{(i)}\right)$ is odd, thus we have $\mathrm{Deg}(F_i) = n$ and $D_c(F_i) = 1$.

4 Concluding Remarks

In this paper, we construct two classes of (n, n)-functions with perfect diffusion property and optimal algebraic degree. These functions provide complete diffusion after iterations. The enumeration results for the constructed functions show that there are many (n, n)-functions which have perfect diffusion property.

The functions constructed in Theorems 2 and 4 represent a theoretical interest, which may have weak resistance to different cryptanalysis. Further improvements in the design of (n, n)-functions with perfect diffusion property are of interest. In addition, the RS (n, n)-functions defined in this paper may be worth discussing in the future for their efficient evaluations and short representations.

Acknowledgments. The authors would like to thank the anonymous referees for their helpful comments.

References

1. Bard, G.V.: Algebraic Cryptanalysis. Springer, New York (2009)
2. Biham, E., Shamir, A.: Differential cryptanalysis of DES-like cryptosystems. J. Cryptol. **4**(1), 3–72 (1991)
3. Carlet, C.: Boolean functions for cryptography and error correcting codes. In: Crama, Y., Hammer, P. (eds.) Boolean Models and Methods in Mathematics, Computer Science, and Engineering, pp. 257–397. Cambridge University Press, London (2010)
4. Carlet, C.: Vectorial Boolean functions for cryptography. In: Crama, Y., Hammer, P. (eds.) Boolean Models and Methods in Mathematics, Computer Science, and Engineering, pp. 398–469. Cambridge University Press, London (2010)
5. Courtois, N.T., Pieprzyk, J.: Cryptanalysis of block ciphers with overdefined systems of equations. In: Zheng, Y. (ed.) ASIACRYPT 2002. LNCS, vol. 2501, pp. 267–287. Springer, Heidelberg (2002)
6. Daemen, J.: Cipher and hash function design strategies based on linear and differential cryptanalysis. Ph.D. thesis, Catholic University of Louvain (1995)
7. Feistel, H.: Cryptography and computer privacy. Sci. Am. **228**(5), 15–23 (1973)
8. Fischer, S., Meier, W.: Algebraic immunity of S-boxes and augmented functions. In: Biryukov, A. (ed.) FSE 2007. LNCS, vol. 4593, pp. 366–381. Springer, Heidelberg (2007)
9. Forré, R.: Methods and instruments for designing S-boxes. J. Cryptol. **2**(3), 115–130 (1990)
10. Kam, J.B., Davida, G.I.: Structured design of substitution-permutation encryption networks. IEEE Trans. Comput. **C–28**(10), 747–753 (1979)
11. Lidl, R., Niederreiter, H.: Finite Fields. Cambridge University Press, New York (1997)
12. Matsui, M.: Linear cryptanalysis method for DES cipher. In: Helleseth, T. (ed.) EUROCRYPT 1993. LNCS, vol. 765, pp. 386–397. Springer, Heidelberg (1994)
13. Maximov, A.: Classes of plateaued rotation symmetric Boolean functions under transformation of Walsh spectra. Cryptology ePrint Archive, Report 2004/354 (2004). https://eprint.iacr.org/2004/354

14. Pieprzyk, J., Qu, C.X.: Fast hashing and rotation-symmetric functions. J. Univers. Comput. Sci. **5**(1), 20–31 (1999)
15. Preneel, B., Bosselaers, A., Rijmen, V., et al.: Comments by the NESSIE project on the AES finalists (2000). http://www.nist.gov/aes
16. Shannon, C.E.: Communication theory of secrecy systems. Bell Syst. Tech. J. **28**(4), 656–715 (1949)

Information Theory

Shannon Entropy Versus Renyi Entropy from a Cryptographic Viewpoint

Maciej Skórski[✉]

Cryptology and Data Security Group, University of Warsaw, Warsaw, Poland
maciej.skorski@mimuw.edu.pl

Abstract. We provide a new inequality that links two important entropy notions: Shannon Entropy H_1 and collision entropy H_2. Our formula gives the *worst possible* amount of collision entropy in a probability distribution, when its Shannon Entropy is fixed. While in practice it is easier to evaluate Shannon entropy than other entropy notions, it is well known in folklore that it does not provide a good estimate of randomness quality from a cryptographic viewpoint, except very special settings. Our results and techniques put this in a quantitative form, allowing us to precisely answer the following questions:

(a) How accurately does Shannon entropy estimate uniformity? Concretely, if the Shannon entropy of an n-bit source X is $n - \epsilon$, where ϵ is a small number, can we conclude that X is close to uniform? This question is motivated by uniformity tests based on entropy estimators, like Maurer's Universal Test.

(b) How much randomness can we extract having high Shannon entropy? That is, if the Shannon entropy of an n-bit source X is $n - O(1)$, how many almost uniform bits can we retrieve, at least? This question is motivated by the folklore upper bound $O(\log(n))$.

(c) Can we use high Shannon entropy for key derivation? More precisely, if we have an n-bit source X of Shannon entropy $n - O(1)$, can we use it as a secure key for some applications, such as square-secure applications? This is motivated by recent improvements in key derivation obtained by Barak et al. (CRYPTO'11) and Dodis et al. (TCC'14), which consider keys with some entropy deficiency.

Our approach involves convex optimization techniques, which yield the shape of the "worst" distribution, and the use of the Lambert W function, by which we resolve equations coming from Shannon Entropy constraints. We believe that it may be useful and of independent interests elsewhere, particularly for studying Shannon Entropy with constraints.

Keywords: Shannon entropy · Renyi entropy · Smooth renyi entropy · Min-entropy · Lambda w function

A full version of this paper is available at https://eprint.iacr.org/2014/967.pdf.

M. Skórski — This work was partly supported by the WELCOME/2010-4/2 grant founded within the framework of the EU Innovative Economy Operational Programme.

J. Groth (Ed.): IMACC 2015, LNCS 9496, pp. 257–274, 2015.
DOI: 10.1007/978-3-319-27239-9_16

1   Introduction

1.1   Entropy Measures

Entropy, as a measure of randomness contained in a probability distribution, is a fundamental concept in information theory and cryptography. There exist many entropy definitions and they are not equally good for all applications. While the most famous (and most liberal) Shannon Entropy [Sha48], which quantifies the encoding length, is extremely useful in information theory, more conservative measures, like min-entropy (which quantifies unpredictability) or collision entropy (which bounds collision probability between samples), are necessary in cryptographic applications, like extracting randomness [NZ96, HILL99, RW05] or key derivation [DY13, BDK+11]. Any misunderstanding about what is a suitable entropy notion may be a serious problem not only of a theoretical concern, because it leads to vulnerabilities due to overestimating security. In fact, when entropy is overestimated, security of real-world applications can be broken [DPR+13]. Standards [BK12, AIS11] recommend to use more conservative entropy metrics in practical designs, but in the other hand Shannon entropy is easier to evaluate [AIS11] (in particular when the distribution of the randomness source is not exactly known) and moreover Shannon entropy estimators have already been relatively well studied and are being used in practice [Mau92, Cor99, BL05, LPR11].

1.2   Motivations and Goals of this Work

The aim of this paper is to provide sharp separation results between Shannon entropy and Renyi entropy (focusing on collision entropy and min-entropy). Under certain conditions, for example when consecutive bits of a given random variable are independent (produced by a memoryless source), they are comparable [RW05, Hol11] (this observation is closely related to a result in information theory known as the Asymptotic Equipartition Property [Cac97]). Such a simplifying assumption is used to argue about provable security of true random number generators [BKMS09, VSH11, LPR11], and may be enforced in certain settings, for example when certifying devices in a laboratory [BL05]. But in general (especially from a theoretical viewpoint) neither min-entropy (being of fundamental importance for general randomness extraction [RW05, Sha11]) nor collision entropy, useful for key derivation [DY13, BDK+11, Shi15], randomness extraction [HILL99], and random number generating [BKMS09, BST03]) *cannot* be well estimated by Shannon entropy. Still, in practice Shannon entropy remains an important tool for testing cryptographic quality of randomness [AIS11]. In this paper we address the natural question

> How bad is Shannon entropy as an estimate of cryptographic quality of randomness?

and answer it in a series of bounds, focusing on three important cryptographic applications, which require entropy estimation: (a) uniformity testing, (b) general randomness extraction and (c) key derivation.

1.3   Our Results and Techniques

Brief Summary. We investigate in details the gap between Shannon Entropy and Renyi Entropy (focusing on smooth collision entropy and smooth min-entropy) in a given entropy source. We impose no restrictions on the source and obtain general and tight bounds, identifying the worst case. Our results are mostly negative, in the sense that the gap may be very big, so that even almost full Shannon Entropy does not guarantee that the given distribution is close to uniform or that it may used to derive a secure key. This agrees with folklore. However, to the best of our knowledge, our analysis for the first time provides a comprehensive and detailed study of this problem, establishing tight bounds. Moreover, our techniques may be of independent interests and can be extended to compare Renyi entropy of different orders.

Results and Corollaries. *Bounding Renyi Entropy by Shannon Entropy.* Being interested in establishing a bound on the amount of extractable entropy in terms of Shannon Entropy only, we ask the following question

Q: Suppose that the Shannon Entropy $H_1(X)$ of an n-bit random variable X is at least k. What is the best lower bound on the collision entropy $H_2(X)$?

We give a complete answer to this question in Sect. 3.1. It is briefly summarized in Table 1 below.

Table 1. Minimal collision entropy given Shannon entropy constraints.

Domain of X	$H_1(X)$	Region	Max. ℓ_2-distance to uniform	Min. value of $H_2(X)$
$\{0,1\}^n$	$n - \Delta$	$2^n \Delta \geqslant 13$	$\Theta\left(\frac{\Delta}{\log(2^n \Delta)}\right)$	$n - \log_2\left(1 + \Theta\left(2^n \Delta^2 \log^{-2}(2^n \Delta)\right)\right)$
		$2^n \Delta \leqslant 13$	$O\left(\Delta\right)$	$n - \log_2\left(1 + O\left(2^n \Delta^2\right)\right)$

The Shape of the Worst-case Distribution. Interestingly, the description of the "worst" distribution X is pretty simple: it is a combination of a one-point heavy mass with a flat distribution outside. In fact, it has been already observed in the literature that such a shape provides good separations for Shannon Entropy [Cac97]. However, as far as we know, our paper is the first one which provides a full proof that this shape is really best possible.

Infeasibility of Uniformity Tests Based on Entropy Estimators. If an n-bit random variable X satisfies $H_1(X) = n$ then it must be uniform. It might be tempting to think that a very small entropy gap $\Delta = n - H_1(X)$ (when entropy is very "condensed") implies closeness to the uniform distribution. Clearly, this is a necessary condition. For example, standards for random number generating [AIS11] require the Shannon entropy of raw bits to be at least 0.997 per bit on average.

Q: Suppose that the Shannon Entropy $H_1(X)$ of an n-bit random variable X is at least $n - \Delta$, where $\Delta \approx 0$. What is the best upper bound on the distance between X and the uniform distribution U_n?

There are popular statistical randomness tests [Mau92, Cor99] which are based on the fact that very small Δ is necessary to a very small statistical distance. Theoretically, they can detect any deviation at any confidence level. In this paper we quantify what is well known in folklore, namely that this approach *cannot be provable secure and efficient* at the same time. Based on the results summarized in Table 1, we prove that for the statistical distance (the ℓ_1 distance) the gap Δ can be as small as ϵ but still the source is ϵ/n-far from the uniform distribution. Putting this statement around, to guarantee ϵ-closeness we need to estimate the entropy up to a tiny margin $n\epsilon$. This shows that an application of entropy estimators to test sequences of truly random bits may be problematic, because estimating entropy within such a small margin is computationally inefficient. Having said this, we stress that entropy estimators like Maurer's Universal Test [Mau92] are powerful tools capable of discovering most of defects which appear within a broader margin of error.

Large Gap Between Shannon and Smooth Collision Entropy. Many constructions in cryptography require min-entropy. However, the weaker notion of collision entropy found also many applications, especially for problems when one deals with imperfect randomness. The collision entropy of a distribution X constitutes a lower bound on the number of extractable almost-uniform bits, according to the Leftover Hash Lemma [HILL99, RW05]. Moreover, the recent improvements in key derivation [DY13, BDK+11] show that for some applications we can use high collision entropy to generate secure keys, wasting much less entropy comparing to extractors-based techniques (see Sect. 2.5). For example, consider the one-time MAC with a 160-bit key over $GF(2^{80})$, where the key is written as (a, b) and the tag for a message x is $ax + b$. The security is $\epsilon = 2^{-80}$ when the key is uniform [DY13]. We also know that it is $\epsilon = 2^{-70}$-secure when the key has $150 = 160 - 10$ bits of collision entropy. Suppose that a Shannon entropy estimator indicates 159 bits of entropy. Is our scheme secure? This discussion motivates the following question

Q: Suppose that the Shannon Entropy $H_1(X)$ of a random variable $X \in \{0, 1\}^n$ is at least $n - \Delta$ where $\Delta \leqslant 1$. What is the best lower bound on $H_2(X)$? Does it help if we consider only $H_2(X')$ where X' is close to X?

As a negative result, we demonstrate that the gap between the Shannon Entropy and Renyi Entropy could be almost as big as the length of the entropy source output (that is almost maximal possible). Moreover, smoothing entropy, even with weak security requirements, does not help. For example, we construct a 256-bit string of more than 255 bits of Shannon Entropy, but only 19 bits of (smooth) Renyi entropy. This is just an illustrative example, we provide a more general analysis in Corollary 4 in Sect. 4.

Large Gap Between Shannon and Extractable Entropy. Min entropy gives only a lower bound on extractable entropy. However, its smooth version can be used to establish an upper bound on the amount of almost random bits, of required quality, that can be extracted from a given source [RW05].

Q: Suppose that the Shannon Entropy $H_1(X)$ of a random variable $X \in \{0,1\}^n$ is at least $n - \Delta$ where $\Delta < 1$. How many bits that are close to uniform can be extracted from X?

Again, analogously to the previous result, we provide a separation between Shannon and extractable entropy, where the gap is almost as big as the length of the random variable. For example, we construct a 256-bit string of more than 255.5 bits of Shannon Entropy, but only 10 bits of extractable entropy, even if we allow them to be of very weak quality, not really close to uniform! This is just an illustrative example, we provide a more precise and general statement. To our knowledge, the concrete tight bounds we provide are new, though a similar "extreme" numerical example can be found in [Cac97]. The separation is again a straightforward application of ideas behind the proof of the results in Table 1

Converting Shannon Entropy into Renyi Entropy. Even though the gap in our separations are almost as big as the length of the source output, there might be small amount of Renyi Entropy in every distribution of high Shannon Entropy.

Q: Suppose that the Shannon Entropy of an n-bit random variable X is at least $n - \Delta$ where $\Delta \geqslant 1$. Does X have some non-trivial amount of collision entropy?

This question may be relevant in settings, when one would like to check whether some (not really big though) collision entropy is present in the source. For example, there are necessary conditions on security of message authentication codes in terms of collision entropy [Shi15]. We establish a simple and tight bound on this amount: it is about $2\log_2 n - 2\log_2 \Delta$. For example, in the concrete case of a 256-bit string of Shannon Entropy 255 we find that the necessary amount of Renyi entropy is 15. We also establish an interesting rule of thumb: for much more than one bit of Renyi entropy in the output of a source, its Shannon Entropy must be bigger than the half of its length. Again, besides this numerical example we provide detailed and general bounds.

Techniques. To prove our main technical results, we use standard convex optimization techniques combined with some calculus which allows us to deal with implicit equations. In particular, we demonstrate that the Lambert-W function is useful in studying Shannon Entropy constraints.

1.4 Organization of the Paper

We start with necessary definitions and explanations of basic concepts in Sect. 2. Our main result is discussed in Sect. 3. Further applications are given in Sect. 4. We end with the conclusion in Sect. 5. The proofs of main results, which are technical and complicated a bit, appear in Sect. 5.

2 Preliminaries

2.1 Basic Notions

By U_S we denote the uniform distribution over a set S, and U_n is a shortcut for the uniform n-bit distribution. The probability mass function of a random variable X is denoted by P_X.

2.2 Quantifying Closenes of Distributions

The closeness of two distributions X, Y over the same domain Ω is most commonly measured by the so called statistical or variational distance $\mathrm{SD}(X; Y)$. It is defined as the half of the ℓ_1-distance between the probability mass functions $\mathrm{SD}(X; Y) = \frac{1}{2} \mathrm{d}_1(P_X; P_Y) = \frac{1}{2} \sum_x |\Pr[X = x] - \Pr[Y = x]|$. In this paper we use also the ℓ_2-distance between probability distributions, defined as $\mathrm{d}_2(P_X; P_Y) = \sqrt{\sum_x (\Pr[X = x] - \Pr[Y = x])^2}$. These two ℓ_p distances are related by $\mathrm{d}_2(\cdot) < \mathrm{d}_1(\cdot) \leqslant \sqrt{|\Omega|} \cdot \mathrm{d}_2(\cdot)$. In information theory the closeness of two distributions is often measures using so called *divergences*. The Kullback-Leibler divergence between X and Y is defined as $\mathrm{KL}(X \parallel Y) = -\sum_x P_X(x) \log \frac{P_X(x)}{P_Y(x)}$, and the Renyi divergence of order 2 equals $D_2(X \parallel Y) = \sum_x \frac{(P_X(x) - P_Y(x))^2}{P_X(x)}$. We have $D_2(X \parallel U_S) = H_2(U_S) - H_2(X) = \log_2(|S|\mathrm{CP}(X))$.

For convenience we define also the collision probability of X as the probability that two independent copies of X collide: $\mathrm{CP}(X) = \sum_x \Pr[X = x]^2$.

2.3 Entropy Definitions

Below we define the three key entropy measures, already mentioned in the introduction. It is worth noting that they all are special cases of a much bigger parametrized family of Renyi entropies. However the common convention in cryptography, where only these three matter, is to slightly abuse the terminology and to refer to collision entropy when talking about Renyi entropy, keeping the names for Shannon and Min-Entropy.

Definition 1 (Entropy Notions). *The Shannon Entropy* $\mathrm{H}(X) = H_1(X)$, *the collision entropy (or Renyi entropy)* $H_2(X)$, *and the Min-Entropy* $\mathrm{H}_\infty(X)$ *of a distribution* X *are defined as follows*

$$H(X) = \sum_x \Pr[X = x] \log \Pr[X = x] \tag{1}$$

$$H_2(X) = -\log\left(\sum_x \Pr[X = x]^2\right) \tag{2}$$

$$\mathrm{H}_\infty(X) = -\log \max_x \Pr[X = x]. \tag{3}$$

Remark 1 (Comparing Different Entropies). It is easy to see that we have

$$\mathrm{H}(X) \geqslant H_2(X) \geqslant \mathrm{H}_\infty(X),$$

with the equality if and only if X is uniform.

2.4   Entropy Smoothing

THE CONCEPT. Entropy Smoothing is a very useful concept of replacing one distribution by a distribution which is very close in the statistical distance (which allows keeping its most important properties, like the amount of extractable entropy) but more convenient for the application at hand (e.g. a better structure, removed singularities, more entropy).

APPLICATIONS OF SMOOTH ENTROPY. The smoothing technique is typically used to *increase entropy* by cutting off big but rare "peaks" in a probability distribution, that is point masses relatively heavy comparing to others. Probably the most famous example is the so called Asymptotic Equipartition Property (AEP). Imagine a sequence X of n independent Bernoulli trials, where 1 appears with probability $p > 1/2$. Among all n-bit sequences the most likely ones are those with 1 in almost all places. In particular $H_\infty(X) = -n \log p$. However, for most of the sequences the number of 1's oscillates around pn (these are so called typical sequences). By Hoeffding's concentration inequality, the number of 1's is at most $pn + h$ with probability $1 - \exp(-2h^2/n)$. For large n and suitably chosen h, the distribution of X approaches a distribution X' of min-entropy $H_\infty(X') \approx -n(p \log p + (1 - p) \log(1 - p)) \approx H(X)$ (the relative error here is of order $O(n^{-1/2})$), much larger than the min-entropy of the original distribution! A quantitative version of this fact was used in the famous construction of a pseudorandom generator from any one-way function [HILL88]. Renner and Wolf [RW04] revisited the smoothing technique in entropy framework and came up with new applications.

Definition 2 (Smooth Entropy, *[RW04]*). *Suppose that $\alpha \in \{1, 2, \infty\}$. We say that the ϵ-smooth entropy of order α of X is at least k if there exists a random variable X' such that $SD(X; X') \leqslant \epsilon$ and $H_\alpha(X') \geqslant k$.*

For shortness, we also say smooth Shannon Entropy, smooth Renyi entropy or smooth min-entropy. We also define the *extractable entropy* of X as follows

Definition 3 (Extractable Entropy, *[RW05]*). *The ϵ-extractable entropy of X is defined to be*

$$H_{\text{ext}}^\epsilon(X) = \max_{\mathcal{U}:\ \exists f \in \Gamma^\epsilon(X \to \mathcal{U})} \log |\mathcal{U}| \qquad (4)$$

where $\Gamma^\epsilon(X \to \mathcal{U})$ consists of all functions f such that $SD(f(X, R); U_\mathcal{U}, R) \leqslant \epsilon$ where R is uniform and independent of X and $U_\mathcal{U}$.

2.5   Randomness Extraction and Key Derivation

Roughly speaking, an extractor is a randomized function which produces an almost uniform string from a longer string but not of full entropy. The randomization here is necessary if one wants an extractor working with all high-entropy sources; the role of that auxiliary randomness is similar to the purpose of catalysts in chemistry.

Definition 4 (Strong Extractors [NZ96]). *A strong (k, ϵ)-extractor is a function* $\mathrm{Ext} : \{0,1\}^n \times \{0,1\}^d \to \{0,1\}^k$ *such that*

$$\mathrm{SD}(\mathrm{Ext}(X, U_d), U_d; U_{k+d}) \leqslant \epsilon. \tag{5}$$

A very simple, efficient and optimal (with respect to the necessarily entropy loss) extractor is based on universal hash functions. Recall that a class $\mathcal{H}$ of functions from n to m bits is universal [CW79] if for any different x, y there are exactly $|\mathcal{H}|/2^m$ functions $h \in \mathcal{H}$ such that $h(x) = h(y)$.

Lemma 1 (Leftover Hash Lemma). *Let $\mathcal{H}$ be a universal class of functions from n to random m bits, let H be chosen from $\mathcal{H}$ at random and let X be an n-bit variable. If $H_2(X) \geqslant k$, then $\mathrm{SD}(H(X), H; U_m, H) \leqslant \frac{1}{2} \cdot 2^{\frac{m-k}{2}}$.*

By Lemma 1 and the properties of the statistical distance we obtain

Corollary 1 (Bound on Extractable Entropy, [RW05]). *We have $H_\infty^\epsilon(X) \geqslant H_{\mathrm{ext}}^\epsilon(X) \geqslant H_2^{\epsilon/2}(X) - 2\log(1/\epsilon) - 1$.*

Note that to extract k bits ϵ-close to uniform we need to invest $k + 2\log(1/\epsilon)$ bits of (collision) entropy; the loss of $2\log(1/\epsilon)$ bits here is optimal [RTS00]. While there are many other extractors, the Leftover Hash Lemma is particularly often used in the TRNG design [BST03, BKMS09, VSH11] because it is simple, efficient, and provable secure. Extractors based on the LHL are also very important in key derivation problems [BDK+11]. Note that the LHL uses only collision entropy, weaker than min-entropy.

To get an illustrative example, note that deriving a key which is ϵ-close to uniform with $\epsilon = 2^{-80}$ requires losing $L = 2\log(1/\epsilon) = 160$ bits of entropy. Sometimes we can't afford to lose so much. In special cases, in particular for so called *square-friendly applications* [BDK+11, DY13] we can get an improvement over Corollary 1. In particular, for these applications (which include message authentication codes or digital signatures), we can apply X of collision entropy $k < m$, still achieving some non-trivial security.

Theorem 1 (Beating the $2\log(1/\epsilon)$ Entropy Loss for Some Applications. [BDK+11]). *Let P be an ϵ-secure and σ-square-secure application (when keyed with U_m). Let $\mathcal{H}$ be a universal class of functions from n to random m bits, let H be chosen from $\mathcal{H}$ at random. Then for any X of length n and collision-entropy k, the application P keyed with $H(X)$ given H is ϵ'-secure when $\epsilon' \leqslant \epsilon + \sqrt{\sigma} \cdot 2^{\frac{m-k}{2}}$.*

In particular, when $\sigma = \epsilon$ we get around of the RT-bound, achieving $\epsilon' \approx 2\epsilon$ with only $k = m + \log(1/\epsilon)$. This way we save $\log(1/\epsilon) \approx 80$ bits.

3   Main Result

In this section we calculate what is the minimal collision entropy in a distribution having a certain amount of Shannon entropy. First, by means of convex optimization, we show in Sect. 3.1 that the uniform distribution with one extra heavy mass is the "worst" shape. Next, using some facts about Lambert W function, in Sect. 3.2 we solve the corresponding implicit equation and derive a closed-form answer.

3.1 Maximizing Collisions Given Shannon Entropy

Below we answer the posted question on the best bound on H_2 in terms of H_1. The "worst case" distribution, which minimizes the gap, is pretty simple: it is a combination of a one-point mass at some point and a uniform distribution outside.

Theorem 2. *Let X be a random variable with values in a d-element set. If $H(X) = k$, then*

$$H_2(X) \geqslant -\log_2 \left(b^2 + \frac{(1-b)^2}{d-1} \right) \tag{6}$$

where b is the unique solution to

$$H(b) + (1-b)\log_2(d-1) = k \tag{7}$$

under the restriction $b \geqslant \frac{1}{d}$ ($H(b)$ denotes the entropy of a bit equal 1 with probability b). The bound in Eq. (6) is best possible.

Remark 2 (The Implicit Equation in Theorem 2). The number b is defined nondirectly depending on d and k. In Sect. 3.2, we will show how to accurately approximate the solution of this equation.

The proof of Theorem 2 appears in Appendix A. The main idea is to write down the posted question as a constrained optimization problem and apply standard Lagrange multipliers techniques.

3.2 Closed-Form Bounds for Solutions

Below we present a tight formula approximating the solution to Eq. (7). We will substitute it to Eq. (6) in order to obtain a closed-form expression.

Lemma 2 (The solution for Moderate Gaps). *Let b be the solution to Eq. (7) and let $\Delta = \log_2 d - k$ be the entropy gap. Suppose $d\Delta \geqslant 13$. Then we have*

$$\frac{0.84\Delta}{\log_2(d\Delta) - 1.52} \leqslant b \leqslant \frac{1.37\Delta}{\log_2(d\Delta) - 1.98} \tag{8}$$

The proof is referred to Appendix B. The main idea is to solve Eq. (8) approximately using the so called Lambert W function, that matches Shannon-like expressions of the form $y \log y$. Here we discuss the lemma and its applications.

Remark 3 (Establishing Tighter Constants). The inspection of the proof shows that the numerical constants in Lemma 2 can be made sharper, if needed. Under the mild assumption that $\Delta^{-1} = 2^{o(\log_2 d)}$ one can get

$$b = \frac{(1 + o(1))\Delta}{\log_2(d\Delta) - \log_2 e - \log_2 \log_2 e + o(1)} \qquad (9)$$

The gap between 1.52 and 1.98 is self-improving, in the sense that knowing in advance a better upper bound on b one can make it closer to 0. In turn, the gap between 0.84 and 1.37 can be made closer to 0 by choosing in the proof a more accurate approximation for the Lambert W function.

Now we are ready to compute minimal collision entropy given Shannon Entropy.

Corollary 2 (Minimal Collision Entropy, General Case). *Let X^* minimizes $H_2(X)$ subject to $H(X) \geqslant n - \Delta$ where X takes its values in a given d-element set. If $d\Delta \geqslant 13$ then*

$$\frac{0.55\Delta}{\log_2(d\Delta)} \leqslant d_2(X^*; U) \leqslant \frac{3.24\Delta}{\log_2(d\Delta)}, \qquad (10)$$

where U is uniform over the domain of X. If $d\Delta < 13$ then

$$d_2(X^*; U) < 0.88\Delta. \qquad (11)$$

The collision entropy is obtained as $H_2(X^*) = -\log_2\left(\frac{1}{d} + d_2(X^*; U)^2\right)$.

Proof (Proof of Corollary 2). We will consider two cases.

Case I: $d\Delta \geqslant 13$. By Lemma 2 we get

$$\frac{0.84\Delta}{\log_2(d\Delta)} \leqslant b \leqslant \frac{2.95\Delta}{\log_2(d\Delta)} \qquad (12)$$

By the last inequality and the fact that $x \to \frac{x}{\log_2 x}$ is increasing for $x \geqslant e$ we get

$$bd \geqslant \frac{0.84 d\Delta}{\log_2(d\Delta)} \geqslant 2.95$$

Let $b_0 = \frac{1}{d}$. By the last inequality we get $b - b_0 \geqslant 0.66b$. Since

$$b^2 + \frac{(1-b)^2}{d-1} = b_0 + \frac{d}{d-1} \cdot (b - b_0)^2,$$

by the identity $d_2(X; U)^2 = \sum_x \Pr[X = x]^2 - \frac{1}{d}$ and the definition of collision entropy we get

$$d_2(X^*, U)^2 = CP(X^*) - b_0 = \frac{d}{d-1} \cdot (b - b_0)^2.$$

Note that $d\Delta \geqslant 13$ implies $d\log_2 d \geqslant 13$ (because $\Delta \leqslant \log_2 d$) and hence $d > 5$. By this inequality and $b - b_0 \geqslant 0.66b$ we finally obtain

$$0.43b^2 \leqslant d_2(X^*; U)^2 \leqslant 1.2b^2 \tag{13}$$

and the result for the case $d\Delta \geqslant 13$ follows by combining Eqs. (12) and (13).

Case II: $d\Delta < 13$. We do a trick to "embed" our problem into a higher dimension. If $\boldsymbol{p} \in \mathbb{R}^d$ is the distribution of X, define $\boldsymbol{p}' \in \mathbb{R}^{d+1}$ by $p'_i = (1 - \gamma)p'_i$ for $i \leqslant d$ and $p'_{d+1} = \gamma$. It is easy to check that $H_1(\boldsymbol{p}') = -(1 - \gamma)\log_2(1 - \gamma) - \gamma\log_2\gamma + (1 - \gamma)H_1(\boldsymbol{p})$. Setting $\gamma = \frac{1}{1+2^{H_1(\boldsymbol{p})}}$ we get

$$
\begin{aligned}
H_1(\boldsymbol{p}') - H_1(\boldsymbol{p}) &= -(1 - \gamma)\log_2(1 - \gamma) - \gamma\log_2\gamma - \gamma H_1(\boldsymbol{p}) \\
&\quad - (1 - \gamma)\log_2(1 - \gamma) - \gamma\log_2\left(2^{H_1(\boldsymbol{p})}\gamma\right) \\
&= \log_2\frac{2^{H_1(\boldsymbol{p})} + 1}{2^{H_1(\boldsymbol{p})}} \\
&\geqslant \log_2\frac{d + 1}{d} \\
&\geqslant (1 - b)\log_2\frac{d}{d - 1}
\end{aligned}
$$

where the first inequality follows by $H_1(\boldsymbol{p}) \leqslant \log_2 d$, and the second inequality follows because $b \geqslant \frac{1}{d}$ implies that is suffices to prove $\log_2\frac{d+1}{d} \geqslant \left(1 - \frac{1}{d}\right)\log_2\frac{d}{d-1}$ or equivalently that $d\log_2\frac{d+1}{d} \geqslant (d - 1)\log_2\frac{d}{d-1}$; this is true because the map $u \to u\log_2(1 + u^{-1})$ is increasing in u for $u > 0$ (we skip an easy argument, which simply checks the derivative). Since $H_1(\boldsymbol{p}') - H_1(\boldsymbol{p}) = 0$ for $\gamma = 0$ and since $H_1(\boldsymbol{p}') - H_1(\boldsymbol{p}) \geqslant (1 - b)\log_2\frac{d}{d-1} > (1 - b)\log_2\frac{d+1}{d}$ for $\frac{1}{1+2^{H_1(\boldsymbol{p})}}$ for $\geqslant (1 - b)\log_2\frac{d}{d-1}$, by continuity we conclude that there exists $\gamma = \gamma_b$, between 0 and $\frac{1}{1+2^{H_1(\boldsymbol{p})}}$, such that $\boldsymbol{p}'$ satisfies

$$(1 - b)\log_2\frac{d + 1}{d} = H_1(\boldsymbol{p}') - H_1(\boldsymbol{p}).$$

Adding this Eq. (7) by sides, we conclude that also b solves 7 with the dimension d replaced by $d' = d + 1$ and the constraint k replaced by $k' = H_1(\boldsymbol{p}')$. By $H_1(\boldsymbol{p}') - H_1(\boldsymbol{p}) \geqslant \log_2\frac{d+1}{d}$ we conclude that $\Delta' = \log_2(d + 1) - H_1(\boldsymbol{p}') \leqslant \log_2 d - H_1(\boldsymbol{p}) = \Delta$ so the entropy gap is even smaller. After a finite number of step, we end with $\Delta' \leqslant \Delta$, the same b and $d'\Delta' \geqslant 13$. Then by the first case we get that the squared distance is at most $O(\Delta'^2) = O(\Delta^2)$.

4 Applications

4.1 Negative Results

The first result we provide is motivated by uniformness testing based on Shannon entropy. We hope that n-bit distribution with entropy $n - \Delta$ where $\Delta \approx 0$, that is with an extremely small entropy defficency, is close to uniform. We show that for this to be true, Δ has to be negligible.

Corollary 3 (Shannon Entropy Estimators are Inefficient as Uniformity Tests). *Suppose that $n \gg 1$ and $\epsilon > 2^{-0.9n}$. Then there exists a distribution $X \in \{0,1\}^n$ such that $H_1(X) \geqslant n - \epsilon$ but $\mathrm{SD}(X; U_n) = \Omega(\epsilon/n)$.*

Remark 4. Note that typically one estimates Shannon Entropy within an additive error $O(1)$. However here, to prove that the distribution is ϵ-close to uniform, one has to estimate the entropy with an error $O(n\epsilon)$, which is much tighter! The best known bounds on the running time for an additive error $O(\epsilon)$ are polynomial in ϵ [AOST14,Hol06][1]. With ϵ secure (meaning small) enough for cryptographic purposes, such a precision is simply not achievable within reasonable time.

Proof (Proof of Corollary 3). Take $d = 2^n$ in Corollary 2 and $\Delta = \epsilon$. Suppose that $\Delta = \Omega(2^{-0.9n})$. We have $d_2(X; U_n) = \Theta(\Delta n^{-1})$. In the other hand we have the trivial inequality $d_2(X; U_n) \leqslant 4 \cdot \mathrm{SD}(X; U_n)$ (which is a consequence of standard facts about ℓ_p-norms) and the result follows.

Corollary 4 (Separating Smooth Renyi Entropy and Shannon Entropy). *For any n, δ such that $2^{-n} < \delta < \frac{1}{6}$, there exists a distribution $X \in \{0,1\}^n$ such that $\mathrm{H}(X) \geqslant (1 - 2\delta)n + \log_2(1 - 2^{-n})$, $H_2(X) \leqslant 2\log_2(1/\delta) - 2$ and $\mathrm{H}_2^\epsilon(X) \leqslant H_2(X) + 1$ for every $\epsilon \leqslant \delta$. For a concrete setting consider $n = 256$ and $\delta = 2^{-10}$. We have $\mathrm{H}(X) > 255$ but $H_2(X) \leqslant 18$ and $\mathrm{H}_2^\epsilon(X) \leqslant 19$ for every $\epsilon < 2^{-9}$!*

Proof. We use a distribution of the same form as the optimal distribution as for problem (15). Denote $N = 2^n$ and define $\boldsymbol{p}_i = \frac{1-2\delta}{N-1}$ for $i = 2, \ldots, N$, and $\boldsymbol{p}_1 = 2\delta$. It is easy to see that $\mathrm{H}(\boldsymbol{p}) \geqslant (1 - 2\delta)n + \log_2(1 - 2^{-n})$ and $H_2(\boldsymbol{p}) < \log(1/\delta) - 2$. Consider now arbitrary distribution $\boldsymbol{p}'$ such that $\mathrm{SD}(\boldsymbol{p}; \boldsymbol{p}') \leqslant \epsilon$. We have $\boldsymbol{p}'_i = \boldsymbol{p}_i + \epsilon_i$ where $\sum_i \epsilon_i = 0$ and $\sum_i |\epsilon_i| = 2\epsilon$. Note that

$$\sum_{i>1} \boldsymbol{p}'^2_i - \sum_{i>1} \boldsymbol{p}_i^2 > 2\sum_{i>1} \boldsymbol{p}_i \epsilon_i$$

$$> -\frac{2(1 - 2\delta)\epsilon}{N - 1}$$

$$= -\frac{2\epsilon}{1 - 2\delta} \cdot \sum_{i>1} \boldsymbol{p}_i^2,$$

and $\boldsymbol{p}'^2_1 - \boldsymbol{p}_1^2 \geqslant -\delta^2 = -\frac{1}{2}\boldsymbol{p}_1^2$. Thus, for $2\epsilon + \delta < \frac{1}{2}$ it follows that $\sum_{i \geqslant 1} \boldsymbol{p}'^2_i \geqslant (1 - \frac{1}{2}) \sum_{i \geqslant 1} \boldsymbol{p}_i^2$ and the result follows.

Corollary 5 (Separating Extractable Entropy and Shannon Entropy). *For any $n \geqslant 1$, $\epsilon \in (0,1)$ and $\delta > 2^{-n}$, there exists a random variable $X \in \{0,1\}^n$ such that $\mathrm{H}(X) \geqslant (1 - \epsilon - \delta)n + \log_2(1 - 2^{-n})$ but $\mathrm{H}_{\mathrm{ext}}^\epsilon(X) \leqslant \log(1/\delta)$. For a concrete setting consider $n = 256$ and $\delta = 2^{-10}$. We have $\mathrm{H}(X) > 255.5$ but $\mathrm{H}_{\mathrm{ext}}^\epsilon(X) \leqslant 10$ for every $\epsilon < 2^{-10}$!*

---

[1] More precisely they require $\mathrm{poly}(\epsilon^{-1})$ independent samples.

Proof (Proof of Corollary 5). We use a distribution of the same form as the optimal distribution as for problem (15). Fix ϵ, δ (we can assume $\epsilon + \delta < 1$) and denote $N = 2^n$. We define $p_i = \frac{1-\epsilon-\delta}{N-1}$ for $i = 2, \ldots, N$, and $p_1 = \epsilon + \delta$. Note that $p_i < \delta$ for $i \neq 1$. It follows then that $\mathrm{H}_\infty^\epsilon(\boldsymbol{p}) \leqslant \log(1/\epsilon)$. In the other hand, note that $\boldsymbol{p}$ is a convex combination of the distribution uniform over the first $N-1$ points (with the weight $1 - \epsilon - \delta$ and a point mass at N (with the weight $\epsilon + \delta$. It follows that Shannon Entropy of $\boldsymbol{p}$ is at least $(1 - \epsilon - \delta) \cdot \log_2(N-1)$.

4.2   Positive Results

Now we address the question what happens when $\Delta > 1$. This is motivated by settings where keys with entropy deficiency can be applied (cf. Theorem 1 and related references).

Corollary 6 (Collision Entropy When the Shannon Gap is Moderate).
Let $k \leqslant n-1$ and let $X^* \in \{0,1\}^n$ minimizes $H_2(X)$ subject to $\mathrm{H}(X) \geqslant k$ where $X \in \{0,1\}^n$. Then

$$2 \log_2 n - 2 \log_2(n - k) \leqslant H_2(X^*) \leqslant 2 \log_2 n - 2 \log_2(n + 1 - k) + 1. \qquad (14)$$

For instance, if $k = 255$ then $15 < H_2(X^*) < 16$.

Proof (Proof of Corollary 6). Let b be the solution to Eq. (7) (here we have $d = 2^n$). Since $0 \leqslant H(b) \leqslant 1$ we have $\frac{k}{\log_2(d-1)} \geqslant 1 - b \geqslant \frac{k-1}{\log_2(d-1)}$. We improve the left-hand side inequality a little bit

Claim. We have $1 - \frac{k-1}{\log_2 d} \geqslant b \geqslant 1 - \frac{k}{\log_2 d}$.

Proof (Proof of Sect. 4.2). Since $b \geqslant \frac{1}{d}$ we have $\log_2(d - 1) - \log(1 - b) \geqslant \log_2 d$ and therefore

$$\begin{aligned} k &= -b \log_2 b - (1 - b) \log_2(1 - b) + (1 - b) \log_2(d - 1) \\ &\geqslant -b \log_2 b + (1 - b) \log_2 d \end{aligned}$$

from which it follows that $1 - b \leqslant \frac{k}{\log_2 d}$. The left part is already proved.

The result now easily follows by observing that $\frac{(1-b)^2}{d-1} \geqslant b^2$ holds true for $b \leqslant \frac{-1+\sqrt{d-1}}{d-2} \leqslant \frac{1}{2}$, also for $d = 2$. This is indeed satisfied by Sect. 4.2 and $k \leqslant \log_2 d - 1$.

4.3   Bounds in Terms of the Renyi Divergence

Our Corollary 2 gives a bound on the ℓ_2-distance between X and U. Note that

$$d_2(X; U)^2 = \mathrm{CP}(X) - d^{-1} = d^{-1}(d\mathrm{CP}(X) - 1) = d^{-1}\left(2^{D_2(X\|U)} - 1\right)$$

and thus our bounds can be expressed in terms of the Renyi divergence D_2. Since we find the distribution X with possibly minimal entropy, this gives an *upper bound* on the divergence in terms the Shannon entropy.

5 Conclusion

Our results put in a quantitative form the well-accepted fact that Shannon Entropy does not have good cryptographic properties, unless additional strong assumptions are imposed on the entropy source. The techniques we applied may be of independent interests.

Acknowledgment. The author thanks anonymous reviewers for their valuable comments.

A Proof of Theorem 2

Proof (Proof of Theorem 2). Consider the corresponding optimization problem

$$\underset{\boldsymbol{p}\in\mathbb{R}^d}{\text{minimize}} \quad -\log_2\left(\sum_{i=1}^{d} \boldsymbol{p}_i^2\right)$$

$$\text{subject to} \quad 0 < \boldsymbol{p}_i, \ i = 1,\ldots,d.$$

$$\sum_{i=1}^{d} \boldsymbol{p}_i - 1 = 0 \tag{15}$$

$$\sum_{i=1}^{d} -\boldsymbol{p}_i \log_2 \boldsymbol{p}_i = k$$

The Lagrangian associated to (15) is given by

$$L(\boldsymbol{p},(\lambda_1,\lambda_2)) = -\log_2\left(\sum_{i=1}^{d}\boldsymbol{p}_i^2\right) - \lambda_1\left(\sum_{i=1}^{d}\boldsymbol{p}_i - 1\right) - \lambda_2\left(-\sum_{i=1}^{d}\boldsymbol{p}_i\log_2\boldsymbol{p}_i - k\right) \tag{16}$$

Claim. The first and second derivative of the Lagrangian (16) are given by

$$\frac{\partial L}{\partial \boldsymbol{p}_i} = -2\log_2 e \cdot \frac{\boldsymbol{p}_i}{\boldsymbol{p}^2} - \lambda_1 + \lambda_2\log_2 e + \lambda_2\log_2 \boldsymbol{p}_i \tag{17}$$

$$\frac{\partial^2 L}{\partial \boldsymbol{p}_i \boldsymbol{p}_j} = 4\log_2 e \cdot \frac{\boldsymbol{p}_i \boldsymbol{p}_j}{(\boldsymbol{p}^2)^2} + [i=j]\cdot\left(-\frac{2\log_2 e}{\boldsymbol{p}^2} + \frac{\lambda_2\log_2 e}{\boldsymbol{p}_i}\right) \tag{18}$$

Claim. Let $\boldsymbol{p}^*$ be a non-uniform optimal point to 15. Then it satisfies $\boldsymbol{p}_i^* \in \{a,b\}$ for every i, where a,b are some constant such that

$$-\frac{2\log_2 e}{\boldsymbol{p}^{*2}} + \frac{\lambda_2\log_2 e}{a} > 0 > -\frac{2\log_2 e}{\boldsymbol{p}^{*2}} + \frac{\lambda_2\log_2 e}{b} \tag{19}$$

Proof (Proof of Appendix A). At the optimal point $\boldsymbol{p}$ we have $\frac{\partial L}{\partial \boldsymbol{p}_i} = 0$ which means

$$-2\log_2 e \cdot \frac{\boldsymbol{p}_i}{\boldsymbol{p}^2} - \lambda_1 + \lambda_2 \log_2 e + \lambda_2 \log_2 \boldsymbol{p}_i = 0, \quad i = 1, \ldots, d. \qquad (20)$$

Think of $\boldsymbol{p}^2$ as a constant, for a moment. Then the left-hand side of Eq. (20) is of the form $-c_1 \boldsymbol{p}_i + c_2 \log_2 \boldsymbol{p}_i + c_3$ with some positive constant c_1 and real constants c_2, c_3. Since the derivative of this function equals $-c_1 + \frac{c_2}{\boldsymbol{p}_i}$, the left-hand side is either decreasing (when $c_2 \leqslant 0$) or concave (when $c_2 > 0$). For the non-uniform solution the latter must be true (because otherwise $\boldsymbol{p}_i$ for $i = 1, \ldots, d$ are equal). Hence the Eq. (20) has at most two solutions $\{a, b\}$, where $a < b$ and both are not dependent on i. Moreover, its left-hand side has the maximum between a and b, thus we must have $-c_1 + \frac{c_2}{a} > 0 > -c_1 + \frac{c_2}{b}$. Expressing this in terms of λ_1, λ_2 we get Eq. (19).

Claim. Let $\boldsymbol{p}^*$ and a, b be as in Appendix A. Then $\boldsymbol{p}_i = a$ for all but one index i.

Proof (Proof of Appendix A). The tangent space of the constraints $\sum_{i=1}^d \boldsymbol{p}_i - 1 = 0$ and $-\sum_{i=1}^d \boldsymbol{p}_i \log_2 \boldsymbol{p}_i - k = 0$ at the point $\boldsymbol{p}$ is the set of all vectors $h \in \mathbb{R}^d$ satisfying the following conditions

$$
\begin{aligned}
\sum_{i=1}^d h_i &= 0 \\
\sum_{i=1}^d -(\log_2 \boldsymbol{p}_i + \log_2 e)h_i &= 0
\end{aligned}
\qquad (21)
$$

Intuitively, the tangent space includes all infinitesimally small movements that are consistent with the constraints. Let $D^2 L = \left(\frac{\partial^2 L}{\partial \boldsymbol{p}_i \boldsymbol{p}_j} \right)_{i,j}$ be the second derivative of L. It is well known that the necessary second order condition for the minimizer $\boldsymbol{p}$ is $h^T(D^2)Lh \geqslant 0$ for all vectors in the tangent space (21). We have

$$h^T \cdot (D^2 L) \cdot h = 4\log_2 e \cdot \frac{\left(\sum_{i=}^d \boldsymbol{p}_i h_i\right)^2}{(\boldsymbol{p}^2)^2} + \sum_{i=1}^d \left(-\frac{2\log_2 e}{\boldsymbol{p}^2} + \frac{\lambda_2 \log_2 e}{\boldsymbol{p}_i} \right) h_i^2.$$

Now, if the are two different indexes i_1, i_2 such that $\boldsymbol{p}_{i_1}^* = \boldsymbol{p}_{i_2}^* = b$, we can define $h_{i_1} = -\delta$, $h_{i_2} = \delta$ and $h_i = 0$ for $i \notin \{i_1, i_2\}$. Then we get

$$h^T \cdot (D^2 L) \cdot h = 2 \left(-\frac{2\log_2 e}{\boldsymbol{p}^2} + \frac{\lambda_2 \log_2 e}{b} \right) \delta^2$$

which is negative according to Eq. (19). Thus we have reached a contradiction.

Finally, taking into account the case of possibly uniform $\boldsymbol{p}^*$ and combining it with the last claim we get

Claim. The optimal point $\boldsymbol{p}^*$ satisfies $\boldsymbol{p}_{i_0}^* = b$ and $\boldsymbol{p}_i^* = \frac{1-b}{d-1}$ for $i \neq i_0$, for some $b \geqslant \frac{1}{d}$. Then we have $\mathrm{H}(\boldsymbol{p}^*) = \mathrm{H}(b) + (1-b)\log_2(d-1)$ and $H_2(\boldsymbol{p}^*) = -\log_2\left(b^2 + \frac{(1-b)^2}{d-1}\right)$.

It remains to take a closer look at Eq. (7). It defines b as an *implicit function* of k and d. Its uniqueness is a consequence of the following claim

Claim. The function $f(b) = H(b) + (1 - b)\log_2(d - 1)$ is strictly decreasing and concave for $b \geqslant \frac{1}{d}$.

Proof (Proof of Appendix A). The derivative equals $\frac{\partial f}{\partial b} = -\log_2 \frac{b}{1-b} - \log_2(d-1)$ and hence, for $\frac{1}{d} < b < 1$, is at most $-\log_2 \frac{\frac{1}{d}}{1 - \frac{1}{d}} - \log_2(d-1) = 0$. The second derivative is $\frac{\partial^2 f}{\partial b^2} = -\frac{\log_2 e}{b(1-b)}$. Thus, the claim follows.
The statement follows now by Appendices A and B.

B   Proof of Lemma 2

Proof. Let $\Delta = \log_2 d - k$ be the gap in the Shannon Entropy. Note that from Eq. (7) and the inequality $-2 \leqslant d(\log_2(d-1) - \log_2 d) \leqslant -\log_2 e$ it follows that

$$-b\log_2 b - (1 - b)\log_2(1 - b) - b\log_2 d = -\Delta + C_1(d) \cdot d^{-1}$$

where $\log_2 e \leqslant C_1 \leqslant 2$. Note that $f\left(\frac{1}{2}\right) = -1 + \frac{1}{2}\log_2(d - 1) < \log_2 d - 1$. Since $\Delta \leqslant 1$ implies $f(b) \geqslant \log_2 d - 1$, by Appendix A we conclude that $b < \frac{1}{2}$. Next, observe that $1 \leqslant \frac{-(1-b)\log_2(1-b)}{b} \leqslant \log_2 e$, for $0 < b < \frac{1}{2}$. This means that $-(1 - b)\log_2(1 - b) = -b\log_2 C_2(d)$ where $\frac{1}{e} \leqslant C_2(d) \leqslant \frac{1}{2}$. Now we have

$$-b\log_2(C_2(d) \cdot d \cdot b) = -\Delta + C_1(d) \cdot d^{-1}.$$

Let $y = C_2(d)\cdot d\cdot b$. Our equation is equivalent to $y \ln y = C_3(d)\cdot d\cdot\Delta - C_1(d)C_3(d)$. where $C_3 = C_2/\log_2 e$. Using the Lambert-W function, which is defined as $W(x)\cdot e^{W(x)} = x$, we can solve this equations as

$$b = \frac{e^{W(C_3(d)d\Delta - C_3(d)C_1(d))}}{C_2(d)d}. \tag{22}$$

For $x \geqslant e$ we have the well-known approximation for the Lambert W function [HH08]

$$\ln x - \ln\ln x < W(x) \leqslant \ln x - \ln\ln x + \ln(1 + e^{-1}). \tag{23}$$

Provided that $C_3(d)d\Delta - C_3(d)C_1(d) \geqslant 1$, which is satisfied if $d\Delta \geqslant 6$, we obtain

$$b = \frac{C_3(d)d\Delta - C_3(d)C_1(d)}{C_3(d)d \cdot \log_2\left(C_3(d)d\Delta - C_3(d)C_1(d)\right)} \cdot C_4(d) \tag{24}$$

where $1 \leqslant C_4(d) \leqslant 1 + e^{-1}$. Since the function $x \to \frac{x}{\log_2 x}$ is increasing for $x \geqslant e$ and since for $d\Delta \geqslant 13$ we have $C_3(d)d\Delta - C_3(d)C_1(d) \geqslant e$, from Eq. (24) we get

$$b \leqslant \frac{C_3(d)d\Delta}{C_3(d)d \cdot \log_2\left(C_3(d)d\Delta\right)} \cdot C_4(d) = \frac{C_4(d)\Delta}{\log_2\left(C_3(d)d\Delta\right)} \tag{25}$$

from which the right part of Eq. (8) follows by the inequalities on C_3 and C_4. For the lower bound, note that for $d\Delta \geqslant 13$ we have $C_3(d)d\Delta - C_3(d)C_1(d) \geqslant C_3(d)d\Delta \cdot \frac{11}{13}$ because it reduces to $C_1(d) \leqslant 2$, and that $C_3(d)d\Delta \cdot \frac{11}{13} \geqslant 13 \cdot \frac{1}{e\log_2 e} \cdot \frac{11}{13} > e$. Therefore, by Eq. (24) and the mononicity of $\frac{x}{\log_2 x}$ we get

$$b \geqslant \frac{\frac{11}{13}C_3(d)d\Delta}{C_3(d)d \cdot \log_2\left(\frac{11}{13}C_3(d)d\Delta\right)} \cdot C_4(d) = \frac{\frac{11}{13}C_4(d)\Delta}{\log_2\left(\frac{11}{13}C_3(d)d\Delta\right)}, \qquad (26)$$

from which the left part of Eq. (8) follows by the inequalities on C_3 and C_4.

References

[AIS11] A proposal for: Functionality classes for random number generators1, Technical report AIS 30, Bonn, Germany, September 2011. http://tinyurl.com/bkwt2wf

[AOST14] Acharya, J., Orlitsky, A., Suresh, A.T., Tyagi, H.: The complexity of estimating renyi entropy, CoRR abs/1408.1000 (2014)

[BDK+11] Barak, B., Dodis, Y., Krawczyk, H., Pereira, O., Pietrzak, K., Standaert, F.-X., Yu, Y.: Leftover hash lemma, revisited. In: Rogaway, P. (ed.) CRYPTO 2011. LNCS, vol. 6841, pp. 1–20. Springer, Heidelberg (2011)

[BK12] Barker, E.B., Kelsey, J.M.: Sp 800–90a recommendation for random number generation using deterministic random bit generators, Technical report, Gaithersburg, MD, United States (2012)

[BKMS09] Bouda, J., Krhovjak, J., Matyas, V., Svenda, P.: Towards true random number generation in mobile environments. In: Jøsang, A., Maseng, T., Knapskog, S.J. (eds.) NordSec 2009. LNCS, vol. 5838, pp. 179–189. Springer, Heidelberg (2009)

[BL05] Bucci, M., Luzzi, R.: Design of testable random bit generators. In: Rao, J.R., Sunar, B. (eds.) CHES 2005. LNCS, vol. 3659, pp. 147–156. Springer, Heidelberg (2005)

[BST03] Barak, B., Shaltiel, R., Tromer, E.: True random number generators secure in a changing environment. In: Walter, C.D., Koç, Ç.K., Paar, C. (eds.) CHES 2003. LNCS, vol. 2779, pp. 166–180. Springer, Heidelberg (2003)

[Cac97] Cachin, C.: Smooth entropy and rényi entropy. In: Fumy, W. (ed.) EUROCRYPT 1997. LNCS, vol. 1233, pp. 193–208. Springer, Heidelberg (1997)

[Cor99] Coron, J.-S.: On the security of random sources. In: Imai, H., Zheng, Y. (eds.) PKC 1999. LNCS, vol. 1560, p. 29. Springer, Heidelberg (1999)

[CW79] Carter, J.L., Wegman, M.N.: Universal classes of hash functions. J. Comput. Syst. Sci. 18(2), 143–154 (1979)

[DPR+13] Dodis, Y., Pointcheval, D., Ruhault, S., Vergniaud, D., Wichs, D.: Security analysis of pseudo-random number generators with input: /dev/random is not robust. In: Proceedings of the 2013 ACM SIGSAC Conference on Computer and Communications Security, CCS 2013, pp. 647–658. ACM, New York (2013)

[DY13] Dodis, Y., Yu, Y.: Overcoming weak expectations. In: Sahai, A. (ed.) TCC 2013. LNCS, vol. 7785, pp. 1–22. Springer, Heidelberg (2013)

[HH08] Hoorfar, A., Hassani, M.: Inequalities on the lambert w function and hyperpower function. J. Inequal. Pure Appl. Math. 9(2), 07–15 (2008)

[HILL88] Hstad, J., Impagliazzo, R., Levin, L.A., Luby, M.: Pseudo-random generation from one-way functions. In: Proceedings of the 20TH STOC, pp. 12–24 (1988)

[HILL99] Hastad, J., Impagliazzo, R., Levin, L.A., Luby, M.: A pseudorandom generator from any one-way function. SIAM J. Comput. **28**(4), 1364–1396 (1999)

[Hol06] Holenstein, T.: Pseudorandom generators from one-way functions: a simple construction for any hardness. In: Halevi, S., Rabin, T. (eds.) TCC 2006. LNCS, vol. 3876, pp. 443–461. Springer, Heidelberg (2006)

[Hol11] Holenstein, T.: On the randomness of repeated experiment

[LPR11] Lauradoux, C., Ponge, J., Röck, A.: Online Entropy Estimation for Non-Binary Sources and Applications on iPhone. Rapport de recherche, Inria (2011)

[Mau92] Maurer, U.: A universal statistical test for random bit generators. J. Cryptology **5**, 89–105 (1992)

[NZ96] Nisan, N., Zuckerman, D.: Randomness is linear in space. J. Comput. Syst. Sci. **52**(1), 43–52 (1996)

[RTS00] Radhakrishnan, J., Ta-Shma, A.: Bounds for dispersers, extractors, and depth-two superconcentrators. SIAM J. Discrete Math. **13**, 2000 (2000)

[RW04] Renner, R., Wolf, S.: Smooth renyi entropy and applications. In: Proceedings of the International Symposium on Information Theory, ISIT 2004, p. 232. IEEE (2004)

[RW05] Renner, R.S., Wolf, S.: Simple and tight bounds for information reconciliation and privacy amplification. In: Roy, B. (ed.) ASIACRYPT 2005. LNCS, vol. 3788, pp. 199–216. Springer, Heidelberg (2005)

[Sha48] Shannon, C.E.: A mathematical theory of communication. Bell Syst. Tech. J. **27**(3), 379–423 (1948)

[Sha11] Shaltiel, R.: An introduction to randomness extractors. In: Aceto, L., Henzinger, M., Sgall, J. (eds.) ICALP 2011, Part II. LNCS, vol. 6756, pp. 21–41. Springer, Heidelberg (2011)

[Shi15] Shikata, J.: Design and analysis of information-theoretically secure authentication codes with non-uniformly random keys. IACR Cryptology ePrint Arch. **2015**, 250 (2015)

[VSH11] Voris, J., Saxena, N., Halevi, T.: Accelerometers and randomness: perfect together. In: Proceedings of the Fourth ACM Conference on Wireless Network Security, WiSec 2011, pp. 115–126. ACM, New York (2011)

Leakage Resilience

Continuous After-the-Fact Leakage-Resilient eCK-Secure Key Exchange

Janaka Alawatugoda[1,4](✉), Douglas Stebila[1,2], and Colin Boyd[3]

[1] School of Electrical Engineering and Computer Science,
Queensland University of Technology, Brisbane, Australia
{janaka.alawatugoda,stebila}@qut.edu.au
[2] School of Mathematical Sciences, Queensland University of Technology,
Brisbane, Australia
[3] Department of Telematics, Norwegian University of Science and Technology,
Trondheim, Norway
colin.boyd@item.ntnu.no
[4] Department of Computer Engineering, University of Peradeniya,
Peradeniya, Sri Lanka

Abstract. Security models for two-party authenticated key exchange (AKE) protocols have developed over time to capture the security of AKE protocols even when the adversary learns certain secret values. Increased granularity of security can be modelled by considering partial leakage of secrets in the manner of models for leakage-resilient cryptography, designed to capture side-channel attacks. In this work, we use the strongest known partial-leakage-based security model for key exchange protocols, namely continuous after-the-fact leakage eCK (CAFL-eCK) model. We resolve an open problem by constructing the first concrete two-pass leakage-resilient key exchange protocol that is secure in the CAFL-eCK model.

Keywords: Key exchange protocols, Side-channel attacks, Security models, Leakage-resilience, After-the-fact leakage

1 Introduction

During the past two decades side-channel attacks have become a familiar method of attacking cryptographic systems. Examples of information which may leak during executions of cryptographic systems, and so allow side-channel attacks, include timing information [6,8,18], electromagnetic radiation [15], and power consumption [21]. Leakage may reveal partial information about the secret parameters which have been used for computations in cryptographic systems. In order to abstractly model leakage attacks, cryptographers have proposed the notion of *leakage-resilient* cryptography [1,4,7,13,14,16,17,20]. In this notion the information that leaks is not fixed, but instead chosen adversarially, so as to model any possible physical leakage function. A variety of different cryptographic primitives have been developed in recent years. As one of the most widely used

© Springer International Publishing Switzerland 2015
J. Groth (Ed.): IMACC 2015, LNCS 9496, pp. 277–294, 2015.
DOI: 10.1007/978-3-319-27239-9_17

cryptographic primitives, it is important to analyze the leakage resilience of key exchange protocols.

Earlier key exchange security models, such as the Bellare–Rogaway [5], Canetti–Krawczyk [9], and extended Canetti–Krawczyk (eCK) [19] models, aim to capture security against an adversary who can fully compromise some, but not all, secret values. This is not a very granular form of leakage, and thus is not suitable for modelling side-channel attacks in key exchange protocols enabled by partial leakage of secret keys. This motivates the development of leakage-resilient key exchange security models [3,4,11,22,23]. Among them the generic security model proposed by Alawatugoda et al. [3] in 2014 facilitates more granular leakage.

Alawatugoda et al. [3] proposed a *generic leakage-security model* for key exchange protocols, which can be instantiated as either a *bounded* leakage variant or as a *continuous* leakage variant. In the bounded leakage variant, the total amount of leakage is bounded, whereas in the continuous leakage variant, each protocol execution may reveal a fixed amount of leakage. Further, the adversary is allowed to obtain the leakage even after the session key is established for the session under attack (after-the-fact leakage). In Sect. 3 we review the continuous leakage instantiation of the security model proposed by Alawatugoda et al.

Alawatugoda et al. [3] also provided a generic construction for a protocol which is proven secure in their generic leakage-security model. However, when it comes to a concrete construction, the proposed generic protocol can only be instantiated in a way that is secure in the *bounded* version of the security model. Until now there are no suitable cryptographic primitives which can be used to instantiate the generic protocol in the continuous leakage variant of the security model.

Our aim is to propose a concrete protocol construction which can be proven secure in the continuous leakage instantiation of the security model of Alawatugoda et al. Thus, we introduce the first concrete construction of *continuous* and *after-the-fact* leakage-resilient key exchange protocol.

Bounded Leakage and Continuous Leakage. Generally, in models assuming bounded leakage there is an upper bound on the amount of leakage information for the entire period of execution. The security guarantee only holds if the leakage amount is below the prescribed bound. Differently, in models allowing continuous leakage the adversary is allowed to obtain leakage over and over for a polynomial number of iterations during the period of execution. Naturally, there is a bound on the amount of leakage that the adversary can obtain in each single iteration, but the total amount of leakage that the adversary can obtain for the entire period of execution is unbounded.

After-the-Fact Leakage. The concept of after-the-fact leakage has been applied previously to encryption primitives. Generally, leakage which happens after the challenge is given to the adversary is considered as after-the-fact leakage. In key exchange security models, the challenge to the adversary is to distinguish the session key of a chosen session, usually called the *test session*, from a random session key [5,9,19], After-the-fact leakage is the leakage which happens after the test session is established.

Our Contribution. Alawatugoda et al. [3] left the construction of a continuous after-the-fact leakage-resilient eCK secure key exchange protocol as an open problem. In this paper, we construct such a protocol (protocol P2) using existing leakage-resilient cryptographic primitives. We use leakage-resilient storage schemes and their refreshing protocols [12] for this construction.

Table 1 compares the proposed protocol P2, with the NAXOS protocol [19], the Moriyama-Okamoto (MO) protocol [22] and the generic Alawatugoda et al. [3] protocol instantiation, by means of computation cost, security model and the proof model.

Table 1. Security and efficiency comparison of leakage-resilient key exchange protocols

Protocol	Initiator cost	Responder cost	Leakage Feature	After-the-fact	Proof model
NAXOS [19]	4 Exp	4 Exp	None	None	Random oracle
MO [22]	8 Exp	8 Exp	Bounded	No	Standard
Alawatugoda et al. [3]	12 Exp	12 Exp	Bounded	Yes	Standard
Protocol P2 (this paper)	6 Exp	6 Exp	Continuous	Yes	Random oracle

In protocol P2, the secret key is encoded into two equal-sized parts of some chosen size, and the leakage bound from each of the two parts is 15 % of the size of a part, per occurrence. Since this is a continuous leakage model the total leakage amount is unbounded. More details of the leakage tolerance of protocol P2 may be found in Sect. 5.3.

2   Preliminaries

We discuss the preliminaries which we use for the protocol constructions.

2.1   Diffie–Hellman Problems

Let $\mathcal{G}$ be a group generation algorithm and $(\mathbb{G}, q, g) \leftarrow \mathcal{G}(1^k)$, where $\mathbb{G}$ is a cyclic group of prime order q and g is an arbitrary generator.

Definition 1 (Computational Diffie–Hellman (CDH) Problem). *Given an instance (g, g^a, g^b) for $a, b \xleftarrow{\$} \mathbb{Z}_q$, the CDH problem is to compute g^{ab}.*

Definition 2 (Decision Diffie–Hellman (DDH) Problem). *Given an instance (g, g^a, g^b, g^c) for $a, b \xleftarrow{\$} \mathbb{Z}_q$ and either $c \xleftarrow{\$} \mathbb{Z}_q$ or $c = ab$, the DDH problem is to distinguish whether $c = ab$ or not.*

Definition 3 (Gap Diffie–Hellman (GDH) Problem). *Given an instance (g, g^a, g^b) for $a, b \xleftarrow{\$} \mathbb{Z}_q$, the GDH problem is to find g^{ab} given access to an oracle $\mathcal{O}$ that solves the DDH problem.*

2.2   Leakage-Resilient Storage

We review the definitions of leakage-resilient storage according to Dziembowski et al. [12]. The idea behind their construction is to split the storage of elements into two parts using a randomized encoding function. As long as leakage is then limited from each of its two parts then no adversary can learn useful information about an encoded element. The key observation of Dziembowski et al. is then to show how such encodings can be *refreshed* in a leakage-resilient way so that the new parts can be re-used. To construct a continuous leakage-resilient primitive the relevant secrets are split, used separately, and then refreshed between any two usages.

Definition 4 (Dziembowski-Faust Leakage-Resilient Storage Scheme).
For any $m, n \in \mathbb{N}$, the storage scheme $\Lambda_{\mathbb{Z}_q^*}^{n,m} = (\text{Encode}_{\mathbb{Z}_q^*}^{n,m}, \text{Decode}_{\mathbb{Z}_q^*}^{n,m})$ efficiently stores elements $s \in (\mathbb{Z}_q^*)^m$ where:

- $\text{Encode}_{\mathbb{Z}_q^*}^{n,m}(s) : s_L \xleftarrow{\$} (\mathbb{Z}_q^*)^n \setminus \{(0^n)\}$, then $s_R \leftarrow (\mathbb{Z}_q^*)^{n \times m}$ such that $s_L \cdot s_R = s$
  and outputs (s_L, s_R).
- $\text{Decode}_{\mathbb{Z}_q^*}^{n,m}(s_L, s_R) :$ outputs $s_L \cdot s_R$.

In the model we expect an adversary to see the results of a leakage function applied to s_L and s_R. This may happen each time computation occurs.

Definition 5 (λ-limited Adversary). *If the amount of leakage obtained by the adversary from each of s_L and s_R is limited to $\lambda = (\lambda_1, \lambda_2)$ bits in total respectively, the adversary is known as a λ-limited adversary.*

Definition 6 (($\lambda_\Lambda, \epsilon_1$)-secure leakage-resilient storage scheme). *We say $\Lambda = (\text{Encode}, \text{Decode})$ is $(\lambda_\Lambda, \epsilon_1)$-secure leakage-resilient, if for any $s_0, s_1 \xleftarrow{\$} (\mathbb{Z}_q^*)^m$ and any λ_Λ-limited adversary $\mathcal{C}$, the leakage from $\text{Encode}(s_0) = (s_{0L}, s_{0R})$ and $\text{Encode}(s_1) = (s_{1L}, s_{1R})$ are statistically ϵ_1-close. For an adversary-chosen leakage function $\mathbf{f} = (f_1, f_2)$, and a secret s such that $\text{Encode}(s) = (s_L, s_R)$, the leakage is denoted as $(f_1(s_L), f_2(s_R))$.*

Lemma 1 ([12]). *Suppose that $m < n/20$. Then $\Lambda_{\mathbb{Z}_q^*}^{n,m} = (\text{Encode}_{\mathbb{Z}_q^*}^{n,m}, \text{Decode}_{\mathbb{Z}_q^*}^{n,m})$ is (λ, ϵ)-secure for some ϵ and $\lambda = (0.3 \cdot n \log q, 0.3 \cdot n \log q)$.*

The encoding function can be used to design different leakage resilient schemes with bounded leakage. The next step is to define how to *refresh* the encoding so that a continuous leakage is also possible to defend against.

Definition 7 (Refreshing of Leakage-Resilient Storage). *Let $(L', R') \leftarrow \text{Refresh}_{\mathbb{Z}_q^*}^{n,m}(L, R)$ be a refreshing protocol that works as follows:*

- *Input : (L, R) such that $L \in (\mathbb{Z}_q^*)^n$ and $R \in (\mathbb{Z}_q^*)^{n \times m}$.*
- *Refreshing R :*
    1. *$A \xleftarrow{\$} (\mathbb{Z}_q^*)^n \setminus \{(0^n)\}$ and $B \leftarrow$ non singular $(\mathbb{Z}_q^*)^{n \times m}$ such that $A \cdot B = (0^m)$.*

2. $M \leftarrow$ non-singular $(\mathbb{Z}_q^*)^{n \times n}$ such that $L \cdot M = A$.
3. $X \leftarrow M \cdot B$ and $R' \leftarrow R + X$.
- Refreshing L :
  1. $\tilde{A} \xleftarrow{\$} (\mathbb{Z}_q^*)^n \setminus \{(0^n)\}$ and $\tilde{B} \leftarrow$ non singular $(\mathbb{Z}_q^*)^{n \times m}$ such that $\tilde{A} \cdot \tilde{B} = (0^m)$.
  2. $\tilde{M} \leftarrow$ non-singular $(\mathbb{Z}_q^*)^{n \times n}$ such that $\tilde{M} \cdot R' = \tilde{B}$.
  3. $Y \leftarrow \tilde{A} \cdot \tilde{M}$ and $L' \leftarrow L + Y$.
- Output : (L', R')

Let $\Lambda = (\text{Encode}, \text{Decode})$ be a $(\boldsymbol{\lambda}_\Lambda, \epsilon_1)$-secure leakage-resilient storage scheme and Refresh be a refreshing protocol. We consider the following experiment Exp, which runs Refresh for ℓ rounds and lets the adversary obtain leakage in each round. For refreshing protocol Refresh, a $\boldsymbol{\lambda}_{\text{Refresh}}$-limited adversary $\mathcal{B}$, $\ell \in \mathbb{N}$ and $s \xleftarrow{\$} (\mathbb{Z}_q^*)^m$, we denote the following experiment by $\text{Exp}_{(\text{Refresh}, \Lambda)}(\mathcal{B}, s, \ell)$:

1. For a secret s, the initial encoding is generated as $(s_L^0, s_R^0) \leftarrow \text{Encode}(s)$.
2. For $j = 1$ to ℓ run $\mathcal{B}$ against the jth round of the refreshing protocol.
3. Return whatever $\mathcal{B}$ outputs.

We require that the adversary $\mathcal{B}$ outputs a single bit $b \in \{0, 1\}$ upon performing the experiment Exp using $s \xleftarrow{\$} \{s_0, s_1\} \in (\mathbb{Z}_q^*)^m$. Now we define leakage-resilient security of a refreshing protocol.

Definition 8 (($\ell, \boldsymbol{\lambda}_{\text{Refresh}}, \epsilon_2$)-**secure Leakage-Resilient Refreshing Protocol**). *For a* ($\boldsymbol{\lambda}_\Lambda, \epsilon_1$)-*secure Leakage-Resilient Storage Scheme* $\Lambda = (\text{Encode}, \text{Decode})$ *with message space* $(\mathbb{Z}_q^*)^m$, *Refresh is* ($\ell, \boldsymbol{\lambda}_{\text{Refresh}}, \epsilon_2$)-*secure leakage-resilient, if for every* $\boldsymbol{\lambda}_{\text{Refresh}}$-*limited adversary* $\mathcal{B}$ *and any two secrets* $s_0, s_1 \in (\mathbb{Z}_q^*)^m$, *the statistical distance between* $\text{Exp}_{(\text{Refresh}, \Lambda)}(\mathcal{B}, s_0, \ell)$ *and* $\text{Exp}_{(\text{Refresh}, \Lambda)}(\mathcal{B}, s_1, \ell)$ *is bounded by* ϵ_2.

Theorem 1 ([12]). *Let* $m/3 \leq n, n \geq 16$ *and* $\ell \in \mathbb{N}$. *Let* n, m *and* $\mathbb{Z}_q^*$ *be such that* $\Lambda_{\mathbb{Z}_q^*}^{n,m}$ *is* ($\boldsymbol{\lambda}, \epsilon$)-*secure leakage-resilient storage scheme (Definitions 4 and 6). Then the refreshing protocol* $\text{Refresh}_{\mathbb{Z}_q^*}^{n,m}$ *(Definition 7) is a* ($\ell, \boldsymbol{\lambda}/2, \epsilon'$)-*secure leakage-resilient refreshing protocol for* $\Lambda_{\mathbb{Z}_q^*}^{n,m}$ *(Definition 8) with* $\epsilon' = 2\ell q(3q^m \epsilon + mq^{-n-1})$.

3   Continuous After-the-Fact Leakage eCK Model and the eCK Model

In 2014 Alawatugoda et al. [3] proposed a new security model for key exchange protocols, namely the generic after-the-fact leakage eCK (($\cdot$)AFL-eCK) model which, in addition to the adversarial capabilities of the eCK model [19], is equipped with an adversary-chosen, efficiently computable, adaptive leakage function $\mathbf{f}$, enabling the adversary to obtain the leakage of long-term secret keys of protocol principals. Therefore the ($\cdot$)AFL-eCK model captures all the attacks captured by the eCK model, and captures the partial leakage of long-term secret keys due to side-channel attacks.

The eCK Model. In the eCK model, in sessions where the adversary does not modify the communication between parties (passive sessions), the adversary is allowed to reveal both ephemeral secrets, long-term secrets, or one of each from two different parties, whereas in sessions where the adversary may forge the communication of one of the parties (active sessions), the adversary is allowed to reveal the long-term or ephemeral secret of the other party. The security challenge is to distinguish the real session key from a random session key, in an adversary-chosen protocol session.

Generic After-the-Fact Leakage eCK Model. The generic $(\cdot)$AFL-eCK model can be instantiated in two different ways which leads to two security models. Namely, *bounded* after-the-fact leakage eCK (BAFL-eCK) model and *continuous* after-the-fact leakage eCK (CAFL-eCK) model. The BAFL-eCK model allows the adversary to obtain a bounded amount of leakage of the long-term secret keys of the protocol principals, as well as reveal session keys, long-term secret keys and ephemeral keys. Differently, the CAFL-eCK model allows the adversary to continuously obtain arbitrarily large amount of leakage of the long-term secret keys of the protocol principals, enforcing the restriction that the amount of leakage per observation is bounded.

Below we revisit the definitions of the CAFL-eCK model, and we also recall the definitions of the eCK model as a comparison to the CAFL-eCK definitions.

3.1 Partner Sessions in the CAFL-eCK Model

Definition 9 (Partner Sessions in the CAFL-eCK Model). *Two oracles $\Pi_{U,V}^s$ and $\Pi_{U',V'}^{s'}$ are said to be partners if all of the following hold:*

1. *both $\Pi_{U,V}^s$ and $\Pi_{U',V'}^{s'}$ have computed session keys;*
2. *messages sent from $\Pi_{U,V}^s$ and messages received by $\Pi_{U',V'}^{s'}$ are identical;*
3. *messages sent from $\Pi_{U',V'}^{s'}$ and messages received by $\Pi_{U,V}^s$ are identical;*
4. *$U' = V$ and $V' = U$;*
5. *Exactly one of U and V is the initiator and the other is the responder.*

The protocol is said to be correct if two partner oracles compute identical session keys.

The definition of partner sessions is the same in the eCK model.

3.2 Leakage in the CAFL-eCK Model

A realistic way in which side-channel attacks can be mounted against key exchange protocols seems to be to obtain the leakage information from the protocol computations which use the secret keys. Following the previously used premise in other leakage models that "only computation leaks information", leakage is modelled where any computation takes place using secret keys. In normal protocol models, by issuing a Send query, the adversary will get a protocol message which is computed according to the normal protocol computations. Sending an adversary-chosen, efficiently computable adaptive leakage function with the Send query thus reflects the concept "only computation leaks information".

A tuple of t adaptively chosen efficiently computable leakage functions $\mathbf{f} = (f_{1j}, f_{2j}, \ldots, f_{tj})$ are introduced; j indicates the jth leakage occurrence and the size t of the tuple is *protocol-specific*. A key exchange protocol may use more than one cryptographic primitive where each primitive uses a distinct secret key. Hence, it is necessary to address the leakage of secret keys from each of those primitives. Otherwise, some cryptographic primitives which have been used to construct a key exchange protocol may be stateful and the secret key is encoded into number of parts. The execution of a stateful cryptographic primitive is split into a number of sequential stages and each of these stages uses one part of the secret key. Hence, it is necessary to address the leakage of each of these encoded parts of the secret key.

Note that the adversary is restricted to obtain leakage from each key part independently: the adversary cannot use the output of f_{1j} as an input parameter to the f_{2j} and so on. This prevents the adversary from observing a connection between each part.

3.3   Adversarial Powers of the CAFL-eCK Model

The adversary $\mathcal{A}$ controls the whole network. $\mathcal{A}$ interacts with a set of oracles which represent protocol instances. The following query allows the adversary to run the protocol.

– Send($U, V, s, m, \mathbf{f}$) query: The oracle $\Pi_{U,V}^s$, computes the next protocol message according to the protocol specification and sends it to the adversary $\mathcal{A}$, along with the leakage $\mathbf{f}(sk_U)$. $\mathcal{A}$ can also use this query to activate a new protocol instance as an initiator with blank m.

In the eCK model Send query is same as the above except the leakage function $\mathbf{f}$.

The following set of queries allow the adversary $\mathcal{A}$ to compromise certain session specific ephemeral secrets and long-term secrets from the protocol principals.

– SessionKeyReveal(U, V, s) query: $\mathcal{A}$ is given the session key of the oracle $\Pi_{U,V}^s$.
– EphemeralKeyReveal(U, V, s) query: $\mathcal{A}$ is given the ephemeral keys (per-session randomness) of the oracle $\Pi_{U,V}^s$.
– Corrupt(U) query: $\mathcal{A}$ is given the long-term secrets of the principal U. This query does not reveal any session keys or ephemeral keys to $\mathcal{A}$.

SessionKeyReveal, EphemeralKeyReveal and Corrupt (Long-term key reveal) queries are the same in the eCK model.

Once the oracle $\Pi_{U,V}^s$ has accepted a session key, asking the following query the adversary $\mathcal{A}$ attempt to distinguish it from a random session key. The Test query is used to formalize the notion of the semantic security of a key exchange protocol.

- Test(U, s) query: When $\mathcal{A}$ asks the Test query, the challenger first chooses a random bit $b \xleftarrow{\$} \{0, 1\}$ and if $b = 1$ then the actual session key is returned to $\mathcal{A}$, otherwise a random string chosen from the same session key space is returned to $\mathcal{A}$. This query is only allowed to be asked once across all sessions.

The Test query is the same in the eCK model.

3.4 Freshness Definition of the CAFL-eCK Model

Definition 10 ($\boldsymbol{\lambda}$ – CAFL-eCK-freshness). *Let $\boldsymbol{\lambda} = (\lambda_1, \ldots, \lambda_t)$ be a vector of t elements (same size as $\mathbf{f}$ in Send query). An oracle $\Pi^s_{U,V}$ is said to be $\boldsymbol{\lambda}$ – CAFL-eCK-fresh if and only if:*

1. *The oracle $\Pi^s_{U,V}$ or its partner, $\Pi^{s'}_{V,U}$ (if it exists) has not been asked a* SessionKeyReveal.
2. *If the partner $\Pi^{s'}_{V,U}$ exists, none of the following combinations have been asked:*
   (a) Corrupt(U) *and* EphemeralKeyReveal(U, V, s).
   (b) Corrupt(V) *and* EphemeralKeyReveal(V, U, s').
3. *If the partner $\Pi^{s'}_{V,U}$ does not exist, none of the following combinations have been asked:*
   (a) Corrupt(V).
   (b) Corrupt(U) *and* EphemeralKeyReveal(U, V, s).
4. *For each* Send$(U, \cdot, \cdot, \cdot, \mathbf{f})$ *query, size of the output of $|f_{ij}(sk_{U\,i})| \leq \lambda_i$.*
5. *For each* Send$(V, \cdot, \cdot, \cdot, \mathbf{f})$ *queries, size of the output of $|f_{ij}(sk_{V\,i})| \leq \lambda_i$.*

The eCK-freshness is slightly different from the $\boldsymbol{\lambda}$ – CAFL-eCK-freshness by stripping off points 4 and 5.

3.5 Security Game and Security Definition of the CAFL-eCK Model

Definition 11 ($\boldsymbol{\lambda}$ – CAFL-eCK Security Game). *Security of a key exchange protocol in the CAFL-eCK model is defined using the following security game, which is played by the adversary $\mathcal{A}$ against the protocol challenger.*

- ***Stage 1:*** *$\mathcal{A}$ may ask any of* Send, SessionKeyReveal, EphemeralKeyReveal *and* Corrupt *queries to any oracle at will.*
- ***Stage 2:*** *$\mathcal{A}$ chooses a $\boldsymbol{\lambda}$ – CAFL-eCK-fresh oracle and asks a* Test *query. The challenger chooses a random bit $b \xleftarrow{\$} \{0, 1\}$, and if $b = 1$ then the actual session key is returned to $\mathcal{A}$, otherwise a random string chosen from the same session key space is returned to $\mathcal{A}$.*
- ***Stage 3:*** *$\mathcal{A}$ continues asking* Send, SessionKeyReveal, EphemeralKeyReveal *and* Corrupt *queries. $\mathcal{A}$ may not ask a query that violates the $\boldsymbol{\lambda}$ – CAFL-eCK-freshness of the test session.*
- ***Stage 4:*** *At some point $\mathcal{A}$ outputs the bit $b' \leftarrow \{0, 1\}$ which is its guess of the value b on the test session. $\mathcal{A}$ wins if $b' = b$.*

The eCK security game is same as the above, except that in Stage 2 and Stage 3 eCK-fresh oracles are chosen instead of λ − CAFL-eCK-fresh oracles. $Succ_{\mathcal{A}}$ is the event that the adversary $\mathcal{A}$ wins the security game in Definition 11.

Definition 12 (λ − CAFL-eCK-security). *A protocol π is said to be λ − CAFL-eCK secure if there is no adversary $\mathcal{A}$ that can win the λ − CAFL-eCK security game with significant advantage. The advantage of an adversary $\mathcal{A}$ is defined as* $\mathrm{Adv}_{\pi}^{\lambda-\mathrm{CAFL-eCK}}(\mathcal{A}) = |2\Pr(Succ_{\mathcal{A}}) - 1|$.

3.6   Practical Interpretation of Security of CAFL-eCK Model

We review the relationship between the CAFL-eCK model and real world attack scenarios.

- **Active adversarial capabilities:** Send queries address the powers of an active adversary who can control the message flow over the network. In the previous security models, this property is addressed by introducing the *send* query.
- **Side-channel attacks:** Leakage functions are embedded with the Send query. Thus, assuming that the leakage happens when computations take place in principals, a wide variety of side-channel attacks such as timing attacks, EM emission based attacks, power analysis attacks, which are based on *continuous leakage of long-term secrets* are addressed. This property is not addressed in the earlier security models such as the BR models, the CK model, the eCK model and the Moriyama-Okamoto model.
- **Malware attacks:** EphemeralKeyReveal queries cover the malware attacks which steal stored ephemeral keys, given that the long-term keys may be securely stored separately from the ephemeral keys in places such as smart cards or hardware security modules. Separately, Corrupt queries address malware attacks which steal the long-term secret keys of protocol principals. In the previous security models, this property is addressed by introducing the *ephemeral-key reveal, session-state reveal* and *corrupt* queries.
- **Weak random number generators:** Due to weak random number generators, the adversary may correctly determine the produced random number. EphemeralKeyReveal query addresses situations where the adversary can get the ephemeral secrets. In the previous security models, this property is addressed by introducing the *ephemeral-key reveal query* or the *session-state reveal* query.

4   Protocol P1: Simple eCK-Secure Key Exchange

The motivation of LaMacchia et al. [19] in designing the eCK model was that an adversary should have to compromise both the long-term and ephemeral secret keys of a party in order to recover the session key. In this section we look at construction paradigms of eCK-secure key exchange protocols, because our aim is to construct a CAFL-eCK-secure key exchange protocol using a eCK-secure key exchange protocol as the underlying primitive.

In the NAXOS protocol, [19], this is accomplished using what is now called the "NAXOS trick": a "pseudo" ephemeral key $\widetilde{esk}$ is computed as the hash of the long-term key lsk and the actual ephemeral key esk: $\widetilde{esk} \leftarrow H(esk, lsk)$. The value $\widetilde{esk}$ is never stored, and thus in the eCK model the adversary must learn both esk and lsk in order to be able to compute $\widetilde{esk}$. The initiator must compute $\widetilde{esk} = H(esk, lsk)$ twice: once when sending its Diffie–Hellman ephemeral public key $g^{\widetilde{esk}}$, and once when computing the Diffie–Hellman shared secrets from the received values. This is to avoid storing a single value that, when compromised, can be used to compute the session key.

Moving to the leakage-resilient setting requires rethinking the NAXOS trick. Alawatugoda et al. [3] have proposed a generic construction of an after-the-fact leakage eCK $((\cdot)\text{AFL-eCK})$-secure key exchange protocol, which uses a leakage-resilient NAXOS trick. The leakage-resilient NAXOS trick is obtained using a decryption function of an after-the-fact leakage-resilient public key encryption scheme. A concrete construction of a BAFL-eCK-secure protocol is possible since there exists a bounded after-the-fact leakage-resilient public key encryption scheme which can be used to obtain the required leakage-resilient NAXOS trick, but it is not possible to construct a CAFL-eCK-secure protocol since there is no continuous after-the-fact leakage-resilient scheme available. Therefore, an attempt to construct a CAFL-eCK-secure key exchange protocol using the leakage-resilient NAXOS approach is not likely at this stage.

4.1   Description of Protocol P1

Our aim is to construct an eCK-secure key exchange protocol which does not use the NAXOS trick, but combines long-term secret keys and ephemeral secret keys to compute the session key, in a way that guarantees eCK security of the protocol. The protocol P1 shown in Table 2 is a Diffie–Hellman-type [10] key agreement protocol. Let $\mathbb{G}$ be a group of prime order q and generator g. After exchanging the public values both principals compute a Diffie–Hellman-type shared secret, and then compute the session key using a random oracle H. We use the random oracle because otherwise it is not possible to perfectly simulate the interaction between the adversary and the protocol, in a situation where the simulator does not know a long-term secret key of a protocol principal.

In order to compute the session key, protocol P1 combines four components $(Z_1 \leftarrow B^a, Z_3 \leftarrow Y^a, Z_4 \leftarrow Y^x, Z_2 \leftarrow B^x)$ using the random oracle function H. These four components cannot be recovered by the attacker without both the ephemeral and long-term secret keys of at least one protocol principal, which allows a proof of eCK security.

Though the design of protocol P1 is quite straightforward, we could not find it given explicitly in the literature: most work on the design of eCK-secure protocols seeks to create more efficient protocols than this naive protocol, but the naive protocol is more appropriate for building into a leakage-resilient protocol.

Table 2. Protocol P1

Alice (Initiator)	Bob (Responder)
Initial Setup	
$a \xleftarrow{\$} \mathbb{Z}_q^*, A \leftarrow g^a$	$b \xleftarrow{\$} \mathbb{Z}_q^*, B \leftarrow g^b$
Protocol Execution	
$x \xleftarrow{\$} \mathbb{Z}_q^*, X \leftarrow g^x \quad \xrightarrow{Alice, X}$	$y \xleftarrow{\$} \mathbb{Z}_q^*, Y \leftarrow g^y$
$\xleftarrow{Bob, Y}$	
$Z_1 \leftarrow B^a, Z_2 \leftarrow B^x$	$Z_1' \leftarrow A^b, Z_2' \leftarrow X^b$
$Z_3 \leftarrow Y^a, Z_4 \leftarrow Y^x$	$Z_3' \leftarrow A^y, Z_4' \leftarrow X^y$
$K \leftarrow \mathrm{H}(Z_1, Z_2, Z_3, Z_4, Alice, X, Bob, Y)$	$K \leftarrow \mathrm{H}(Z_1', Z_2', Z_3', Z_4', Alice, X, Bob, Y)$
K is the session key	

Leakage-Resilient Rethinking of Protocol P1. Moving to the leakage-resilient setting requires rethinking the exponentiation computation in a leakage-resilient manner. Since there exist leakage-resilient encoding schemes and leakage-resilient refreshing protocols for them (Definitions 4 and 7) our aim is computing the required exponentiations in a leakage-resilient manner using the available leakage-resilient storage and refreshing schemes. For now we look at the eCK security of protocol P1, and later in Sect. 5 we will look at the leakage-resilient modification to protocol P1 in detail.

4.2   Security Analysis of Protocol P1

Theorem 2. *If* H *is modeled as a random oracle and* $\mathbb{G}$ *is a group of prime order* q *and generator* g, *where the gap Diffie–Hellman* (GDH) *problem is hard, then protocol* P1 *is secure in the* eCK *model.*

Let $\mathcal{U} = \{U_1, \ldots, U_{N_P}\}$ *be a set of* N_P *parties. Each party* U_i *owns at most* N_S *number of protocol sessions. Let* $\mathcal{A}$ *be an adversary against protocol* P1. *Then,* $\mathcal{B}$ *is an algorithm which is constructed using the adversary* $\mathcal{A}$, *against the* GDH *problem such that the advantage of* $\mathcal{A}$ *against the* eCK-*security of protocol* P1, $\mathrm{Adv}_{\mathrm{P1}}^{\mathrm{eCK}}$ *is:*

$$\mathrm{Adv}_{\mathrm{P1}}^{\mathrm{eCK}}(\mathcal{A}) \leq \max\left(N_P^2 N_S^2 \left(\mathrm{Pr}_{g,q}^{\mathrm{GDH}}(\mathcal{B})\right), N_P^2 \left(\mathrm{Pr}_{g,q}^{\mathrm{GDH}}(\mathcal{B})\right), \quad N_P^2 N_S \left(\mathrm{Pr}_{g,q}^{\mathrm{GDH}}(\mathcal{B})\right), \right.$$

$$\left. N_P^2 N_S \left(\mathrm{Pr}_{g,q}^{\mathrm{GDH}}(\mathcal{B})\right), N_P^2 N_S \left(\mathrm{Pr}_{g,q}^{\mathrm{GDH}}(\mathcal{B})\right), N_P^2 \left(\mathrm{Pr}_{g,q}^{\mathrm{GDH}}(\mathcal{B})\right) \right).$$

Proof Sketch: Let **A** denote the event that $\mathcal{A}$ wins the eCK challenge. Let **H** denote the event that $\mathcal{A}$ queries the random oracle H with $(\mathrm{CDH}(A^*, B^*), \mathrm{CDH}(B^*, X^*), \mathrm{CDH}(A^*, Y^*), \mathrm{CDH}(X^*, Y^*), initiator, X, responder, Y)$, where A^*, B^* are the long-term public-keys of the two partners to the test session, and X^*, Y^* are their ephemeral public keys for this session. Note that when $A = g^a, B = g^b$, $\mathrm{CDH}(A, B) = g^{ab}$; also *initiator* is the initiator of the session and *responder* is the responder of the session.

$$\mathrm{Pr}(\mathbf{A}) \leq \mathrm{Pr}(\mathbf{A} \wedge \mathbf{H}) + \mathrm{Pr}(\mathbf{A} \wedge \bar{\mathbf{H}}).$$

Without the event **H** occurring, the session key given as the answer to the Test query is random-looking to the adversary, and therefore $\Pr(\mathbf{A}|\bar{\mathbf{H}}) = \frac{1}{2}$. $\Pr(\mathbf{A} \wedge \bar{\mathbf{H}}) = \Pr(\mathbf{A}|\bar{\mathbf{H}}) \Pr(\bar{\mathbf{H}})$, and therefore $\Pr(\mathbf{A} \wedge \bar{\mathbf{H}}) \leq \frac{1}{2}$. Hence,

$$\Pr(\mathbf{A}) \leq \frac{1}{2} + \Pr(\mathbf{A} \wedge \mathbf{H}),$$

that is $\Pr(\mathbf{A} \wedge \mathbf{H}) = \mathrm{Adv}_{\mathrm{P1}}^{\mathrm{eCK}}(\mathcal{A})$. Henceforth, the event $(\mathbf{A} \wedge \mathbf{H})$ is denoted as $\mathbf{A}^*$.

Note 1. Let $\mathcal{B}$ be an algorithm against a GDH challenger. $\mathcal{B}$ receives $L = g^\ell, W = g^w$ as the GDH challenge and $\mathcal{B}$ has access to a DDH oracle, which outputs 1 if the input is a tuple of $(g^\alpha, g^\beta, g^{\alpha\beta})$. $\Omega : \mathbb{G} \times \mathbb{G} \to \mathbb{G}$ is a random function known only to $\mathcal{B}$, such that $\Omega(\varPhi, \varTheta) = \Omega(\varTheta, \varPhi)$ for all $\varPhi, \varTheta \in \mathbb{G}$. $\mathcal{B}$ will use $\Omega(\varPhi, \varTheta)$ as $\mathrm{CDH}(\varPhi, \varTheta)$ in situations where $\mathcal{B}$ does not know $\log_g \varPhi$ and $\log_g \varTheta$. Except with negligible probability, $\mathcal{A}$ will not recognize that $\Omega(\varPhi, \varTheta)$ is being used as $\mathrm{CDH}(\varPhi, \varTheta)$.

We construct the algorithm $\mathcal{B}$ using $\mathcal{A}$ as a sub-routine. $\mathcal{B}$ receives $L = g^\ell, W = g^w$ as the GDH challenge. We consider the following mutually exclusive events, under two main cases:

1. A partner to the test session exists: the adversary is allowed to corrupt both principals or reveal ephemeral keys from both sessions of the test session.
   (a) Adversary corrupts both the owner and partner principals to the test session - Event $\mathbf{E_{1a}}$
   (b) Adversary corrupts neither owner nor partner principal to the test session - Event $\mathbf{E_{1b}}$
   (c) Adversary corrupts the owner to the test session, but does not corrupt the partner to the test session - Event $\mathbf{E_{1c}}$
   (d) Adversary corrupts the partner to the test session, but does not corrupt the owner to the test session - Event $\mathbf{E_{1d}}$
2. A partner to the test session does not exist: the adversary is not allowed to corrupt the intended partner principal to the test session.
   (a) Adversary corrupts the owner to the test session - Event $\mathbf{E_{2a}}$
   (b) Adversary does not corrupt the owner to the test session - Event $\mathbf{E_{2b}}$

In any other situation the test session is no longer fresh. If event $\mathbf{A}^*$ happens at least one of the following event should happen.

$$[(\mathbf{E_{1a}} \wedge \mathbf{A}^*), (\mathbf{E_{1b}} \wedge \mathbf{A}^*), (\mathbf{E_{1c}} \wedge \mathbf{A}^*), (\mathbf{E_{1d}} \wedge \mathbf{A}^*), (\mathbf{E_{2a}} \wedge \mathbf{A}^*), (\mathbf{E_{2b}} \wedge \mathbf{A}^*)]$$

Hence,

$$\mathrm{Adv}_{\mathrm{P1}}^{\mathrm{eCK}} \leq \max\Big(\Pr(\mathbf{E_{1a}} \wedge \mathbf{A}^*), \Pr(\mathbf{E_{1b}} \wedge \mathbf{A}^*), \Pr(\mathbf{E_{1c}} \wedge \mathbf{A}^*),$$
$$\Pr(\mathbf{E_{1d}} \wedge \mathbf{A}^*), \Pr(\mathbf{E_{2a}} \wedge \mathbf{A}^*), \Pr(\mathbf{E_{2b}} \wedge \mathbf{A}^*)\Big).$$

Complete security analysis of each event is available in the full version of this paper [2].   □

5   Protocol P2: A Leakage-Resilient Version of P1

Protocol P1 is an eCK-secure key exchange protocol (Theorem 2). The eCK model considers an environment where partial information leakage does not take place. Following the concept that only computation leaks information, we now assume that the leakage of long-term secret keys happens when computations are performed using them. Then, instead of the *non-leakage* eCK model which we used for the security proof of protocol P1, we consider the CAFL-eCK model which additionally allows the adversary to obtain continuous leakage of long-term secret keys.

Our idea is to perform the computations which use long-term secret keys (exponentiation operations) in such a way that the resulting leakage from the long-term secrets should not leak sufficient information to reveal them to the adversary. To overcome that challenge we use a leakage-resilient storage scheme and a leakage-resilient refreshing protocol, and modify the architecture of protocol P1, in such a way that the secret keys s are encoded into two portions s_L, s_R, Exponentiations are computed using two portions s_L, s_R instead of directly using s, and the two portions s_L, s_R are being refreshed continuously. Thus, we add leakage resiliency to the eCK-secure protocol P1 and construct protocol P2 such that it is leakage-resilient and eCK-secure.

Obtaining Leakage Resiliency by Encoding Secrets. In this setting we encode a secret s using an Encode function of a leakage-resilient storage scheme $\Lambda = (\text{Encode}, \text{Decode})$. So the secret s is encoded as $(s_L, s_R) \leftarrow \text{Encode}(s)$. As mentioned in the Definition 2.4.1 the leakage-resilient storage scheme randomly chooses s_L and then computes s_R such that $s_L \cdot s_R = s$. We define the tuple leakage parameter $\boldsymbol{\lambda} = (\lambda_1, \lambda_2)$ as follows: $\boldsymbol{\lambda}$-limited adversary $\mathcal{A}$ sends a leakage function $\mathbf{f} = (f_{1j}, f_{2j})$ and obtains at most λ_1, λ_2 amount of leakage from each of the two encodings of the secret s respectively: $f_{1j}(s_L)$ and $f_{2j}(s_R)$.

As mentioned in Definition 7, the leakage-resilient storage scheme can continuously refresh the encodings of the secret. Therefore, after executing the refreshing protocol it outputs new random-looking encodings of the same secret. So for the $\boldsymbol{\lambda}$-limited adversary again the situation is as before. Thus, refreshing the encodings will help to obtain leakage resilience over a number of protocol executions.

The computation of exponentiations is also split into two parts. Let $\mathbb{G}$ be a group of prime order q with generator g. Let $s \xleftarrow{\$} \mathbb{Z}_q^*$ be a long-term secret key and $E = g^e$ be a received ephemeral value. Then, the value Z needs to be computed as $Z \leftarrow E^s$. In the leakage-resilient setting, in the initial setup the secret key is encoded as $s_L, s_R \leftarrow \text{Encode}_{\mathbb{Z}_q^*}^{n,1}(s)$. So the vector $s_L = (s_{L1}, \cdots, s_{Ln})$ and the vector $s_R = (s_{R1}, \cdots, s_{Rn})$ are such that $s = s_{L1}s_{R1} + \cdots + s_{Ln}s_{Rn}$. Then the computation of E^s can be performed as two component-wise computations as follows: compute the intermediate vector $T \leftarrow E^{s_L} = (E^{s_{L1}}, \cdots, E^{s_{Ln}})$ and then compute the element $Z \leftarrow T^{s_R} = E^{s_{L1}s_{R1}}E^{s_{L2}s_{R2}} \dots E^{s_{L1}s_{R1}} = E^{s_{L1}s_{R1} + \cdots + s_{Ln}s_{Rn}} = E^s$.

5.1 Description of Protocol P2

Using the above ideas, by encoding the secret using a leakage-resilient storage scheme, and refreshing the encoded secret using a refreshing protocol, it is possible to hide the secret from a λ-limited adversary. Further, it is possible to successfully compute the exponentiation using the encoded secrets. We now use these primitives to construct a CAFL-eCK-secure key exchange protocol, using an eCK-secure key exchange protocol as an underlying primitive.

Let $\Lambda_{\mathbb{Z}_q^*}^{n,1} = (\text{Encode}_{\mathbb{Z}_q^*}^{n,1}, \text{Decode}_{\mathbb{Z}_q^*}^{n,1})$ be the leakage-resilient storage scheme which is used to encode secret keys and $\text{Refresh}_{\mathbb{Z}_q^*}^{n,1}$ be the $(\ell, \lambda, \epsilon)$-secure leakage-resilient refreshing protocol of $\Lambda_{\mathbb{Z}_q^*}^{n,1}$.

As we can see, the obvious way of key generation (initial setup) in a protocol principal of this protocol is as follows: first pick $a \xleftarrow{\$} \mathbb{Z}_q^*$ as the long-term secret key, then encode the secret key as $(a_L^0, a_R^0) \leftarrow \text{Encode}_{\mathbb{Z}_q^*}^{n,1}(a)$, then compute the long-term public key $A = g^a$ using the two encodings (a_L^0, a_R^0), and finally erase a from the memory. The potential threat to that key generation mechanism is that even though the long-term secret key a is erased from the memory, it might not be properly erased and can be leaked to the adversary during the key generation. In order to avoid such a vulnerability, we randomly picks two values $a_L^0 \xleftarrow{\$} (\mathbb{Z}_q^*)^n \setminus \{(0^n)\}$, $a_R^0 \xleftarrow{\$} (\mathbb{Z}_q^*)^{n \times 1} \setminus \{(0^{n \times 1})\}$ and use them as the encodings of the long-term secret key a of a protocol principal. As explained earlier, we use a_L^0, a_R^0 to compute the corresponding long-term public key A in two steps as $a' \leftarrow g^{a_L^0}$ and $A \leftarrow a'^{a_R^0}$. Thus, it is possible to avoid exposing the un-encoded secret key a at any point of time in the key generation and hence avoid leaking directly from a at the key generation step. Further, the random vector a_L^0 is multiplied with the random vector a_R^0, such that $a = a_L^0 \cdot a_R^0$, which will give a random integer a in the group $\mathbb{Z}_q^*$. Therefore, this approach is same as picking $a \xleftarrow{\$} \mathbb{Z}_q^*$ at first and then encode, but in the reverse order. During protocol execution both a_L^0, a_R^0 are continuously refreshed and refreshed encodings a_L^j, a_R^j are used to exponentiation computations.

Table 3 shows protocol P2. In this setting leakage of a long-term secret key does not happen directly from the long-term secret key itself, but from the two encodings of the long-term secret key (the leakage function $\mathbf{f} = (f_{1j}, f_{2j})$ directs to the each individual encoding). During the exponentiation computations and the refreshing operation collectively at most $\boldsymbol{\lambda} = (\lambda_1, \lambda_2)$ leakage is allowed to the adversary from each of the two portions independently. Then, the two portions of the encoded long-term secret key are refreshed and in the next protocol session another λ-bounded leakage is allowed. Thus, continuous leakage is allowed.

5.2 Security Analysis of Protocol P2

Theorem 3. *If the underlying refreshing protocol* $\text{Refresh}_{\mathbb{Z}_q^*}^{n,1}$ *is* $(\ell, \lambda, \epsilon)$-*secure leakage-resilient refreshing protocol of the leakage-resilient storage scheme* $\Lambda_{\mathbb{Z}_q^*}^{n,1}$

Table 3. Concrete construction of Protocol P2

Alice (Initiator)	Bob (Responder)
Initial Setup	
$a_L^0 \xleftarrow{\$} (\mathbb{Z}_q^*)^n \setminus \{(0^n)\}, a_R^0 \xleftarrow{\$} (\mathbb{Z}_q^*)^{n \times 1} \setminus \{(0^{n \times 1})\}$	$b_L^0 \xleftarrow{\$} (\mathbb{Z}_q^*)^n \setminus \{(0^n)\}, b_R^0 \xleftarrow{\$} (\mathbb{Z}_q^*)^{n \times 1} \setminus \{(0^{n \times 1})\}$
$a' \leftarrow g^{a_L^0}, A \leftarrow (a')^{a_R^0}$	$b' \leftarrow g^{b_L^0}, B \leftarrow (b')^{b_R^0}$

Alice (Initiator)	Bob (Responder)
Protocol Execution	
$x \xleftarrow{\$} \mathbb{Z}_q^*, X \leftarrow g^x$ $\xrightarrow{Alice,X}$ $\xleftarrow{Bob,Y}$	$y \xleftarrow{\$} \mathbb{Z}_q^*, Y \leftarrow g^y$
$T_1 \leftarrow B^{a_L^j}, Z_1 \leftarrow T_1^{a_R^j}$	$T_3 \leftarrow A^{b_L^j}, Z_1' \leftarrow T_3^{b_R^j}$
$Z_2 \leftarrow B^x$	$T_4 \leftarrow X^{b_L^j}, Z_2' \leftarrow T_4^{b_R^j}$
$T_2 \leftarrow Y^{a_L^j}, Z_3 \leftarrow T_2^{a_R^j}$	$Z_3' \leftarrow A^y$
$Z_4 \leftarrow Y^x$	$Z_4' \leftarrow X^y$
$(a_L^{j+1}, a_R^{j+1}) \leftarrow \mathrm{Refresh}_{\mathbb{Z}_q^*}^{n,1}(a_L^j, a_R^j)$	$(b_L^{j+1}, b_R^{j+1}) \leftarrow \mathrm{Refresh}_{\mathbb{Z}_q^*}^{n,1}(b_L^j, b_R^j)$
$K \leftarrow \mathrm{H}(Z_1, Z_2, Z_3, Z_4, Alice, X, Bob, Y)$	$K \leftarrow \mathrm{H}(Z_1', Z_2', Z_3', Z_4', Alice, X, Bob, Y)$
K is the session key	

and the underlying key exchange protocol P1 is eCK-secure key exchange protocol, then protocol P2 is λ-CAFL-eCK-secure.

Let $\mathcal{A}$ be an adversary against the key exchange protocol P2. Then the advantage of $\mathcal{A}$ against the CAFL-eCK-security of protocol P2 is:

$$\mathrm{Adv}_{P2}^{\lambda-\mathrm{CAFL\text{-}eCK}}(\mathcal{A}) \leq N_P \left(\mathrm{Adv}_{P1}^{\mathrm{eCK}}(\mathcal{A}) + \epsilon \right).$$

Proof. The proof proceeds by a sequence of games.

- **Game 1.** This is the original game.
- **Game 2.** Same as Game 1 with the following exception: before $\mathcal{A}$ begins, an identity of a random principal $U^* \xleftarrow{\$} \{U_1, \ldots, U_{N_P}\}$ is chosen. Challenger expects that the adversary will issue the Test for a session which involves the principal U^* ($\Pi_{U^*, \cdot}$ or $\Pi_{\cdot, U^*}$). If not the challenger aborts the game.
- **Game 3.** Same as Game 2 with the following exception: challenger picks a random $s \xleftarrow{\$} \mathbb{Z}_q^*$ and uses encodings of s to simulate the adversarial leakage queries $\mathbf{f} = (f_{1j}, f_{2j})$ of the principal U^*.

We now analyze the adversary's advantage of distinguishing each game from the previous game. Let $\mathrm{Adv}_{\mathrm{Game}\ x}(\mathcal{A})$ denote the advantage of the adversary $\mathcal{A}$ winning Game x.

Game 1 is the original game. Hence,

$$\mathrm{Adv}_{\mathrm{Game}\ 1}(\mathcal{A}) = \mathrm{Adv}_{P2}^{\lambda-\mathrm{CAFL\text{-}eCK}}(\mathcal{A}). \tag{1}$$

Game 1 and Game 2. The probability of Game 2 to be halted due to incorrect choice of the test session is $1 - \frac{1}{N_P}$. Unless the incorrect choice happens, Game 2 is identical to Game 1. Hence,

$$\mathrm{Adv}_{\mathrm{Game}\ 2}(\mathcal{A}) = \frac{1}{N_P} \mathrm{Adv}_{\mathrm{Game}\ 1}(\mathcal{A}). \tag{2}$$

Game 2 and Game 3. We construct an algorithm $\mathcal{B}$ against a leakage-resilient refreshing protocol challenger of $\text{Refresh}_{\mathbb{Z}_q^*}^{n,1}$, using the adversary $\mathcal{A}$ as a subroutine.

The $(\ell, \boldsymbol{\lambda}, \epsilon)$-$\text{Refresh}_{\mathbb{Z}_q^*}^{n,1}$ refreshing protocol challenger chooses $s_0, s_1 \xleftarrow{\$} \mathbb{Z}_q^*$ and sends them to the algorithm $\mathcal{B}$. Further, the refreshing protocol challenger randomly chooses $s \xleftarrow{\$} \{s_0, s_1\}$ and uses s as the secret to compute the leakage from encodings of s. Let $\boldsymbol{\lambda} = (\lambda_1, \lambda_2)$ be the leakage bound and the refreshing protocol challenger continuously refresh the two encodings of the secret s.

When the algorithm $\mathcal{B}$ gets the challenge of s_0, s_1 from the refreshing protocol challenger, $\mathcal{B}$ uses s_0 as the secret key of the protocol principal U^* and computes the corresponding public key. For all other protocol principals $\mathcal{B}$ sets secret/public key pairs by itself. Using the setup keys, $\mathcal{B}$ computes answers to all the queries from $\mathcal{A}$ and simulates the view of CAFL-eCK challenger of protocol P2. $\mathcal{B}$ computes the leakage of secret keys by computing the adversarial leakage function $\mathbf{f}$ on the corresponding secret key (encodings of secret key), except the secret key of the protocol principal U^*. In order to obtain the leakage of the secret key of U^*, algorithm $\mathcal{B}$ queries the refreshing protocol challenger with the adversarial leakage function $\mathbf{f}$, and passes that leakage to $\mathcal{A}$.

If the secret s chosen by the refreshing protocol challenger is s_0, the leakage of the secret key of U^* simulated by $\mathcal{B}$ (with the aid of the refreshing protocol challenger) is the real leakage. Then the simulation is identical to Game 2. Otherwise, the leakage of the secret key of U^* simulated by $\mathcal{B}$ is a leakage of a random value. Then the simulation is identical to Game 3. Hence,

$$|\text{Adv}_{\text{Game 2}}(\mathcal{A}) - \text{Adv}_{\text{Game 3}}(\mathcal{A})| \leq \epsilon. \tag{3}$$

Game 3. Since the leakage is computed using a random s value, the adversary $\mathcal{A}$ will not get any advantage due to the leakage. Therefore, the advantage $\mathcal{A}$ will get is same as the advantage that $\mathcal{A}$ has against eCK challenger of protocol P1. Because both P1 and P2 are effectively doing the same computation, regardless of the protocol P2, and with no useful leakage the CAFL-eCK model is same as the eCK model. Hence,

$$\text{Adv}_{\text{Game 3}}(\mathcal{A}) = \text{Adv}_{\text{P1}}^{\text{eCK}}(\mathcal{A}). \tag{4}$$

We find,

$$\text{Adv}_{\text{P2}}^{\boldsymbol{\lambda}-\text{CAFL-eCK}}(\mathcal{A}) \leq N_P \left(\text{Adv}_{\text{P1}}^{\text{eCK}}(\mathcal{A}) + \epsilon \right). \qquad \square$$

5.3   Leakage Tolerance of Protocol P2

The order of the group $\mathbb{G}$ is q. Let $m = 1$ in the leakage-resilient storage scheme $\Lambda_{\mathbb{Z}_q^*}^{n,1}$. According to the Lemma 1, if $m < n/20$, then the leakage parameter for the leakage-resilient storage scheme is $\boldsymbol{\lambda}_\Lambda = (0.3n \log q, 0.3n \log q)$. Let $n = 21$, then $\boldsymbol{\lambda}_\Lambda = (6.3 \log q, 6.3 \log q)$ bits. According to the Theorem 1, if $m/3 \leq n$ and $n \geq 16$, the refreshing protocol $\text{Refresh}_{\mathbb{Z}_q^*}^{n,1}$ of the leakage-resilient storage scheme

$\Lambda_{\mathbb{Z}_q^*}^{n,1}$ is tolerant to (continuous) leakage up to $\lambda_{\text{Refresh}} = \lambda_\Lambda/2 = (3.15 \log q, 3.15 \log q)$ bits, per occurrence.

When a secret key s (of size $\log q$ bits) of protocol P2 is encoded into two parts, the left part s_L will be $n \cdot \log q = 21 \log q$ bits and the right part s_R will be $n \cdot 1 \cdot \log q = 21 \log q$ bits. For a tuple leakage function $\mathbf{f} = (f_{1j}, f_{2j})$ (each leakage function $f_{(.)}$ for each of the two parts s_L and s_R), there exists a tuple leakage bound $\lambda = (\lambda, \lambda)$ for each leakage function $f_{(.)}$, such that $\lambda = 3.15 \log q$ bits, per occurrence, which is $\frac{3.15 \log q}{21 \log q} \times 100\% = 15\%$ of the size of a part. The overall leakage amount is unbounded since continuous leakage is allowed.

6 Conclusion

In this paper we answered that open problem of constructing a concrete CAFL-eCK secure key exchange protocol by using a leakage-resilient storage scheme and its refreshing protocol. The main technique used to achieve after-the-fact leakage resilience is encoding the secret key into two parts and only allowing the independent leakage from each part. As future work it is worthwhile to investigate whether there are other techniques to achieve after-the-fact leakage resilience, rather than encoding the secret into parts. Moving to the standard model is another possible research direction. Strengthening the security model, by not just restricting to the independent leakage from each part, would be a more challenging research direction.

Acknowledgements. This research was supported in part by Australian Research Council (ARC) Discovery Project grant DP130104304.

References

1. Akavia, A., Goldwasser, S., Vaikuntanathan, V.: Simultaneous hardcore bits and cryptography against memory attacks. In: Reingold, O. (ed.) TCC 2009. LNCS, vol. 5444, pp. 474–495. Springer, Heidelberg (2009)
2. Alawatugoda, J., Boyd, C., Stebila, D.: Continuous after-the-fact leakage-resilient eck-secure key exchange. IACR Cryptology ePrint Archive 2015:335 (2015)
3. Alawatugoda, J., Stebila, D., Boyd, C.: Modelling after-the-fact leakage for key exchange. In: ASIACCS (2014)
4. Alwen, J., Dodis, Y., Wichs, D.: Leakage-resilient public-key cryptography in the bounded-retrieval model. In: Halevi, S. (ed.) CRYPTO 2009. LNCS, vol. 5677, pp. 36–54. Springer, Heidelberg (2009)
5. Bellare, M., Rogaway, P.: Entity authentication and key distribution. In: Stinson, D.R. (ed.) CRYPTO 1993. LNCS, vol. 773, pp. 232–249. Springer, Heidelberg (1994)
6. Bernstein, D.J.: Cache-timing attacks on AES. Technical report (2005). http://cr.yp.to/antiforgery/cachetiming-20050414.pdf
7. Brakerski, Z., Kalai, Y.T., Katz, J., Vaikuntanathan, V.: Overcoming the hole in the bucket: public-key cryptography resilient to continual memory leakage. IACR Cryptology ePrint Archive, Report 2010/278 (2010)

8. Brumley, D., Boneh, D.: Remote timing attacks are practical. In: USENIX Security Symposium, pp. 1–14 (2003)

9. Canetti, R., Krawczyk, H.: Analysis of key-exchange protocols and their use for building secure channels. In: Pfitzmann, B. (ed.) EUROCRYPT 2001. LNCS, vol. 2045, pp. 453–474. Springer, Heidelberg (2001)

10. Diffie, W., Hellman, M.: New directions in cryptography. IEEE Trans. Inf. Theory **22**, 644–654 (1976)

11. Dodis, Y., Haralambiev, K., López-Alt, A., Wichs, D.: Efficient public-key cryptography in the presence of key leakage. In: Abe, M. (ed.) ASIACRYPT 2010. LNCS, vol. 6477, pp. 613–631. Springer, Heidelberg (2010)

12. Dziembowski, S., Faust, S.: Leakage-resilient cryptography from the inner-product extractor. In: Lee, D.H., Wang, X. (eds.) ASIACRYPT 2011. LNCS, vol. 7073, pp. 702–721. Springer, Heidelberg (2011)

13. Dziembowski, S., Pietrzak, K.: Leakage-resilient cryptography. In: IEEE Symposium on Foundations of Computer Science, pp. 293–302 (2008)

14. Faust, S., Kiltz, E., Pietrzak, K., Rothblum, G.N.: Leakage-resilient signatures. IACR Cryptology ePrint Archive, Report 2009/282 (2009)

15. Hutter, M., Mangard, S., Feldhofer, M.: Power and EM attacks on passive 13.56 MHz RFID devices. In: Paillier, P., Verbauwhede, I. (eds.) CHES 2007. LNCS, vol. 4727, pp. 320–333. Springer, Heidelberg (2007)

16. Katz, J., Vaikuntanathan, V.: Signature schemes with bounded leakage resilience. In: Matsui, M. (ed.) ASIACRYPT 2009. LNCS, vol. 5912, pp. 703–720. Springer, Heidelberg (2009)

17. Kiltz, E., Pietrzak, K.: Leakage resilient ElGamal encryption. In: Abe, M. (ed.) ASIACRYPT 2010. LNCS, vol. 6477, pp. 595–612. Springer, Heidelberg (2010)

18. Kocher, P.C.: Timing attacks on implementations of Diffie-Hellman, RSA, DSS, and other systems. In: Koblitz, N. (ed.) CRYPTO 1996. LNCS, vol. 1109, pp. 104–113. Springer, Heidelberg (1996)

19. LaMacchia, B., Lauter, K., Mityagin, A.: Stronger security of authenticated key exchange. In: Susilo, W., Liu, J.K., Mu, Y. (eds.) ProvSec 2007. LNCS, vol. 4784, pp. 1–16. Springer, Heidelberg (2007)

20. Malkin, T., Teranishi, I., Vahlis, Y., Yung, M.: Signatures resilient to continual leakage on memory and computation. In: Ishai, Y. (ed.) TCC 2011. LNCS, vol. 6597, pp. 89–106. Springer, Heidelberg (2011)

21. Messerges, T., Dabbish, E., Sloan, R.: Examining smart-card security under the threat of power analysis attacks. IEEE Trans. Comput. **51**, 541–552 (2002)

22. Moriyama, D., Okamoto, T.: Leakage resilient eCK-secure key exchange protocol without random oracles. In: ASIACCS, pp. 441–447 (2011)

23. Yang, G., Mu, Y., Susilo, W., Wong, D.S.: Leakage resilient authenticated key exchange secure in the auxiliary input model. In: Deng, R.H., Feng, T. (eds.) ISPEC 2013. LNCS, vol. 7863, pp. 204–217. Springer, Heidelberg (2013)

A Leakage Resilient MAC

Daniel P. Martin[1]([⊠]), Elisabeth Oswald[1], Martijn Stam[1], and Marcin Wójcik[2]

[1] Department of Computer Science, University of Bristol, Bristol, UK
{dan.martin,elisabeth.oswald,martijn.stam}@bris.ac.uk
[2] The Computer Laboratory, University of Cambridge, Cambridge, UK
marcin.wojcik@cl.cam.ac.uk

Abstract. We put forward the first practical message authentication code (MAC) which is provably secure against continuous leakage under the Only Computation Leaks Information (OCLI) assumption. Within the context of continuous leakage, we introduce a novel modular proof technique: while most previous schemes are proven secure directly in the face of leakage, we reduce the (leakage) security of our scheme to its non-leakage security. This modularity, while known in other contexts, has two advantages: it makes it clearer which parts of the proof rely on which assumptions (i.e. whether a given assumption is needed for the leakage or the non-leakage security) and it also means that, if the security of the non-leakage version is improved, the security in the face of leakage is improved 'for free'. We conclude the paper by discussing implementations; one on a popular core for embedded systems (the ARM Cortex-M4) and one on a high end processor (Intel i7), and investigate some performance and security aspects.

Keywords: Leakage resilience · Message authentication code · Provable security · Side channels · Implementation

1 Introduction

Side channel leakage (*e.g.* via timing, power or EM side channels) enables the extraction of secret data out of cryptographic devices, as initially demonstrated by Kocher (*et al.*) in 1996 and 1999 [17,18]. The engineering community reacted quickly by developing a variety of countermeasures that are commonly described as masking and hiding (see [20]). Such countermeasures intend to *reduce* the overall exploitable leakage via techniques that are cheap to implement.

Initially with hesitance, but more lately with much enthusiasm, the theory community picked up on the fact that schemes are needed which can tolerate some leakage. Complementary to the engineering approach, the aim is to design schemes which do not reduce leakage but cope with it, normally via updating the keys. The most compelling property of this approach is that the security

M. Wójcik—This work was conducted while the author was at the University of Bristol.

© Springer International Publishing Switzerland 2015
J. Groth (Ed.): IMACC 2015, LNCS 9496, pp. 295–310, 2015.
DOI: 10.1007/978-3-319-27239-9_18

definitions intrinsically incorporate leakage and hence security proofs then hold even in the presence of leakage. The main drawback of having theoretical backing of security seems to be that the resulting schemes are typically considerably less efficient than other schemes. A prime example of such a scheme is the stream cipher by Dziembowski and Pietrzak [5].

Despite the fact that almost all real word cryptographic protocols require some form of authentication, there is a distinct gap in the literature when it comes to leakage resilient message authentication codes (MACs). Hazay *et al.* [14] produce a MAC from minimal assumptions (existence of a one way function). While only relying on minimal assumptions is an advantage from a theoretical perspective, the scheme has a major drawback in that it only allows a bounded amount of leakage (this bound relates to the total leakage of the device). This makes the scheme unsuitable for practice. In his Master's thesis, Schipper [30] discusses a MAC construction in yet another security model. However unfortunately this MAC is also undesirable for practice as the number of AES calls used by verification grows logarithmically in the number of tag queries. Pereira *et al.* [28] create a leakage resilient MAC in the simulatable leakage model, following on from the work of Standaert *et al.* [33]. However due to the use of components which are not allowed to leak, and that the simulator given has been shown to be insecure by Longo *et al.* [19], it is not clear what practical guarantees it will provide when implemented.

1.1 Our Contribution

Inspired by the bilinear ElGamal cryptosystem by Kiltz and Pietrzak [15], we propose a MAC scheme that is secure within the continuous leakage model, using the Only Computation Leaks Information assumption (discussed in Sect. 2). To our knowledge this is the first MAC scheme to be given within this model, which has become one of the more desirable models due to its closer link with practical side channel scenarios.

In Sect. 3 we give our basic MAC construction and prove it secure in the random oracle model without leakage. Unlike previous work (where schemes have to be completely re-proven when considering leakage), we can construct our proof when considering leakage by a reduction to the non-leaky version (see Sect. 4). This is the first proof to achieve such a clean reduction, which has several advantages. Firstly it shows more clearly how much the leakage is impacting on the security of a scheme. This also implies if the security of the basic MAC construction is tightened, the security of the MAC construction with leakage is tightened 'for free'. This manifests itself (as seen in the theorem statement) by having the leakage security bound in terms of the security without leakage. Secondly it becomes clearer which further assumptions are required to prove security when assuming leakage: for example the basic MAC construction requires a Random Oracle assumption, while the Generic Group Model is required when leakage is added.

In Sect. 5 we discuss an implementation of our leakage resilient MAC when instantiated over a suitable, pairing supporting, elliptic curve using a well known library (MIRACL). We show that in practice (by compiling our implementation on two very different platforms, an embedded ARM core and a high end INTEL processor) we are reasonably efficient and the cost of providing provable leakage resilience, is not nearly as high as often believed.

In the full version [21] we compare our MAC to the other leakage resilient authentication schemes. We show that compared to the majority of other provably secure schemes we are considerably more efficient. The only scheme which is comparable with regards to efficiency is a signature scheme [12]. We also further elaborate on the leakage of the Jacobi symbol and provide more detailed/formal proofs for the security of the MAC.

1.2 Related Work

Kiltz and Pietrzak [15] combine two techniques that are commonly used within both communities to build a key encapsulation mechanism on top of a key update scheme. The first technique is masking (or secret sharing as it is known by the theoretical community), which involves splitting the key into two parts and then working on each share separately. The second technique is frequent rekeying. Unlike other proposals (e.g. [16] or [1]), which are stateful (and thus need to be synchronised) or ones which needs to transmit a clue [23] to 'synchronise' parties, the proposal by Kiltz and Pietrzak [15] can leverage the algebraic properties of the underlying system such that the resulting system requires no synchronisation. This is achieved by changing the representation of the shares rather than changing the secret itself. Using the same techniques, Galindo and Vivek [12], and Tang et al. [34] create leakage resilient signature schemes. These constructions are proven secure in the continuous leakage model using the OCLI assumption [24] (see also Sect. 2).

Dodis and Pietrzak [4] create a leakage resilient PRF where the leakage functions are chosen non-adaptively before any queries to the PRF are made. Faust et al. [6] construct a simpler leakage resilient PRF, which is achieved at the expense of having to make both the input to the PRF and the leakage non-adaptive. All known PRFs in the continual leakage model have the restriction of being non-adaptive (in the leakage), while MACs do not have this restriction. This shows a separation between PRFs and MACs which does not exist in the non-leakage model but PRFs will still serve as an interesting comparison.

2 Modelling Leakage

In this section we discuss what assumptions we make when modelling leakage. Clearly some restrictions are required on the leakage, otherwise the adversary will be able to win because he can just ask for the key. One of the first decisions to be made is how to define a bound for the leakage (i.e. how many bits about a secret does the adversary get via some side channel). For instance, one could define there to be an overall bound, i.e. the adversary gets at most a certain number of bits, irrespective of how often the construction is actually called (this is called bounded leakage in the literature). Another option would be to impose a per call bound. In this latter case, each call to the construction delivers at most a certain number of bits, while the overall leakage remains unbounded. This type of model is called continuous leakage model and fits best to real world leakage such as power or EM traces.

Whilst some previous works [5,7] make an *a priori* assumption about the computational complexity of the leakage function, we opted for a concrete security statement. This means that the adversarial advantage is explicitly bounded in the complexity of the leakage function as expressed in the number of queries to the generic group oracles (see Sect. 2.2).

Finally we need to restrict the scope of the leakage function because otherwise (given our choices of assumptions above) no security would be possible (because of the infamous 'future computation attack' [15]). We discuss our choice of how to restrict the leakage function in the following.

2.1 Only Computation Leaks Information

Micali and Reyzin [24] introduced the Only Computation Leaks Information (OCLI) assumption. It states that data leakage only occurs on data that is currently being computed on and that data at rest will not leak. Whilst this assumption might not strictly hold in practice (it has been shown to be invalid for some technologies an gate level [29]), it sufficiently captures the behaviour of many state of the art devices.

Application of the OCLI assumption requires splitting a large computation into smaller components that each only operate on a subset of the data available, thus restricting the scope of what can be leaked on. OCLI will be modelled in this paper by splitting a function F into two parts F° and $F^{\bullet}$. The part of the sensitive/exploitable input S used by F° will be denoted S° while the parts of the sensitive input used by $F^{\bullet}$ will be denoted $S^{\bullet}$. Without OCLI, a leakage query could potentially leak on both shares jointly, and thus reveal information about S. However due to OCLI, any leakage query can only ever leak on S° and $S^{\bullet}$ independently, but never jointly on both.

Concretely, in our model the adversary may adaptively (per function call) choose leakage functions $l^{\circ}, l^{\bullet}$ which will leak up to λ bits (this is a security parameter) on F° and $F^{\bullet}$ respectively. The adversary also gets the output $l^{\circ}(S^{\circ}, x^{\circ}, r^{\circ})$ and $l^{\bullet}(S^{\bullet}, x^{\bullet}, r^{\bullet})$ where $x^{\circ}, x^{\bullet}$ is the input to the functions and $r^{\circ}, r^{\bullet}$ is the randomness that they use.

Note that while the leakage functions l° and $l^{\bullet}$ can be chosen adaptively from query to query, they do have to be chosen at the same time for a single query. This restriction—that the leakage function $l^{\bullet}$ is not allowed to depend on the leakage obtained by l°—is quite common in the literature [12,15], and reflects the abilities of a real world adversary (they can't change the measurement set-up mid measurement).

If this leakage process is iterated multiple times an index is used to specify which iteration we are on, for example we use $l_i^{\circ}, l_i^{\bullet}, S_i^{\circ}, S_i^{\bullet}, r_i^{\circ}, r_i^{\bullet}$.

2.2   Bilinear Generic Group Model

We briefly recall the definition of bilinear groups and of bilinear maps, where we adhere to asymmetric pairings (see Galbraith *et al.* [9] for an overview). Let $\mathbb{G}_1, \mathbb{G}_2$, and $\mathbb{G}_3$ be cyclic groups all of prime order p with generators g_1, g_2, and g_3, respectively. A bilinear map is a function $e : \mathbb{G}_1 \times \mathbb{G}_2 \to \mathbb{G}_3$ with the following properties; bilinearity states that $\forall u \in \mathbb{G}_1, v \in \mathbb{G}_2, a, b \in \mathbb{Z}_p : e(u^a, v^b) = e(u, v)^{ab}$, while non-degeneracy $e(g_1, g_2) \neq 1$, stops the construction of trivial maps. From this point onwards we define the generator g_3 of $\mathbb{G}_3$ to be $e(g_1, g_2)$.

The generic group model [22, 26, 32] is well established to prove the security of protocols involving elliptic curves. Its goal is to restrict the adversary in such a way that structure of the underlying group cannot be exploited (beyond what follows from the group axioms). This is achieved by representing each element within the group as a random string and providing oracles for the various group operations. As a consequence, given only a representation of a group element, the only ability the adversary has is to check equality (*i.e.* the adversary must use an oracle to perform any required group operations).

In the Generic Bilinear Group (GBG) model each of the three groups (or two when using a symmetric pairing) has its own randomised encoding. Each of these encodings will be represented by an injective encoding function $\xi_1 : \mathbb{Z}_p \to \Xi_1$, $\xi_2 : \mathbb{Z}_p \to \Xi_2, \xi_3 : \mathbb{Z}_p \to \Xi_3$ for $\mathbb{G}_1, \mathbb{G}_2, \mathbb{G}_3$ respectively, where Ξ_1, Ξ_2, Ξ_3 are sets of bitstrings. The adversary has access to the following 4 oracles:

- $\mathcal{O}_1(\xi_1(a), \xi_1(b)) = \xi_1(a + b \mod p)$
- $\mathcal{O}_2(\xi_2(a), \xi_2(b)) = \xi_2(a + b \mod p)$
- $\mathcal{O}_3(\xi_3(a), \xi_3(b)) = \xi_3(a + b \mod p)$
- $\mathcal{O}_e(\xi_1(a), \xi_2(b)) = \xi_3(a \cdot b \mod p)$

for all $a, b \in \mathbb{Z}_p$. Each of the 4 oracles will return $\bot$ if either of the inputs is not a invalid encoding of an underlying group element. $\mathcal{O}_1, \mathcal{O}_2, \mathcal{O}_3$ perform the group operations of $\mathbb{G}_1, \mathbb{G}_2, \mathbb{G}_3$ respectively, while $\mathcal{O}_e$ performs the pairing operation. To work with these groups an adversary only needs to be given $\xi_1(1)$ and $\xi_2(1)$ (corresponding to the generators of $\mathbb{G}_1$ and $\mathbb{G}_2$ respectively) plus access to the four oracles, from which any group element can be computed.

Leaking on generic group elements only reveals information about their representation. In some proofs (without leakage) that use the generic group model, the representation of group elements can be chosen in such a way that even sampling a random group element is hard (for an adversary). This is typically achieved by representing group elements as 'long' random strings. When leakage is included in proofs, such a strategy would not make sense because it would imply that only 'large' amounts of leakage[1] would strengthen the adversary. We instantiate the generic group model using compact representations instead. By setting $\Xi_i = \{0, 1\}^n$ where $n = \lceil \log p \rceil$ we get the unique representations required. This gives the adversary the ability to sample group elements efficiently and directly.

---

[1] Typically one would need to leak significantly more than $\log p$ bits, where p would be the size of the group.

proc $KG()$:	**proc** $TAG(K, m)$:	**proc** $VRFY(K, T, m)$:
$K \xleftarrow{\$} \mathbb{G}_1$	$W \leftarrow H(m)$	$W \leftarrow H(m)$
Return K	$T \leftarrow e(K, W)$	$T' \leftarrow e(K, W)$
	Return T	Return $T' = T$

Fig. 1. Our bilinear MAC scheme $\mathcal{M}$

In contrast, Kiltz and Pietrzak [15] (and similarly, Galindo and Vivek [12]) use indirect sampling by raising some generator to a random exponent. They allow leakage on both the random representations, as well as their discrete logarithms (with respect to some generator), in order to model the adversary's ability to leak on the sampling computation itself. Our proof can be seen as more restrictive and our proofs only hold for implementing the sampling directly. We remark that it is possible to sample random elliptic curve points efficiently without performing an exponentiation with an unknown exponent, as discussed in more detail in Sect. 5.

3   A MAC Scheme

We define a MAC as a tuple of algorithms $\mathcal{M} = (KG, TAG, VRFY)$ such that:

$$K \xleftarrow{\$} KG()$$
$$\sigma \xleftarrow{\$} TAG(K, m)$$
$$b \leftarrow VRFY(K, \sigma, m).$$

For correctness we require for all valid keys K that $VRFY(K, TAG(K, m), m) = 1$. We use the standard definition of EUF-CMA security for the rest of this section (it is hard to forge a tag on a message which has not been passed to the tagging oracle before).

We now define our basic MAC construction. Using a hash function $H : \{0, 1\}^* \rightarrow \mathbb{G}_2$ our basic MAC scheme $\mathcal{M} = (KG, TAG, VRFY)$ is defined in Fig. 1. It can be shown to provide EUF-CMA security (Theorem 2). The scheme can be understood as follows; key generation consists of generating a random group element of $\mathbb{G}_1$. Tag generation first hashes the message, then takes the resulting hash as input to a bilinear map, using the secret key as other input. The MAC consists of a message, and its tag. Verification simply reconstructs the tag T and checks the correctness.

Before we provide the proof of the MAC we introduce a new Bilinear Diffie–Hellman problem, which we will use in the reduction to show the security of the MAC. This new DH problem will have its security 'sandwiched' between two other well known DH problems.

3.1   A New Bilinear Diffie–Hellman Problem

In Definition 1 we introduce a bilinear problem, which we coin the target bilinear Diffie–Hellman (TBDH) problem. In Theorem 1 we give a reduction to show if Co-Bilinear Diffie–Hellman (CBDH) is assumed to be a hard problem,[2] then so is the TBDH problem. Similarly, it can be shown that if the standard Diffie–Hellman (CDH) Problem is easy in $\mathbb{G}_3$ then the TBDH Problem is easy.

Definition 1 (Target Bilinear Diffie–Hellman Problem). *Given $\mathbb{G}_1, \mathbb{G}_2$, $\mathbb{G}_3$ with a bilinear map e between them, we say the Target Bilinear Diffie–Hellman (TBDH) Problem is hard if given g_2^x, g_3^y it is hard to compute g_3^{xy}, where x, y are sampled uniformly at random from $\mathbb{Z}_p$. Given an adversary A we define its advantage of winning this game as $\mathbf{Adv}^{tbdh}(A) = \Pr\left[\mathcal{A} = g_3^{xy} : \mathcal{A} \leftarrow A(g_1, g_2, g_2^x, g_3^y)\right]$.*

Before relating the TBDH problem to other Diffie–Hellman problems, we recall the CBDH Problem [35]. The CBDH problem states that given g_2^x, g_2^y, (x, y are sampled uniformly at random from $\mathbb{Z}_p$) find g_3^{xy}.

Theorem 1. *Let A be an adversary against the TBDH Problem, then there exists an adversary B (with approximately the same runtime as A) against the CBDH Problem, such that:*

$$\mathbf{Adv}^{\text{tbdh}}(A) < \mathbf{Adv}^{\text{cbdh}}(B).$$

The element g_2^y can be easily can be converted to the element g_3^y by pairing it with the generator g_1, therefore any adversary who can solve the TBDH problem can be used to solve the CBDH problem. A formal reduction is given in the full version of the paper [21].

Theorem 2. *Let $H : \{0,1\}^* \to \mathbb{G}_2$ be modelled as a random oracle and A be an EUF-CMA adversary against $\mathcal{M}$ who makes q_h queries to the hash function and q_v verification queries, then there exists an adversary B (of similar complexity) against the TBDH problem such that:*

$$\mathbf{Adv}_{\mathcal{M}}^{\text{eufcma}}(A) \leq (q_h + 1)(q_v + 1)\mathbf{Adv}^{\text{tbdh}}(B).$$

Proof Intuition. The proof works by reducing the problem of forging the MAC to the problem of solving the TBDH problem. The reduction constructs tags in such a way that it simulates having the key as g_1^y. If the adversary subsequently forges on a point whose hash is g_2^x, the resulting tag will be the answer to the TBDH problem (g_3^{xy}). To answer an adversary's verification queries, we introduce a slight variation of the TBDH problem called the TBDHwO problem (given in the full version) which gives the adversary access to an oracle $test(C)$ to check if $C = g_3^{xy}$. □

---

[2] It is possible to modify our results to the usual notions of negligible advantages against probabilistic polynomial-time adversaries.

4   A Leakage Resilient MAC

We start this section by introducing the definition of a key update mechanism. Kiltz and Pietrzak [15] implicitly constructed and used a key update mechanism within their KEM. This key update mechanism was then used again in the signature scheme by Galindo and Vivek [12] and the signature scheme by Tang *et al.* [34]. After showing that our definition aligns with the KP key update mechanism, we define what it means for a scheme to be compatible with a key update mechanism. We show this is the case for our MAC given in the previous section and then go on to prove our MAC secure in the face of leakage.

4.1   Key Update Mechanism

We define a key update mechanism as a set of tuples $\mathcal{KU} = (Share, Recombine, U^{\circ}, U^{\bullet})$ such that:

$$(S_0^{\circ}, S_0^{\bullet}) \xleftarrow{\$} Share(K)$$
$$(S_{i+1}^{\circ}, r_u) \xleftarrow{\$} U^{\circ}(S_i^{\circ})$$
$$S_{i+1}^{\bullet} \xleftarrow{\$} U^{\bullet}(S_i^{\bullet}, r_u)$$
$$K_i \leftarrow Recombine(S_i^{\circ}, S_i^{\bullet})$$

For correctness we require that $Recombine(Share(K)) = K$.

We define an equivalence class as follows; we say $(S_i^{\circ}, S_i^{\bullet}) \equiv (S_j^{\circ}, S_j^{\bullet})$ if $Recombine(S_i^{\circ}, S_i^{\bullet}) = Recombine(S_j^{\circ}, S_j^{\bullet})$. Then the final requirement is that the algorithms $U^{\circ}, U^{\bullet}$ preserve the equivalence class of the shares (and thus $\forall i : K_i = K$). Formally we require $(S_i^{\circ}, S_i^{\bullet}) \equiv (S_{i+1}^{\circ}, S_{i+1}^{\bullet})$ where $(S_{i+1}^{\circ}, O_i) \xleftarrow{\$} U^{\circ}(S_i^{\circ}), S_i^{\bullet} \xleftarrow{\$} U^{\bullet}(S_i^{\bullet}, O_i)$.

The KP key update mechanism used within the KEM [15] can be seen to fit within this framework. This is due to the fact that the key is initially split into two shares which multiply together to give back the original key. The first share is updated by multiplying it by a random value, while the second share is updated by multiplying it by the inverse of the random value. This forms our equivalence class and thus when the two shares are multiplied together we will recover the original key, regardless of how many times the shares have been updated. The KP key update mechanism will be used for the remainder of this paper (and denoted $\mathcal{KU}$).

Definition 2 (Key Update Splittable). *We say that a tuple of functions* $(F^{\circ}, F^{\bullet})$ *is a split of F conforming to key update mechanism* $\mathcal{KU}$ *if the following two properties hold. Firstly:*

$$\{F(K, x)\}_{\mathcal{R}} = \{F^*(Share(K), x)\}_{\mathcal{R}^*}$$

where F^* is defined in Fig. 2, the equivalence is over the randomness from sets $\mathcal{R}, \mathcal{R}^*$ used by F, F^* respectively. Secondly, that for all sharings $(S_0^{\circ}, S_0^{\bullet})$ the joint distribution on $(S_1^{\circ}, S_1^{\bullet})$ after F^* has been called once is the same as if $(S_0^{\circ}, S_0^{\bullet})$ had been updated using $(U^{\circ}, U^{\bullet})$.

$$\mathbf{proc}\ F^*(S_i^{\circ}, S_i^{\bullet}, x):$$
$$(S_{i+1}^{\circ}, O) \xleftarrow{\$} F^{\circ}(S_i^{\circ}, x)$$
$$(S_{i+1}^{\bullet}, y) \xleftarrow{\$} F^{\bullet}(S_i^{\bullet}, O)$$
$$\text{Return } y$$

Fig. 2. The algorithm F^*

Claim. The MAC $\mathcal{M}$ given in Sect. 3 is Key Update Splittable conforming to the KP Key Update Mechanism $\mathcal{KU}$.

Proof. $\mathcal{M}$ can be converted into $\mathcal{M}^*$ which is given in Fig. 3. Since we have that:

$$Tag^*(S_i^{\circ}, S_i^{\bullet}, m) = T$$
$$= t^{\circ} \cdot t^{\bullet}$$
$$= e(S_i^{\circ}, H(m)) \cdot e(S_i^{\bullet}, H(m))$$
$$= e(S_i^{\circ} \cdot S_i^{\bullet}, H(m))$$
$$= e(K, H(m))$$
$$= Tag(K, m).$$

Since the MAC uses the key update function to update the key, the distributions will be the same. Hence $\mathcal{M}$ is Key Update Splittable. $\square$

There are three algorithms in our leakage resilient MAC: Key Generation, Tag and Verify. Our security definition only allows leakage on Tag, and we now explain why this is necessary. The Key Generation must not leak because it would leak on the original key. In practice, typical (security) devices would be shipped with their keys preinstalled and only the update would be done on the device. This leaves us to consider whether Tag (EUF-CMA-LT) or Verify (EUF-CMA-LV), or both (EUF-CMA-LTV) are allowed to leak. This question has not been considered before in the continual leakage model in the case of symmetric schemes, as all previous schemes in this model were public-key in which the question simply does not arise.

Making Verify leakage resilient is problematic, due to the leakage being adaptive: assume the adversary takes a random group element and a message and sends both to Verify. In our construction (which follows a typical design) Verify has to calculate the correct tag first, and then compare it against the submitted

proc $KG()$:
$K \xleftarrow{\$} \mathbb{G}_1$
$S_0^{\circ} \xleftarrow{\$} \mathbb{G}_1$
$S_0^{\bullet} \leftarrow K \cdot (S_0^{\circ})^{-1}$
Return $(S_0^{\circ}, S_0^{\bullet})$

proc $TAG^{\circ}(S_i^{\circ}, m)$:
$W \leftarrow H(m)$
$t_i^{\circ} \leftarrow e(S_i^{\circ}, W)$
$r_{i+1} \xleftarrow{\$} \mathbb{G}_1$
$s_{i+1}^{\circ} \leftarrow S_i^{\circ} \cdot r_{i+1}$
Return $(S_{i+1}^{\circ}, r_{i+1}, t_i^{\circ}, W)$

proc $TAG^{\bullet}(S_i^{\bullet}, t_i^{\circ}, W, r_{i+1})$:
$t_i^{\bullet} \leftarrow e(S_i^{\bullet}, W)$
$S_{i+1}^{\bullet} \leftarrow S_i^{\bullet} \cdot r_{i+1}^{-1}$
$T \leftarrow t_i^{\circ} \cdot t_i^{\bullet}$
Return $(S_{i+1}^{\bullet}, T)$

proc $VRFY(K, T, m)$:
$W \leftarrow H(m, w)$
$T' \leftarrow e(K, W)$
Return $(T' = T)$

Fig. 3. Leakage resilient MAC $\mathcal{M}^*$

tag. The adversary can keep submitting the same message until he has completely leaked the tag created for comparison. This tag can then be submitted as a forgery since it was never requested from the Tag oracle. This attack will work against any MAC construction which requires reconstruction of the tag as part of verify (it is not specific to our MAC). We leave it as a question for future research how to construct a leakage resilient Verify theoretically. In practice, Verify will leak and whilst we cannot formally include it in the security proof, we can assume that practical countermeasures can be put in place.

Definition 3 (Existential Unforgability Under Chosen Message Attack with Tag Leakage (EUF-CMA-LT)). Let $\mathcal{M}^* = (\mathcal{KU}, TAG^{\circ}, TAG^{\bullet}, VRFY)$ be a Message Authentication Code. Then Fig. 4 defines the EUF-CMA-LT security game. The advantage of an adversary A winning the game is defined as $\mathbf{Adv}_{\mathcal{M}}^{\text{eufcmalt}}(A) = \Pr[\mathbf{Exp}_{\mathcal{M}}^{\text{eufcmalt}}(A) = 1]$.

Theorem 3. The MAC $\mathcal{M}^*$ is EUF-CMA-LT secure in the Generic Group Model. The advantage of a q-query (to the generic group oracles) adversary who is allowed λ bits of leakage is given by:

$$\mathbf{Adv}_{\mathcal{M}^*}^{\text{eufcmalt}}(A) \leq 2^{4 \cdot \lambda} \cdot \mathbf{Adv}_{\mathcal{M}}^{\text{eufcma}}(B) + \frac{q^2}{p}.$$

Proof Intuition. This proof is given in the Generic Group Model (GGM) and shows that even with the use of leakage the adversary cannot get any elements that they could not get when no leakage was involved. After this has been shown, it is reasonably straightforward to argue that without learning any new elements from the leakage then the leakage can increase the adversary's advantage by at most the number of bits that is leaked on for a single element. By showing that each element is only leaked on four times we get that the advantage can only be increased by at most $2^{4\lambda}$ over the advantage in the game where no leakage is involved. □

experiment $\mathbf{Exp}_{\mathcal{M}}^{\text{eufcmalt}}(A)$:

$K \xleftarrow{\$} KG()$

$(S_0^{\newmoon}, S_0^{\fullmoon}) \xleftarrow{\$} SHARE(K)$

$S \leftarrow \{\}$

$(\sigma^*, m^*) \leftarrow A^{Tag(\cdot), Verify(\cdot, \cdot)}$

if $m^* \in S$ then

    return 0

end if

Return $VRFY(K, \sigma^*, m^*)$

proc $Verify(\sigma, m)$:

$b \leftarrow VRFY(K, \sigma, m)$

Return b

proc $Tag(m, l_i^{\newmoon}, l_i^{\fullmoon})$:

$S \leftarrow S \cup \{m\}$

$(S_{i+1}^{\newmoon}, O_i) \xleftarrow{r_i^{\newmoon}} TAG^{\newmoon}(S_i^{\newmoon}, m)$

$\Lambda_i^{\newmoon} \leftarrow l_i^{\newmoon}(S_i^{\newmoon}, r_i^{\newmoon})$

$(S_{i+1}^{\fullmoon}, \sigma) \xleftarrow{r_i^{\fullmoon}} TAG^{\fullmoon}(S_i^{\fullmoon}, O_i)$

$\Lambda_i^{\fullmoon} \leftarrow l_i^{\fullmoon}(S_i^{\fullmoon}, r_i^{\fullmoon}, O_i)$

Return $(\sigma, \Lambda_i^{\newmoon}, \Lambda_i^{\fullmoon})$

Fig. 4. EUF-CMA-LT experiment

5 Practical Aspects of Our Scheme

For our implementation we selected the Barreto-Naehrig (BN) [2] family of pairing-friendly curves. BN curves are defined over a prime field $\mathbb{F}_p$, with prime order and are given by the equation $E : y^2 = x^3 + b$, with $b \neq 0$ (we select $b = 2$). The common feature of this family is their embedding degree of $k = 12$, which to some extent, dictates the security level achieved on the curve. For our implementation we focused on a security level equivalent to 128-bit and 192-bit AES (this security level is before leakage is considered), for which BN curves are ideally suited.

The prime p is given by polynomial $p(u) = 36u^4 + 36u^3 + 24u^2 + 6u + 1$. For efficiency we set $u = -(2^{62} + 2^{55} + 1)$ and $u = -(2^{190} + 2^{19} + 2^{17} + 2^{15} + 2^{13} + 2^{12} + 2^{11} + 2^9 + 2^8 + 2^7 + 2^5 + 2^4 + 2^3 + 1)$ for 128 and 192 bits security level respectively [27]. That determines the size of the operands in the groups $\mathbb{G}_1$, $\mathbb{G}_2$ and $\mathbb{G}_3$, which are over $\mathbb{F}_p$, $\mathbb{F}_{p^2}$ and $\mathbb{F}_{p^{12}}$ respectively. In case of the 128 bit security level, operations are carried out on operands of length 254, 508 and 3048 bits, whereas for 192 bits security level are carried out on operands of 766, 1532 and 9192 bits.

All algorithms in our scheme were implemented using the MIRACL software library [31], which is a portable C/C++ library that supports a wide-rage of different platforms including embedded ones. The advantage of the selected library is the extensive support for highly efficient pairing operations. In addition we extended and adopted functionality provided by the library to our particular case, whenever required. For the underlying pairing operations, MIRACL uses the well-known Miller algorithm [25]. Furthermore, it also applies the Galbraith-Scott method [10], which allows computing a mapping between the groups $\mathbb{G}_2$ and $\mathbb{G}_3$ efficiently. These mappings are further used to speed up arithmetic computations in $\mathbb{G}_2$ and $\mathbb{G}_3$ by applying Gallant-Lambert-Vanstone method [13] (which works when a suitable group mapping is given). All mentioned optimisation strategies increase efficiency of pairing computations, thus speeding up our proposed scheme and making it suitable for more resource-constrained environments.

We implemented and measured execution times of the schemes on both an embedded platform and a high-end device. For the former case, as a target platform, we selected a popular STM32F4Discovery board, which houses 32-bit ARM Cortex-M4 CPU. For the latter case, we utilised the 64-bit Intel Core i7 CPU. The internal clock of the Cortex-M4 was set to available maximum, *i.e.*, 168MHz, whereas the Intel i7 ran with a 3GHz clock. We ran our benchmarks several times to derive median timings.

Table 1. Performance comparison of random point generation methods.

Device	Cortex-M4		Intel i7	
Operation	Time (ms)		Time (μs)	
	128-bits	192-bits	128-bits	192-bits
Random_Sampler	36	588	96	1159
Try_and_Increment	34	569	76	1119
Random_Scalar	173	2827	389	5186
SWEncoding	30	616	121	1217

5.1   Generating Random Curve Points

Generating random group elements securely is vital for key generation and key updates (which happen in TAG). We found four options (see the full version for algorithmic descriptions [21]) for this purpose, which we now discuss in turn. The first one, the Random_Sampler procedure, randomly selects an x-coordinate and checks if it is on the curve. In case of success, the procedure computes an associated y-coordinate, otherwise randomly selects another x-coordinate. The second key generation procedure, Try_and_Increment is very similar to the first one. It differs only in re-selection of x-coordinate in which the procedure increments x-coordinate by 1 and repeats the assessment of whether a new x is on the curve. The third one, Random_Scalar selects a scalar at random and performs a scalar multiplication using a fixed group generator. The last procedure uses the encoding to BN curve [8], where a random element $t \in \mathbb{F}_p$ is transformed into an element of the curve $E(\mathbb{F}_p)$, which was used by Galindo et al. [11]. Note that when using this encoding one has to perform it twice in order to generate a point distributed uniformly at random [11]. Hence in practice the timings are effectively twice as long.

Performance. For a fair comparison of timings, all procedures were implemented without blinding. Applying blinding to the Random_Sampler,

Try_and_Increment and Random_Scalar methods requires one additional multiplication. A more involved blinding method for the BN encoding have been proposed in [8]. Table 1 shows that Random_Sampler and Try_and_Increment are by far the most efficient (recall that the SWEncoding method needs to be performed twice for uniformly distributed points), and it is clear that this advantage would also hold when blinding is included. This is good news as the Random_Scalar method is not only slow, but also known to be very vulnerable to power analysis attacks [3].

Table 2. Performance comparison

Device	Cortex-M4		Intel i7	
Operation	Time (ms)		Time (μs)	
	128-bits	192-bits	128-bits	192-bits
TAG	2146	30317	7935	68692
$VRFY$	2146	30317	7958	68687
KG^*	72	1126	170	2128
TAG^*	4059	57274	15473	130612
$VRFY^* = VRFY$	2146	30317	7958	68687
$Share$	34	566	94	1082
$TAG^{\ominus}$	2183	30883	7974	69650
$TAG^{\ominus}$	1874	26382	7162	60841
$Recombine$	2	11	<1	<1

Attack Vectors. It has been shown practically that the Random_Scalar method can completely leak the entire secret randomness and hence strictly speaking, it cannot be used for schemes proved secure in the continual leakage model. Security aspects of the SWEncoding scheme have been discussed by Galindo *et al.* [11]. They conclude that the SWEncoding might leak via the Jacobi symbol. The security of the other two methods w.r.t. the continual leakage model has not yet been investigated. Hence we will now discuss the security considerations here.

The Try_and_Increment method leaks information about the number of increments via its overall execution time (*e.g.* via power traces). It is not obvious how this information could be utilised efficiently. However it will contribute to the amount of leakage per call for the λ security bound.

The Random_Sampler method chooses values for x independently of previous choices and hence does not leak any additional information on the x from its high level functionality. The only part which may reveal information about the point is the calculation of the Jacobi symbol.

Since for both the SWEncoding and the Random_Sampler any leakage will be from the Jacobi Symbol we now discuss the leakage that may be available during its computation. The Jacobi computation has a conditional operation which swaps the numerator and denominator when certain conditions are met. See the full version [21] for the results of capturing power traces of the Jacobi symbol and the analysis of if there is any information within them which can be exploited. While it is not clear how much use this information is or how it can be exploited, it is recommended to use blinding (with r^2) to hide the point since if $x^3 + b$ is square, $r^2(x^3 + b)$ will also be square for all random r.

5.2 Performance of the Overall Scheme

Finally we give the performance of the high level functions of the MAC constructions (Table 2). The basic MAC scheme had identical operations in the TAG and VRFY procedures (bar the additional equality check in VRFY which is extremely fast), hence the resulting identical timings.

Switching then to the leakage resilient version, it is clear that the cost essentially doubles for the TAG computation. Since we had to assume VRFY was not leaking (recall that our construction, like other MAC constructions requires the reconstruction of TAG during VRFY, which is seemingly impossible to do securely allowing adaptive adversaries in the continual leakage model), the timings for VRFY in the leakage resilient scheme are the same as for the scheme without leakage.

Acknowledgements. Dan Martin and Elisabeth Oswald have been supported in part by EPSRC via grant EP/I005226/1. Marcin Wójcik has been supported by the EU DG Home Affairs - ISEC (Prevention of and Fight against Crime) / INT (Illegal Use of Internet) programme and his research leading to these results has received funding from the European Union's Seventh Framework Programme (FP7/2007-2013) under grant agreement n° 609094.

References

1. Abdalla, M., Belaïd, S., Fouque, P.-A.: Leakage-resilient symmetric encryption via re-keying. In: Bertoni, G., Coron, J.-S. (eds.) CHES 2013. LNCS, vol. 8086, pp. 471–488. Springer, Heidelberg (2013)
2. Barreto, P.S.L.M., Naehrig, M.: Pairing-friendly elliptic curves of prime order. In: Preneel, B., Tavares, S. (eds.) SAC 2005. LNCS, vol. 3897, pp. 319–331. Springer, Heidelberg (2006)
3. Coron, J.-S.: Resistance against differential power analysis for elliptic curve cryptosystems. In: Koç, Ç.K., Paar, C. (eds.) CHES 1999. LNCS, vol. 1717, pp. 292–302. Springer, Heidelberg (1999)
4. Dodis, Y., Pietrzak, K.: Leakage-resilient pseudorandom functions and side-channel attacks on feistel networks. In: Rabin, T. (ed.) CRYPTO 2010. LNCS, vol. 6223, pp. 21–40. Springer, Heidelberg (2010)
5. Dziembowski, S., Pietrzak, K.: Leakage-resilient cryptography. In: 49th FOCS, pp. 293–302. IEEE Computer Society Press, Philadelphia, 25–28 Oct 2008

6. Faust, S., Pietrzak, K., Schipper, J.: Practical leakage-resilient symmetric cryptography. In: Prouff, E., Schaumont, P. (eds.) CHES 2012. LNCS, vol. 7428, pp. 213–232. Springer, Heidelberg (2012)
7. Faust, S., Rabin, T., Reyzin, L., Tromer, E., Vaikuntanathan, V.: Protecting circuits from leakage: the computationally-bounded and noisy cases. In: Gilbert, H. (ed.) EUROCRYPT 2010. LNCS, vol. 6110, pp. 135–156. Springer, Heidelberg (2010)
8. Fouque, P.-A., Tibouchi, M.: Indifferentiable hashing to barreto–naehrig curves. In: Hevia, A., Neven, G. (eds.) LATINCRYPT 2012. LNCS, vol. 7533, pp. 1–17. Springer, Heidelberg (2012)
9. Galbraith, S.D., Paterson, K.G., Smart, N.P.: Pairings for cryptographers. Discrete Appl. Math. **156**, 3113–3121 (2008)
10. Galbraith, S.D., Scott, M.: Exponentiation in pairing-friendly groups using homomorphisms. In: Galbraith, S.D., Paterson, K.G. (eds.) Pairing 2008. LNCS, vol. 5209, pp. 211–224. Springer, Heidelberg (2008)
11. Galindo, D., Großschädl, J., Liu, Z., Vadnala, P.K., Vivek, S.: Implementation and evaluation of a leakage-resilient ElGamal key encapsulation mechanism. Cryptology ePrint Archive, Report 2014/835 (2014). http://eprint.iacr.org/2014/835
12. Galindo, D., Vivek, S.: A practical leakage-resilient signature scheme in the generic group model. In: Knudsen, L.R., Wu, H. (eds.) SAC 2012. LNCS, vol. 7707, pp. 50–65. Springer, Heidelberg (2013)
13. Gallant, R.P., Lambert, R.J., Vanstone, S.A.: Faster point multiplication on elliptic curves with efficient endomorphisms. In: Kilian, J. (ed.) CRYPTO 2001. LNCS, vol. 2139, pp. 190–200. Springer, Heidelberg (2001)
14. Hazay, C., López-Alt, A., Wee, H., Wichs, D.: Leakage-resilient cryptography from minimal assumptions. In: Johansson, T., Nguyen, P.Q. (eds.) EUROCRYPT 2013. LNCS, vol. 7881, pp. 160–176. Springer, Heidelberg (2013)
15. Kiltz, E., Pietrzak, K.: Leakage resilient ElGamal encryption. In: Abe, M. (ed.) ASIACRYPT 2010. LNCS, vol. 6477, pp. 595–612. Springer, Heidelberg (2010)
16. Kocher, P.: Blind signature systems. U.S. Patent #4,759,063
17. Kocher, P.C.: Timing attacks on implementations of diffie-hellman, RSA, DSS, and other systems. In: Koblitz, N. (ed.) CRYPTO 1996. LNCS, vol. 1109, pp. 104–113. Springer, Heidelberg (1996)
18. Kocher, P.C., Jaffe, J., Jun, B.: Differential power analysis. In: Wiener, M. (ed.) CRYPTO 1999. LNCS, vol. 1666, pp. 388–397. Springer, Heidelberg (1999)
19. Longo, J., Martin, D.P., Oswald, E., Page, D., Stam, M., Tunstall, M.J.: Simulatable leakage: analysis, pitfalls, and new constructions. In: Sarkar, P., Iwata, T. (eds.) ASIACRYPT 2014. LNCS, vol. 8873, pp. 223–242. Springer, Heidelberg (2014)
20. Mangard, S., Oswald, E., Popp, T.: Power Analysis Attacks: Revealing the Secrets of Smart Cards. Springer, Heidelberg (2008)
21. Martin, D.P., Oswald, E., Stam, M.: A leakage resilient MAC. Cryptology ePrint Archive, Report 2013/292 (2013). http://eprint.iacr.org/2013/292
22. Maurer, U.M.: Abstract models of computation in cryptography (invited paper). In: Smart, N.P. (ed.) Cryptography and Coding 2005. LNCS, vol. 3796, pp. 1–12. Springer, Heidelberg (2005)
23. Medwed, M., Standaert, F.-X., Großschädl, J., Regazzoni, F.: Fresh Re-keying: security against side-channel and fault attacks for low-cost devices. In: Bernstein, D.J., Lange, T. (eds.) AFRICACRYPT 2010. LNCS, vol. 6055, pp. 279–296. Springer, Heidelberg (2010)

24. Micali, S., Reyzin, L.: Physically observable cryptography (extended abstract). In: Naor, M. (ed.) TCC 2004. LNCS, vol. 2951, pp. 278–296. Springer, Heidelberg (2004)
25. Miller, V.S.: The Weil pairing, and its efficient calculation. J. Cryptology **17**(4), 235–261 (2004)
26. Nechaev, V.I.: Complexity of a determinate algorithm for the discrete logarithm. Math. Notes **55**(2), 165–172 (1994)
27. Pereira, G.C.C.F., Simplício Jr., M.A., Naehrig, M., Barreto, P.S.L.M.: A family of implementation-friendly BN elliptic curves. Cryptology ePrint Archive, Report 2010/429 (2010). http://eprint.iacr.org/2010/429
28. Pereira, O., Standaert, F.X., Vivek, S.: Leakage-resilient authentication and encryption from symmetric cryptographic primitives. In: ACM CCS 15. ACM Press
29. Renauld, M., Standaert, F.-X., Veyrat-Charvillon, N., Kamel, D., Flandre, D.: A formal study of power variability issues and side-channel attacks for nanoscale devices. In: Paterson, K.G. (ed.) EUROCRYPT 2011. LNCS, vol. 6632, pp. 109–128. Springer, Heidelberg (2011)
30. Schipper, J.: Leakage-Resilient Authentication. Ph.D. thesis, Utrecht University (2010)
31. Scott, M.: Miracl-Multiprecision Integer and Rational Arithmetic C/C++ Library. Shamus Software Ltd., Dublin (2003)
32. Shoup, V.: Lower bounds for discrete logarithms and related problems. In: Fumy, W. (ed.) EUROCRYPT 1997. LNCS, vol. 1233, pp. 256–266. Springer, Heidelberg (1997)
33. Standaert, F.-X., Pereira, O., Yu, Y.: Leakage-resilient symmetric cryptography under empirically verifiable assumptions. In: Canetti, R., Garay, J.A. (eds.) CRYPTO 2013, Part I. LNCS, vol. 8042, pp. 335–352. Springer, Heidelberg (2013)
34. Tang, F., Li, H., Niu, Q., Liang, B.: Efficient leakage-resilient signature schemes in the generic bilinear group model. In: Huang, X., Zhou, J. (eds.) ISPEC 2014. LNCS, vol. 8434, pp. 418–432. Springer, Heidelberg (2014)
35. Yacobi, Y.: A note on the bilinear Diffie-Hellman assumption. Cryptology ePrint Archive, Report 2002/113 (2002). http://eprint.iacr.org/2002/113

Leakage-Resilient Identification Schemes from Zero-Knowledge Proofs of Storage

Giuseppe Ateniese[1], Antonio Faonio[2]($\boxtimes$), and Seny Kamara[3]

[1] Sapienza, University of Rome, Rome, Italy
ateniese@di.uniroma1.it
[2] Aarhus University, Aarhus, Denmark
antfa@cs.au.dk
[3] Microsoft Research, Seattle, USA
senyk@microsoft.com

Abstract. We provide a framework for constructing leakage-resilient identification (ID) protocols in the bounded retrieval model (BRM) from proofs of storage (PoS) that hide partial information about the file. More precisely, we describe a generic transformation from any zero-knowledge PoS to a leakage-resilient ID protocol in the BRM. We then describe a ZK-PoS based on RSA which, under our transformation, yields the first ID protocol in the BRM based on RSA (in the ROM). The resulting protocol relies on a different computational assumption and is more efficient than previously-known constructions.

Keywords: Leakage resilience · Bounded retrieval model · Proof of storage · Identification scheme · Generic transformation · RSA security

1 Introduction

Cryptographic schemes are traditionally designed under the assumption that the adversary cannot learn any information about the secret key. In practice, however, this assumption does not always hold as the adversary could recover information about the key through various means such as side-channel attacks [6,20,21,24,27], memory leakage attacks [17] or by compromising the system on which the keys are stored. These attacks, commonly referred to as *leakage attacks*, have motivated the design of *leakage-resilient* cryptosystems which remain secure even against adversaries that may obtain partial information about the secret state (clearly, under some limitations on the kind of leakage allowed). Several models of leakage-resilience have been proposed and many cryptographic primitives have been realized under gradually stronger models [1,11,13,15,19,23,25]. In what follows we discuss only the most relevant to our work, specifically, we focus on the *bounded retrieval model* (BRM). In this model, there is an absolute upper bound λ on the total amount of information the adversary can recover about the secret key. In the BRM this bound is independent of k, the security parameter, thus security can only be achieved if the key is larger

---

A. Faonio—Supported by European Research Council Starting Grant 279447.

J. Groth (Ed.): IMACC 2015, LNCS 9496, pp. 311–328, 2015.
DOI: 10.1007/978-3-319-27239-9_19

than λ. Since the latter can be very large, we require that the efficiency of the scheme be related only to the security parameter. The BRM model was introduced by Di Crescenzo *et al.* [9] and by Dziembowski [12]. The former showed how to construct password-based key agreement protocols while the latter proposed a symmetric-key authenticated key agreement (AKA) protocol. In this work, we consider the problem of identification in the BRM. More precisely, we are interested in practical identification schemes that support large secret keys and whose efficiency is independent of the key length. The problem was first considered by Alwen *et al.* [1], our contribution provides a new and different perspective, which results in a practical scheme based on RSA.

1.1   Our Contributions

We provide a framework for constructing leakage-resilient ID protocols in the BRM from publicly-verifiable proofs of storage (PoS) that are computationally zero-knowledge (ZK). PoS are interactive protocols allowing a client to verify that a server faithfully stores its file. A PoS is publicly verifiable if anyone with access to the client's public-key can verify the server's storage and it is computationally ZK if, roughly speaking, its verification phase leaks no useful information about the file to a bounded adversary. We show how to construct such a scheme based on the RSA assumption.

PoS were introduced independently by Ateniese *et al.* [2] and Juels and Kaliski [18]. Publicly verifiable PoS were first considered in [2] with extensions and improvements given in [4,28]. We summarize the contributions of this work as follows:

1. **(Generality).** We provide a transformation from any zero-knowledge (ZK) PoS to a BRM identification scheme.
2. **(Efficiency).** Our ZK-PoS-to-BRM-ID transformation is very efficient, leading to BRM-ID schemes that are practical and more efficient than prior work.
3. **(Security).** We show how to build ZK-PoS under standard cryptographic assumptions. In particular, we propose a novel BRM-ID scheme based on the standard RSA assumption in the random oracle model (ROM).

1.2   Related Work

Leakage-resilient identification schemes in the BRM were first considered in [1] which proposed a scheme based on the generalized Okamoto scheme (see Okamoto [26]) and the pairing-based public-key homomorphic linear authenticator of Shacham and Waters [28]. In [1], a transformations is also given from absolute leakage-resilient ID schemes to leakage-resilient signature schemes and AKA protocols. The transformation relies on parallel-repetition and consists in taking n independent copies of the basic relative-leakage scheme. Since n is large, this yields complex and relatively inefficient schemes, thus a more efficient transformation is described by the authors that employs subset selection and reduces both communication and time complexity.

For a detailed comparison between the constructions of [1] and our own, we refer the reader to Sect. 4.1. Here, we just mention that the framework of [1] works only for an extension of the Okamoto ID scheme [26] and is not generalizable. Also, the BRM-ID scheme based on the Okamoto ID scheme relies on BLS signatures [7] and thus on the Gap Diffie-Hellman assumption. For the same level of security, we provide schemes that rely on weaker computational assumptions and that are more efficient in terms of computation.

While zero-knowledge PoS can be designed from general-purpose zero-knowledge proofs by having the server prove knowledge of the file, such an approach would not be efficient. The first practical ZK-PoS scheme was proposed by Wang et al. [29] who extended the pairing-based PoS construction of Shacham and Waters [28] to be zero-knowledge. In comparison, our RSA-based ZK-PoS relies on a weaker computational assumption and, as far as we know, is the first construction to have a full proof of security.

1.3 Overview of Our Technique

At a high level, our framework works as follows. The secret key of the identification protocol is the encoding of a randomly-generated file and the public key is the state information generated by encoding the file together with the public key for the PoS. To identify itself, the prover executes the verification phase of the PoS with the verifier to prove that it indeed holds the file. Note that while (in the context of a BRM leakage attack) the verifier can learn λ bits about the key/file, the properties of the PoS allow us to increase the file size beyond λ without increasing the communication complexity of the verification phase.

One problem with the above approach is that standard PoS do not necessarily hide information about the file from the verifier and, therefore, the ID scheme verifier above could learn the remaining $n - \lambda$ bits of the key from the verification phase. To address this, we need a *zero-knowledge* PoS; that is, a PoS with a verification phase that hides all partial information about the file.

More formally, for the identification scheme we consider the security notion of pre-impersonation leakage-resistance, in which an attacker, in a test stage of the experiment, can interact with an honest prover and leak arbitrary functions of the secret key. We model the latter with a leakage oracle that on input an efficiently computable (and adaptively chosen) function f_i outputs the value $f_i(sk)$. The restriction is that the total length of the leaked information is bounded by some a-priori fixed value λ.

For the PoS, we phrase the soundness definition using the paradigm of "witness-extended emulation" (see Lindell [22]). Intuitively, this guarantees that there exists an expected polynomial time extractor that, for any adversary that convinces the verifier with some probability, outputs the original file with approximately the same probability.

The main intuition is that even after the test stage, an adversary cannot have *full knowledge* of the secret key/file. It follows then by the (knowledge) soundness of the PoS that the adversary cannot convince the verifier. In the intuition above we have not defined the meaning of knowledge of the adversary after the test stage. At first glance, one might consider the average conditional min-entropy of the secret key/file after the test stage. This measure, however, is insufficient for two reasons:

1. The PoS is only *computationally* zero knowledge so, in principle, all the min-entropy of the file could be lost after the test stage.
2. The conditional average min-entropy is not "smooth" with respect to statistically-close distributions. Specifically, given a random variable X and two statistically-close random variables Y and Y', there could be an arbitrary gap between $\widetilde{\mathbf{H}}_\infty(X \mid Y)$ and $\widetilde{\mathbf{H}}_\infty(X \mid Y')$. Therefore, even if we considered the stronger notion of statistical zero-knowledge PoS, we might run into the same problem.

We overcome the above problems by considering a slightly different experiment. In the new experiment the prover oracle is substituted by the simulator guaranteed to exist by the zero knowledge property of the PoS. The crux is that a polynomially-bounded adversary cannot distinguish the two experiments and, therefore, it can convince the verifier with approximately the same probability. Now we can give a meaningful lower bound on the average conditional min-entropy of the secret key/file after the test stage. The adversary cannot guess the original secret with probability roughly more than $2^{-|sk|+\lambda} \leq 2^{-\omega(\log k)}$ so, by soundness of the PoS, it cannot convince the verifier with noticeable probability.

Concretely, the proof proceeds in two steps. First, we establish a lower bound on the conditional average min-entropy of an *encoding* f' of a uniformly random file f when the adversary is given access to a leakage oracle parameterized with f', and the randomness necessary to encode f. We then show that if there exists a probabilistic polynomial time (ppt) adversary $\mathcal{A}$ that succeeds in the pre-impersonation leakage experiment with a noticeable probability, then, by the soundness of the PoS, the lower bound on the average conditional min-entropy mentioned above is violated. This follows because we can simulate the pre-impersonation leakage experiment and then successfully extract from the adversary the file f during the impersonation stage. Furthermore, the experiment provides the information necessary to reconstruct f' from f. This leads to a predictor that guesses the encoded file f' with noticeable probability.

A Comparison. Consider the proof of security of the identification schemes presented in [1]. Briefly, their proof technique relies on a collision resistant hash (CRH) function and the identification scheme is a proof of knowledge of a preimage x (the secret key) for an element y (the public key) in the co-domain of the hash function. The reduction samples a secret key x in the domain of the CRH function h and given the secret key, the reduction can easily reply to all the leakage queries. If the adversary succeeds in the pre-impersonation experiment then the reduction can extract a pre-image x'. Their analysis shows that the uncertainty of x is high even after the test stage and therefore with high probability $x' \neq x$ and $y = h(x') = h(x)$. In comparison with our work, they present a direct reduction to the computational problem of breaking a CRH function.

Our proof has a similar interpretation. Given a successful adversary for the pre-impersonation leakage experiment we define a new adversary for the PoS security experiment. This new adversary "forgets" part of the file (namely it has only λ bits of information about it) and convinces the verifier of the PoS scheme, therefore breaking the knowledge soundness of the proof of storage. However, since we cannot directly argue that a forgetful adversary that convinces the verifier breaks the security of PoS, we formalize it providing the two bounds mentioned before. A similar technique, although based on a different measure of min-entropy, was recently used in the context of fully leakage-resilient signature (see Faonio et al. [14]).

2   Definitions

2.1   Preliminaries

If x is a string, we denote its length by $|x|$; if $\mathbf{X}$ is a set, $|\mathbf{X}|$ represents the number of elements in $\mathbf{X}$. When x is chosen randomly in $\mathbf{X}$, we write $x \leftarrow \mathbf{X}$. When $\mathcal{A}$ is an algorithm, we write $y \leftarrow \mathcal{A}(x)$ to denote a run of $\mathcal{A}$ on input x and output y; if $\mathcal{A}$ is randomized, then y is a random variable and $\mathcal{A}(x;r)$ denotes a run of $\mathcal{A}$ on input x and randomness r; sometimes, when $\mathcal{A}$ is deterministic we write $y := \mathcal{A}(x)$. An algorithm $\mathcal{A}$ is probabilistic polynomial-time (ppt) if it is randomized and for any input $x, r \in \{0,1\}^*$ the computation of $\mathcal{A}(x;r)$ terminates in at most $\mathsf{poly}(|x|)$ steps.

Throughout the paper we let k denote the security parameter. We say that a function $\nu : \mathbb{N} \to \mathbb{R}$ is negligible in the security parameter k if $\nu(k) = k^{-\omega(1)}$. A positive function f is noticeable if there exist a positive polynomial p and a number n_0 such that $f(n) \geq 1/p(n)$ for all $n \geq n_0$.

We start by recalling the notion of conditional min-entropy. We adopt the definition given in [1], where the authors generalize the notion of conditional min-entropy to *interactive* predictors that participate in some randomized experiment $\mathbf{E}$. The (average) conditional min-entropy of random variable X given any randomized experiment $\mathbf{E}$ is defined as follows:

$$\widetilde{\mathbf{H}}_\infty\left(X \mid \mathbf{E}\right) = \max_{\mathcal{B}}\left(-\log \Pr\left[\mathcal{B}()^{\mathbf{E}} = X\right]\right),$$

where the maximum is taken over all predictors without any requirement on efficiency. Note that w.l.o.g. the predictor $\mathcal{B}$ is deterministic, in fact, we can derandomize $\mathcal{B}$ by hardwiring the random coins that maximize his outcome. Sometimes we write $\widetilde{\mathbf{H}}_\infty(X|Y)$ for a random variable Y, in this case we mean the average conditional min-entropy of X given the random experiment that gives Y as input to the predictor.

We recall the definition of δ-indistinguishability for ensembles of distribution, both in the computational and statistical flavors.

Definition 1 (Indistinguishability). *Given a function* $\delta : N \to \mathbb{R}$ *and two distribution ensembles* $\{X_k\}_{k \geq 0}$ *and* $\{Y_k\}_{k \geq 0}$ *such that* $|X_k| \leq p(k)$ *and* $|Y_k| \leq p(k)$ *for a polynomial* $p(k)$*, we say that the ensemble* $\{X_k\}_{k \geq 0}$ *is* δ*-indistinguishable from* $\{Y_k\}_{k \geq 0}$ *if for any non-uniform polynomial time distinguisher* $\mathcal{D}$ *the following holds:*

$$|\Pr\left[\mathcal{D}(1^k, X_k) = 1\right] - \Pr\left[\mathcal{D}(1^k, Y_k) = 1\right]| \leq \delta(k).$$

When we refer to statistical δ*-indistinguishability, the equation above holds for all distinguishers without any bound on the running time.*

2.2 Proofs of Storage

Publicly-verifiable PoS consist of two phases: a setup phase where the client encodes the file and sends it to the server; and a verification phase where a verifier (which may or may not be the original client) engages in an interactive protocol with the server to determine if it indeed possesses the file. The encoding algorithm also outputs a "state information" which represents a pointer to the encoded file and has size independent of the file size. Moreover, we require that knowledge of the state information doesn't help a malicious server to violate the soundness property. Later, we formalize this notion by giving to the adversary oracle access to the encoding algorithm.

We consider PoS in which the verification phase requires three moves (as opposed to two as in previous work [2,4,28]): the server generates the first message a using the public key pk and randomness r; the verifier sends a random challenge c; and the server returns a proof π using pk, the encoded file, the challenge and the randomness used to generate the first message a.

Definition 2 (Proof of Storage). *A publicly-verifiable* proof of storage *(PoS) is a tuple of six* ppt *algorithms* $\Pi = (\mathsf{Gen}, \mathsf{Enc}, \mathsf{Comm}, \mathsf{Chall}, \mathsf{Prove}, \mathsf{Vrfy})$ *such that:*

$(pk, sk) \leftarrow \mathsf{Gen}(1^k)$ *is a probabilistic algorithm that is run by the client to set up the scheme. It takes as input a security parameter, and outputs a public and private key pair* (pk, sk)*.*

$(\boldsymbol{f}, st) \leftarrow \mathsf{Enc}_{sk}(\boldsymbol{f})$ *is a probabilistic algorithm that is run by the client in order to encode the file. It takes as input the secret key* sk*, and a file* $\boldsymbol{f}$ *viewed as an n-dimensional vector over a block space* $\mathbf{B} = \{0, 1\}^{p(k)}$ *for some polynomial* $p(k)$ *(let p be the block size of* Π*). It outputs an encoded file* $\boldsymbol{f}$ *and public state information st in* $\{0, 1\}^{\ell_{st}(k)}$ *(let ℓ_{st} be the state information size of* Π*).*

$a \leftarrow \mathsf{Comm}(pk)$ *is a probabilistic algorithm run by the server to generate the first message. It takes as input the public key and outputs an initial message* a*.*

$c \leftarrow \mathsf{Chall}(pk)$ *is a probabilistic algorithm that takes as input the public key and outputs a challenge* c*.*

$\pi \leftarrow \mathsf{Prove}(pk, \boldsymbol{f}, r, c)$ *is a probabilistic algorithm that takes as input the public key* pk*, an encoded file* $\boldsymbol{f}$*, a string* r*, and a challenge* c*. It outputs a proof* π*.*

$b := \mathsf{Vrfy}(pk, st, a, c, \pi)$ *is a deterministic algorithm that takes as input the public key pk, the state information st, the first message a, a challenge c, and a proof π. It outputs a bit, where '1' indicates acceptance and '0' indicates rejection.*

We say that Π is correct if for all $k \in N$, all (pk, sk) output by $\mathsf{Gen}(1^k)$, all $n \in N$ and $\boldsymbol{f} \in \mathbf{B}^n$, all $(\boldsymbol{f}, st)$ output by $\mathsf{Enc}_{sk}(\boldsymbol{f})$, and all c output by $\mathsf{Chall}(pk)$, it holds that

$$\Pr_{r_c, r_p} \left[\mathsf{Vrfy}\left(pk, st, \mathsf{Comm}(pk; \ r_c), c, \mathsf{Prove}(pk, \boldsymbol{f}, r_c, c; \ r_p)\right) = 1 \right] = 1.$$

An important characteristic of a PoS is *locality* which requires that the running time of the Prove algorithm be polynomial in the security parameter (independent of the parameter n).

Locality effectively captures the server-side efficiency guarantee provided by a PoS and, as we will show in Sect. 3, is what allows us to meet the efficiency requirements of the BRM.

Informally, soundness of a PoS guarantees that if the verifier accepts the proof then the prover indeed has sufficient information to recover the entire original file $\boldsymbol{f}$. As noted in [2,10,18,28], soundness can be formalized using the notion of a knowledge extractor [5,16]. As in [4], we phrase our definition using the paradigm of "witness-extended emulation" [22].

Definition 3 (Soundness for a Publicly-Verifiable PoS). *Let $\Pi = (\mathsf{Gen}, \mathsf{Enc}, \mathsf{Comm}, \mathsf{Chall}, \mathsf{Prove}, \mathsf{Vrfy})$ be a publicly-verifiable PoS. We say that Π is sound with knowledge error $\varepsilon(k)$ if there exists an expected polynomial-time knowledge extractor $\mathcal{K}$ such that for all adversaries $\mathcal{A} = (\mathcal{A}_0, \mathcal{A}_1)$ where $\mathcal{A}_0$ is an oracle ppt algorithm and $\mathcal{A}_1$ is an interactive ppt algorithm involved in the following probabilistic experiment:*

1. **Key Stage:** *The challenger computes $(pk, sk) \leftarrow \mathsf{Gen}(1^k)$. The adversary $\mathcal{A}_0$ takes as input pk and gets oracle access to $\mathsf{Enc}_{sk}(\cdot)$. Eventually, $\mathcal{A}_0$ outputs a tuple $(\boldsymbol{f}, st_{\mathcal{A}})$ and the challenger computes $(\boldsymbol{f}, st) \leftarrow \mathsf{Enc}_{sk}(\boldsymbol{f})$.*
2. **Extraction Stage:** *The extractor $\mathcal{K}$ takes as input pk and st and gets access to the oracle $\mathcal{A}_1(st_{\mathcal{A}}, \boldsymbol{f}, st, \ \cdot \ ; \ \cdot \ )$ modeled as an interactive oracle. Finally $\mathcal{K}$ outputs the tuple $((a, c, \pi), \boldsymbol{f}^*)$.*
3. *The output of the experiment is the tuple $(pk, st, (a, c, \pi), \boldsymbol{f}^*, \boldsymbol{f})$.*

The properties listed below hold:

(i) *The following probability is at most $\varepsilon(k)$:*

$$\Pr \left[\mathsf{Vrfy}(pk, st, a, c, \pi) = 1 \ \bigwedge \ \boldsymbol{f}^* \neq \boldsymbol{f} \right], \tag{1}$$

where the probability is over the outputs of the experiment above.
(ii) *For any pk and st, the distribution (a', c', π') induced by an execution of $\mathcal{A}_1(st_{\mathcal{A}}, \boldsymbol{f}, st)$ with an honest verifier and the distribution (a, c, π) as output by the extractor $\mathcal{K}$ in the experiment above are identically distributed.*

We say that Π is sound if $\varepsilon(k)$ is negligible.

For simplicity, we consider only PoS Π where the function Enc is injective for any sk and for any assignment of the internal randomness. This assumption is made without loss of generality, in fact any PoS scheme can be converted into one with this property by "appending the missing data" in the encoded file. By the soundness property, the procedure is efficient and the average size of the encoded file increases only by a negligible factor[1].

We now turn to our definition of zero-knowledge. Namely, we consider the notion of black-box zero-knowledge which guarantees that there exists a simulator for any adversary and the simulator has only black-box oracle access to the adversary's algorithm.

Definition 4 (Zero-Knowledge). *Let* $\Pi = (\mathsf{Gen}, \mathsf{Enc}, \mathsf{Comm}, \mathsf{Chall}, \mathsf{Prove}, \mathsf{Vrfy})$ *be a publicly-verifiable PoS.* Π *is δ-zero-knowledge (δ-ZK) if there is an expected polynomial time transcript simulator $\mathcal{S}$ such that for all non-uniform polynomial time adversaries $\mathcal{A}$, for any $n \geq 0$, for any $\boldsymbol{f} \in \mathbf{B}^n$ and for any infinite sequence $\mathcal{L} = \{(pk, sk, \boldsymbol{f}, st)\}_{k \geq 0}$ indexed by the security parameter k and where (pk, sk) is output by $\mathsf{Gen}(1^k)$ and $(\boldsymbol{f}, st)$ is output by $\mathsf{Enc}_{sk}(\boldsymbol{f})$, the distribution ensemble*

$$\left\{ (a', c', \pi') \leftarrow \mathcal{S}^{\mathcal{A}}(st, pk, sk) \right\}_{(pk, sk, \boldsymbol{f}, st) \in \mathcal{L}}$$

is $\delta(k)$-indistinguishable from the following distribution ensemble:

$$\left\{ (a, c, \pi) : \begin{array}{l} r \leftarrow \{0, 1\}^*; a := \mathsf{Comm}(pk; r); \\ c \leftarrow \mathcal{A}(pk, st, a); \\ \pi \leftarrow \mathsf{Prove}(pk, \boldsymbol{f}, r, c) \end{array} \right\}_{(pk, sk, \boldsymbol{f}, st) \in \mathcal{L}}.$$

In the definition above, the secret key for the PoS is given as input to the simulator. We could consider a stronger definition where the secret key is given to the distinguisher, but we dismissed this option since a weaker zero-knowledge requirement makes our final compiler more general.

2.3   Identification Protocols

An identification protocol allows a prover $\mathcal{P}$ in possession of a secret key sk to prove its identity to a verifier $\mathcal{V}$ that holds the corresponding public key pk.

We consider 3-move identification protocols where the prover generates the first message α using the public key pk and randomness r; the verifier sends a random challenge β; and the prover then computes a response γ using (pk, sk), the randomness r and the verifier's challenge β. Given the transcript of the protocol, the verifier decides whether to accept or not. The prover algorithm of any identification scheme in the BRM must have efficiency essentially independent of the size of the secret key. This is captured by the following definition.

---

[1] To see this, consider the procedure that first encodes using Enc, then runs internally the extractor with oracle access to the honest prover and, if the extractor fails, appends the original file to the encoding. Since the extractor fails only with negligible probability the average size of the encoded file increases only by a negligible factor.

Definition 5 (Identification Protocol in BRM). *A 3-move identification protocol is a protocol between a ppt prover $\mathcal{P}$ and a ppt verifier $\mathcal{V}$ that consists of five polynomial-time algorithms $\Sigma = (\mathsf{Setup}, \mathsf{Comm}, \mathsf{Chall}, \mathsf{Resp}, \mathsf{Vrfy})$ such that:*

$(pk, sk) \leftarrow \mathsf{Setup}(1^k, 1^s)$ is a probabilistic algorithm that takes as input the security parameter and the key-size parameter and outputs a public and private key pair (pk, sk) such that $|pk| = \mathsf{poly}(k)$ and $|sk| = \mathsf{poly}(k, s)$.

$\alpha \leftarrow \mathsf{Comm}(pk)$ is a probabilistic algorithm run by the prover $\mathcal{P}$ to generate the first message. It takes as input the public key and outputs an initial message α.

$\beta \leftarrow \mathsf{Chall}(pk)$ is a probabilistic algorithm run by the verifier $\mathcal{V}$ that takes as input the public key and outputs a challenge β.

$\gamma \leftarrow \mathsf{Resp}(pk, sk, r, \beta)$ is a probabilistic algorithm that is run by the prover $\mathcal{P}$ to generate the second message. It takes as input the public key pk, the secret key sk, the randomness r, and a challenge β (from some associated challenge space), and outputs a response γ.

$b := \mathsf{Vrfy}(pk, \alpha, \beta, \gamma)$ is a deterministic algorithm run by the verifier $\mathcal{V}$ to decide whether to accept the interaction. It takes as input the first message α, the public key pk, a challenge β, and a response γ. It outputs a bit b, where '1' indicates acceptance and '0' indicates rejection.

The following properties hold:

Correctness. *For all $k \in N$, all $s \in N$, all (pk, sk) output by $\mathsf{Setup}(1^k, 1^s)$, and β output by $\mathsf{Chall}(pk)$, it holds that*

$$\Pr_{r, r'}\left[\mathsf{Vrfy}\left(pk, \mathsf{Comm}(pk; r), \beta, \mathsf{Resp}(pk, sk, r, \beta; r')\right) = 1\right] = 1.$$

Efficiency. *The prover $\mathcal{P}$ has running time $\mathsf{poly}(k, \log s)$. We call the* locality *of the protocol the number of bits of the secret key read as a function of the security parameter k.*

By saying "run the protocol Σ" we refer to the execution of the protocol between $\mathcal{P}$ and $\mathcal{V}$.

As in previous work [1,19], we model leakage attacks by providing the adversary with access to a leakage oracle that returns arbitrary bits of information related to the secret key. Since we are working in the BRM, we require that the oracle returns at most λ bits.

Definition 6 (Leakage Oracle). *A leakage oracle $\mathsf{Leak}_{sk}^{\lambda, k}(\cdot)$ is parameterized by a secret key sk, a security parameter k and a leakage parameter λ. It takes as input a function f (specified as a circuit) and returns $f(sk)$ subject to the restriction that the total output length of all its replies is at most λ, otherwise it outputs $\bot$.*

Roughly speaking, security for identification schemes requires that an adversary should not convince an honest verifier to accept an interaction unless it knows the secret key corresponding to a given public key. In the case of *security against impersonation under active attacks*, this should hold even if the adversary is previously allowed to interact with the honest prover a polynomial number of times. In [1], Alwen *et al.* extend this notion to capture leakage attacks by providing the adversary with a $\mathsf{Leak}_{sk}^{\lambda,k}(\cdot)$ oracle. This leads to two definitions: security against pre-impersonation leakage, where the adversary can only access the oracle *before* interacting with the verifier; and security against anytime leakage, where the adversary can access the oracle even *during* the interaction with the verifier.

Definition 7 (Security Against Pre-impersonation Leakage [1]). *Let Σ be an identification protocol and $\mathcal{A} = (\mathcal{A}_0, \mathcal{A}_1)$ be an adversary. Consider the following experiment:*

1. **Key Stage:** *The challenger computes $(pk, sk) \leftarrow \mathsf{Setup}(1^k, 1^s)$.*
2. **Test Stage:** *The adversary $\mathcal{A}_0$ takes as input pk and gets oracle access to $\mathsf{Leak}_{sk}^{\lambda,k}(\cdot)$ and to an honest prover $\mathcal{P}(sk, pk)$, modeled as an oracle that runs (arbitrarily many) proofs upon request; access to proofs is sequential. Finally $\mathcal{A}_0$ outputs $st_{\mathcal{A}}$.*
3. **Impersonation Stage:** *$\mathcal{A}_1(st_{\mathcal{A}})$ executes Σ as a prover with an honest verifier (running with pk).*
4. *The adversary succeeds if the honest verifier accepts the interaction.*

Σ is $\varepsilon(k)$-secure against pre-impersonation leakage $\lambda(k, s)$ if the success probability of every ppt adversary $\mathcal{A}$ and for infinitely many positive integer s in the above experiment is at most $\varepsilon(k)$. We say that Σ is secure against pre-impersonation leakage $\lambda(k, s)$ if $\varepsilon(k)$ is negligible.

3   From Proofs of Storage to Leakage-Resilient ID Protocols

In this section we show how to transform any computationally ZK publicly-verifiable proof of storage into a leakage-resilient identification protocol in the BRM. The basic idea is to use the file as the secret key of the identification protocol and the state information as its public key. A basic version of this approach would work as follows. The honest prover generates a public and private key pair for the PoS. A file is chosen at random and encoded. The encoded file f' serves as the identification secret key, and the state information st together with the public key of the PoS serves as the public key. To identify itself, the prover executes the verification phase of the PoS with the verifier.

One problem with the above approach is that, in the context of a pre-impersonation leakage attack, the adversary receives access to a $\mathsf{Leak}_{f'}^{\lambda,k}(\cdot)$ oracle and to an honest prover. The effect of the leakage oracle can be mitigated somewhat by increasing the size of the file to be larger than λ. Since the communication complexity of the PoS is effectively constant, this will not degrade the efficiency of the protocol. However, to prevent the adversary's interaction with the honest prover from revealing too much information about the file, we will require the verification phase of the PoS to be zero-knowledge.

Let Π = (Gen, Enc, Comm, Prove, Vrfy) be a PoS with block size $p(k)$. Construct a leakage-resilient ID protocol Σ = (Setup, Comm, Resp, Vrfy) as follows:

- Setup($1^k, 1^s$):
    Set $n = s/p(k)$;
    Compute $(pk', sk') \leftarrow \Pi.\mathsf{Gen}(1^k)$ and sample a file $\boldsymbol{f} \leftarrow \mathbf{B}^n$;
    Compute $(\boldsymbol{f}', st) \leftarrow \Pi.\mathsf{Enc}_{sk}(\boldsymbol{f})$ and
    set $sk = \boldsymbol{f}'$ and $pk = (pk', st)$; Delete sk' and $\boldsymbol{f}$.

- Comm($pk; r$): Output $\alpha := \Pi.\mathsf{Comm}(pk; r)$.

- Chall(pk): Output $\beta \leftarrow \Pi.\mathsf{Chall}(pk)$.

- Resp(pk, sk, r, β): Output $\gamma := \Pi.\mathsf{Prove}(pk', \boldsymbol{f}', r, \beta)$.

- Vrfy($pk, \alpha, \beta, \gamma$): Output $b := \Pi.\mathsf{Vrfy}(pk', st, \alpha, \beta, \gamma)$.

Fig. 1. Transforming a ZK PoS with block size $p(k)$ into a leakage-resilient ID protocol.

The compiler is shown in Fig. 1. If the Prove algorithm of Π is local then the resulting scheme is an identification scheme in the BRM. We recall here a lemma from [1] that we make use of.

Lemma 1. *For any random variable X and for any experiment $\mathbf{E}$ with oracle access to $\mathsf{Leak}_X^\lambda(\cdot)$, consider the experiment $\mathbf{E}'$ which is the same as $\mathbf{E}$ except that the predictor does not have oracle access to $\mathsf{Leak}_X^\lambda(\cdot)$, then $\widetilde{\boldsymbol{H}}_\infty(X \mid \mathbf{E}) \geq \widetilde{\boldsymbol{H}}_\infty(X \mid \mathbf{E}') - \lambda$.*

Let $\mathbf{E}$ be the following randomized experiment:

1. It generates a key pair (pk', sk') for Π, samples a file $\boldsymbol{f}$ uniformly at random, samples random coins ω_{enc} and computes $(\boldsymbol{f}', st) := \mathsf{Enc}_{sk'}(\boldsymbol{f}; \omega_{enc})$.
2. The predictor takes as input $pk = (pk', st)$, sk' and ω_{enc} and gets oracle access to $\mathsf{Leak}_{\boldsymbol{f}'}^{\lambda, k}(\cdot)$.

Lemma 2. *Let ℓ_{st} be the size of the state of Π. Then, $\widetilde{\boldsymbol{H}}_\infty(\boldsymbol{f} \mid \mathbf{E}) \geq |\boldsymbol{f}| - \lambda - \ell_{st}$.*

Proof. Consider the experiment $\mathbf{E}'$ which is the same as $\mathbf{E}$ except that $\mathcal{B}$'s oracle access to $\mathsf{Leak}_{\boldsymbol{f}'}^{\lambda, k}$ is removed. We apply Lemma 1:

$$\widetilde{\boldsymbol{H}}_\infty(\boldsymbol{f}' \mid \mathbf{E}) \geq \widetilde{\boldsymbol{H}}_\infty(\boldsymbol{f}' \mid \mathbf{E}') - \lambda,$$

Consider the experiment $\mathbf{E}''$ which is the same as $\mathbf{E}'$ but where the predictor does not get the state information st as input. We apply Lemma 1:

$$\tilde{\mathbf{H}}_\infty \left(f' \mid \mathbf{E}' \right) \geq \tilde{\mathbf{H}}_\infty \left(f' \mid \mathbf{E}'' \right) - \ell_{st}.$$

Notice that in the experiment $\mathbf{E}''$ the information about f' is limited to sk and ω_{enc} and recall that $\mathsf{Enc}_{sk}(\cdot; \omega_{enc})$ is injective, thus any predictor guesses f' with probability $2^{-|f|}$.

$\square$

In the next lemma we give an upper bound on the average conditional min entropy of f' given the experiment $\mathbf{E}$ that depends on the winning probability of a ppt adversary in the pre-impersonation leakage experiment.

Lemma 3. *Let Π be a δ-ZK PoS with knowledge error ε_Π and let $\varepsilon_\mathcal{A}$ be the probability with which an adversary $\mathcal{A}$ succeeds in the pre-impersonation leakage experiment. If δ is negligible then*

$$\tilde{H}_\infty(f \mid \mathbf{E}) \leq \log(1/\varepsilon_\mathcal{A}) + 2\frac{\varepsilon_\Pi}{\varepsilon_\mathcal{A}} + 1.$$

Proof. Consider the predictor $\mathcal{B}$ that, given the public key $pk = (pk', st)$ and sk', ω_{enc} works as follows during the experiment $\mathbf{E}$:

1. **Setup Stage:** It chooses a string ω for $\mathcal{A}_0$ that maximizes the winning probability of $\mathcal{A}$ in the pre-identification leakage experiment. Let $\mathcal{A}^\omega$ be the algorithm $\mathcal{A}_0$ with the randomness fixed to ω.
2. **Test Stage:** It executes $\mathcal{A}^\omega(pk)$ and answers its leakage queries using its own leakage oracle. At the i-th oracle call of $\mathcal{A}^\omega$ to the prover oracle, it executes the simulator $(a'_i, c'_i, \pi'_i) \leftarrow \mathcal{S}^{\mathcal{A}^\omega_i}(sk)$, where $\mathcal{A}^\omega_i$ is a copy of the adversary $\mathcal{A}^\omega$ where the machine state is set as the machine state of $\mathcal{A}^\omega$ just before the i-th call. The messages a'_i and π'_i are sequentially fed to the adversary $\mathcal{A}^\omega$. Eventually, $\mathcal{A}^\omega$ outputs $st_\mathcal{A}$.
3. **Extraction Stage:** It uses the extractor $\mathcal{K}(pk', st)$, guaranteed to exist by the soundness of Π, with $\mathcal{A}_1(st_\mathcal{A})$ to recover a file f^*. It returns as its output $\mathsf{Enc}_{sk}(f^*; \omega_{enc})$.

$\mathcal{A}^\omega$ is deterministic thus, for all i at the i-th interaction with the prover, $\mathcal{A}^\omega$ will reply with the challenge message c_i equal to the one in the simulated transcript. To bound the probability that the extractor $\mathcal{K}$ outputs the correct file, we first argue that the probability with which $\mathcal{A}_1$ succeeds in the impersonation stage is roughly the same whether it receives its state from a $\mathcal{A}^\omega$ that was executed with oracle access to an honest prover or to a simulator.

Proposition 1. *Let $q(k)$ (resp. $q'(k)$) be an upper bound on the number of queries made by $\mathcal{A}^\omega$ to the prover oracle (resp. leakage oracle). The view of $\mathcal{A}^\omega$ in the Test Stage of the predictor $\mathcal{B}$, as described below,*

$$\left\{ pk, \left(a'_i, c'_i, \pi'_i \right)_{i \in [q(k)]}, \left(f_i(f) \right)_{i \in [q'(k)]} \right\},$$

and the view of $\mathcal{A}^\omega$ in the Test Stage of the pre-impersonation leakage experiment

$$\left\{pk, \left(a_i, c_i, \pi_i\right)_{i \in [q(k)]}, \left(f_i(\boldsymbol{f})\right)_{i \in [q'(k)]}\right\},$$

where, for all $i \in [q(k)]$, the tuple (a_i, c_i, π_i) is a transcript of the interaction between $\mathcal{A}_i^\omega$ and the honest prover, are $(q(k)\delta(k))$-indistinguishable.

The proposition can be proved with a hybrid argument based on the zero-knowledge property of the PoS. Indeed, the zero-knowledge property holds for any non-uniform polynomial-time adversary $\mathcal{A}$.

Recall that $\mathcal{A}^\omega$ at the end of the test stage outputs the state information $st_\mathcal{A}$. The probability that $\mathcal{A}_1(st_\mathcal{A})$ succeeds in the impersonation stage is at least $\varepsilon_\mathcal{A} - q\delta \geq \frac{\varepsilon_\mathcal{A}}{2}$. This holds because δ is negligible in k and by Proposition 1. In fact, if this were not the case, the concatenation of $\mathcal{A}^\omega$ and $\mathcal{A}_1(st_\mathcal{A})$ executing Π as prover with an honest verifier would distinguish the two distributions with noticeable probability.

Now, we can bound the probability that the extractor $\mathcal{K}$ outputs the correct file. From the soundness of Π, the extractor $\mathcal{K}$ outputs a tuple $((a, c, \pi), \boldsymbol{f}^*)$ such that $\mathsf{Vrfy}(pk, a, c, \pi) = 1$ and $\boldsymbol{f}^* \neq \boldsymbol{f}$ with probability at most $\varepsilon_\Pi(k)$. But note that

$$\Pr[\mathsf{Vrfy}(pk, a, c, \pi) = 1 \wedge \boldsymbol{f}^* \neq \boldsymbol{f}]$$
$$> \Pr[\mathsf{Vrfy}(pk, a, c, \pi) = 1] - \Pr[\boldsymbol{f}^* = \boldsymbol{f}] \geq \tfrac{\varepsilon_\mathcal{A}}{2} - \Pr[\boldsymbol{f}^* = \boldsymbol{f}].$$

Hence, it follows that

$$\Pr[\boldsymbol{f}^* = \boldsymbol{f}] \geq \tfrac{\varepsilon_\mathcal{A}}{2} - \varepsilon_\Pi = \tfrac{\varepsilon_\mathcal{A}}{2}\left(1 - \tfrac{2\varepsilon_\Pi}{\varepsilon_\mathcal{A}}\right) > \tfrac{\varepsilon_\mathcal{A}}{2} \cdot 2^{-2\frac{\varepsilon_\Pi}{\varepsilon_\mathcal{A}}} = \varepsilon_\mathcal{A} \cdot 2^{-2\frac{\varepsilon_\Pi}{\varepsilon_\mathcal{A}} - 1},$$

where we used $(1 - x) \geq e^{-x} > 2^{-2x}$. The lemma follows because of Eq. (2) below and by taking the log:

$$2^{-\widetilde{\mathbf{H}}_\infty(\boldsymbol{f}'|\mathbf{E})} \geq \Pr\left[\mathcal{B}^\mathbf{E} = \boldsymbol{f}'\right] \geq \Pr\left[\mathsf{Enc}_{sk}(\boldsymbol{f}^*; \omega_{enc}) = \boldsymbol{f}'\right] = \Pr\left[\boldsymbol{f}^* = \boldsymbol{f}\right]. \quad (2)$$

$\square$

We are now ready to prove our main theorem which establishes the security of our transformation.

Theorem 1. *Let Π be a proof of storage that is sound with knowledge error $\varepsilon_\Pi(k)$, computational $\delta(k)$-zero-knowledge and with state information size $\ell_{st}(k)$. If $\delta(k)$ and $\varepsilon_\Pi(k)$ are negligible in k and if $|f| > \lambda + \ell_{st} + \omega(\log k)$, then Σ as in Fig. 1 is secure against pre-impersonation leakage λ.*

Proof. Let $\varepsilon_\mathcal{A}$ be the pre-impersonation leakage winning probability of an adversary $\mathcal{A}$, since ε_Π and δ are negligible in k, by Lemma 3:

$$\widetilde{\mathbf{H}}_\infty(\boldsymbol{f}'|\mathbf{E}) \leq -\log\left(1/\varepsilon_\mathcal{A}\right) + \mathsf{negl}(k) + 1.$$

It follows then that if $\varepsilon_\mathcal{A}$ is noticeable in k, there exists a constant c such that

$$\widetilde{\mathbf{H}}_\infty(\boldsymbol{f}'|\mathbf{E}) \leq c \cdot \log(k) \quad (3)$$

for infinitely many k. Thus, if $|\boldsymbol{f}| > \lambda + \ell_{st} + \omega(\log k)$, Eq. 3 contradicts Lemma 2.

$\square$

4   A ZK-PoS Based on RSA

We now describe a (statistical) zero-knowledge proof of storage. The scheme, described in Fig. 2, is an extension of the RSA-based construction of Ateniese *et al.* [2]. It relies on a modulus generator Gen_Q that takes as input a security parameter 1^k and outputs a tuple (N, p', q') such that $N = (2p' + 1) \cdot (2q' + 1) = p \cdot q$, where p' and q' are random primes such that $p'q' \in [2^{k-1}, 2^k - 1]$ and p and q are primes.

Abstractly, the scheme can be seen as a witness-indistinguishable Sigma protocol (see Cramer [8]) for the relation:

$$\mathcal{R} = \left\{ \Big((pk, st, c), \ (\tilde{t}, \tilde{f}) \Big) \ \middle| \ \frac{\tilde{t}^e}{\prod_i H(st, i)^{c_i}} \equiv g_1^{\tilde{f}} \mod N \right\},$$

where $pk = (N, g_1, g_2, e, H)$ as defined in Fig. 2, and where the equation that defines the relation $\mathcal{R}$ is essentially the verification procedure of the PoS presented in [2]. We note that for any file $\boldsymbol{f} \in \mathbf{B}^n$ and any challenge $\boldsymbol{c} \in \mathbb{Z}_{2^k}^n$, let $\boldsymbol{f}', st \leftarrow \mathsf{Enc}_{sk}(\boldsymbol{f})$ where $\boldsymbol{f}' = (\boldsymbol{f}, \boldsymbol{t})$, a witness for the instance (pk, st, c) can be derived as

$$\tilde{t} = \prod_i t_i^{c_i} \quad \text{and} \quad \tilde{f} = \sum_i c_i \cdot f_i. \tag{4}$$

The witness indistinguishability property of the Sigma protocol is enough to derive the zero-knowledge property of the PoS. Witness indistinguishability means that the distributions of the transcript for two distinct witnesses are indistinguishable, even when the verifier is malicious. Recall that the simulator of ZK-PoS takes as input the secret key $sk = (N, d, H)$, and thus it can efficiently derive a valid witness (t', f') for the instance (pk, st, c) for any challenge c chosen by the adversary. Specifically, it can encode an uniformly random file (or even a fixed one) using the same state information and compute an honest proof of storage for the challenge c and the encoded file[2]. Notice that we are assuming that the first message of the Sigma protocol is independent from the witness, which is usually true for Sigma protocols.

The locality of the scheme depends on how the challenges are generated. In fact, to make the scheme local it is enough to use "probabilistic checking" and make the server generate a proof for a random subset of the blocks. More concretely, we define a distribution $\mathsf{Sparse}(\mathbb{Z}_{2^k}, n, m)$ by sampling a vector c such that for all $i \in [n]$: (1) with probability m/n the element c_i is chosen uniformly at random from $\mathbb{Z}_{2^k}$; otherwise (2) c_i is set to 0. For locality m the challenge is sampled from the distribution $\mathsf{Sparse}(\mathbb{Z}_{2^k}, n, m)$. This ensures that Prove and Vrfy have locality m *on average*. If the scheme needs to be *always* local, the honest-prover can just discard the challenge if the number of non-zero locations in c is not in the range $\{(1 \pm \varepsilon)m\}$, for a constant ε. The behavior will be indistinguishable from the original scheme with all but negligible probability in k.

Theorem 2. *The scheme described in Fig. 2 is statistical zero-knowledge.*

---

[2] The actual simulator does it implicitly, without sampling the entire file.

$\mathsf{Gen}(1^k)$: Set $\bar{k} = \omega(\log k)$ and generate $(N, p', q') \leftarrow \mathsf{Gen}_Q(1^{k+5\bar{k}})$.
Choose a prime e such that $e > 2^{k+5\bar{k}}$ and d such that $ed = 1$ (mod $p'q'$). Let g_1 and g_2 be generators of the unique cyclic subgroup $\mathcal{Q}_N$ of order $p'q'$ (i.e., the set of quadratic residues modulo N). Let $H : \{0,1\}^* \rightarrow \mathcal{Q}_N$ be a RO. Set $pk = (N, g_1, g_2, e, H)$ and $sk = (N, d, H)$.
The block space $\mathbf{B}$ is $\mathbb{Z}_{2^k}$ and the challenge space $\mathbf{C}$ for n-block long file is $\mathbb{Z}_{2^k}^n \times \mathbb{Z}_{2^k}$.

$\mathsf{Enc}_{sk}(\boldsymbol{f})$:
   1. sample $st \leftarrow \{0,1\}^k$.
   2. for $1 \leq i \leq n$:
      (a) set $r_i := H(st, i)$.
      (b) compute $t_i := \left(r_i \cdot g_1^{f_i} \right)^d \bmod N$.
   3. let $\boldsymbol{t} := (t_1, \ldots, t_n)$
   4. output the encoded file $\boldsymbol{f}' := (\boldsymbol{f}, \boldsymbol{t})$ and state information st.

$\mathsf{Comm}(pk)$:
    sample $z_1 \leftarrow \mathbb{Z}_{2^{k+4\bar{k}}}$ and $z_2 \leftarrow \mathbb{Z}_{2^{k+8\bar{k}}}$ and
    output $a := g_1^{z_1} \cdot g_2^{e \cdot z_2} \bmod N$

$\mathsf{Chall}(pk)$:
    sample $\boldsymbol{c} \leftarrow \mathsf{Sparse}(\mathbb{Z}_{2^k}, n, m)$ and $v \leftarrow \mathbb{Z}_{2^k}$ and
    output $c := (\boldsymbol{c}, v)$.

$\mathsf{Prove}(pk, \boldsymbol{f}', a, c)$:
   1. parse c as $\boldsymbol{c} \in \mathbb{Z}_{2^k}^n$ and $v \in \mathbb{Z}_{2^k}$
   2. sample $\rho \leftarrow \mathbb{Z}_{2^{k+6\bar{k}}}$
   3. compute $\tau := g_2^\rho \cdot \Pi_i t_i^{c_i} \bmod N$
   4. compute $\mu := z_1 + v \cdot \sum_i c_i \cdot f_i$
   5. compute $\sigma := z_2 + v \cdot \rho$
   6. output $\pi := (\tau, \mu, \sigma)$

$\mathsf{Vrfy}(pk, st, \boldsymbol{c}, \pi)$:
   1. for $1 \leq i \leq n$, set $r_i := H(st, i)$
   2. output 1 iff $\mu < 2^{k+5\bar{k}}$ and $a \cdot (\tau^e / \Pi_i r_i^{c_i})^v \stackrel{?}{\equiv} g_1^\mu \cdot g_2^{e \cdot \sigma}$ (mod N)

Fig. 2. A statistical ZK PoS based on RSA with locality parameter m.

Proof. For any adversary $\mathcal{A}$, consider the simulator $\mathcal{S}^{\mathcal{A}}$ that on input the key pair (N, g_1, g_2, e, d, H) samples $a \leftarrow \mathsf{Comm}(pk)$, then executes $(\boldsymbol{c}, v) \leftarrow \mathcal{A}(pk, st, a)$. If $\mathcal{A}$ aborts then the simulator returns the special symbol $\perp$. Otherwise, with the knowledge of the secret key, the simulator computes $v' := v^{-1} \mod p'q'$ and samples an element μ in $\mathbb{Z}_{2^{k+4\bar{k}}}$, an element σ in $\mathbb{Z}_{2^{k+8\bar{k}}}$ and sets

$$\tau := \left(\left(g_1^\mu \cdot g_2^\sigma \cdot a^{-1} \right)^{v'} \cdot \Pi_i r_i^{c_i} \right)^d \mod N$$

where $r_i := H(st, i)$, and outputs the tuple $(st, a, (\boldsymbol{c}, v), (\tau, \mu, \sigma))$.

The output of $\mathcal{S}^{\mathcal{A}}$ is statistically close to a real transcript since a, v and $\boldsymbol{c}$ are distributed exactly as they would be in a real transcript, and since τ, μ, and σ are statistically close to elements from a real transcript. Moreover by definition $v < p'$ and $v < q'$, thus the element $v^{-1} \mod p'q'$ is well defined. □

Theorem 3. *For locality parameter $m = \omega(\log k)$, the scheme described in Fig. 2 is sound if the RSA assumption holds with respect to Gen_Q.*

The proof defines an extraction strategy and analyzes the expected running time and the probability of successfully retrieve the original file. We follow the blueprint of the extractor given in [2], however, there are some extra complications due to the outer sigma protocol and the locality of the scheme. The proof is deferred in the full version [3].

4.1 Efficiency Comparison with Previous Work

We compare the identification scheme derived by applying our transformation to the RSA-based ZK PoS from Sect. 4 with the third (and most efficient) construction of Alwen *et al.* [1].

In the following, we denote our construction by RSA-ID and that of Alwen *et al.* by GDH-ID.

We consider multiplications and additions as constant-time operations and denote by t_e the time for an exponentiation, by t_s the time for an exponentiation with a small (i.e., $o(k)$) exponent, and by t_p the time for a pairing operation. For the same security level, modular exponentiations in RSA groups are more expensive than modular exponentiations in groups for which GDH seems to hold, therefore we distinguish them by using the upper scripts RSA and GDH to indicate in which group the operations are carried out. We can assume that $t_e^{\mathsf{GDH}} < t_e^{\mathsf{RSA}} \ll t_p$.

In GDH-ID, the prover needs $\Omega(\ell \cdot m \cdot t_e^{\mathsf{GDH}})$ work to generate each of its two messages (the first and third) while the verifier needs $\Omega(m \cdot t_e^{\mathsf{GDH}} + t_p)$ time to verify the interaction[3]. For our construction, on the other hand, the prover needs only $O(t_e^{\mathsf{RSA}})$ (i.e., two exponentiations and one multiplication) and $O(t_e^{\mathsf{RSA}} + m \cdot t_s^{\mathsf{RSA}})$ work for the first and third messages, respectively, and the verifier requires only $O(t_e^{\mathsf{RSA}} + m \cdot t_s^{\mathsf{RSA}})$ time to verify the interaction. We also note that while the locality m in RSA-ID can be any function that is $\omega(\log k)$, in GDH-ID m must be at least $\Omega(k)$. In particular, to get approximately $1/2$ tolerance of relative leakage, m must be 12 times larger than k.

---

[3] The integer parameter $\ell \geq 2$ in their construction can be arbitrarily set.

With respect to communication complexity, the third message of GDH-ID requires roughly ℓ times the number of group element as the third message of RSA-ID—though GDH-ID works in smaller groups than RSA-ID for the same security parameter.

There are two negative aspects of RSA-ID compared with GDH-ID: The first is that, for the same security level, RSA groups are bigger than groups for which GDH seems to hold; The second is the ratio between the secret-key size and the leakage tolerated. However, the difference is relevant only when ℓ is $\omega(1)$ and m is $\omega(k)$ in which case the time complexity of GDH-ID becomes much worse than that of RSA-ID.

Acknowledgments. We are grateful to Jonathan Katz for his insightful comments, suggestions, and contributions to this work. G. Ateniese was partially supported by the European Unions Horizon 2020 research and innovation programme under grant agreement No 644666, project "SUNFISH".

References

1. Alwen, J., Dodis, Y., Wichs, D.: Leakage-resilient public-key cryptography in the bounded-retrieval model. In: Halevi, S. (ed.) CRYPTO 2009. LNCS, vol. 5677, pp. 36–54. Springer, Heidelberg (2009)
2. Ateniese, G., Burns, R., Curtmola, R., Herring, J., Kissner, L., Peterson, Z., Song, D.: Provable data possession at untrusted stores. In: CCS (2007)
3. Ateniese, G., Faonio, A., Kamara, S.: Leakage-resilient identification schemes from zero-knowledge proofs of storage. Technical report (2015)
4. Ateniese, G., Kamara, S., Katz, J.: Proofs of storage from homomorphic identification protocols. In: Matsui, M. (ed.) ASIACRYPT 2009. LNCS, vol. 5912, pp. 319–333. Springer, Heidelberg (2009)
5. Bellare, M., Goldreich, O.: On defining proofs of knowledge. In: Brickell, E.F. (ed.) CRYPTO 1992. LNCS, vol. 740, pp. 390–420. Springer, Heidelberg (1993)
6. Boneh, D., Brumley, D.: Remote timing attacks are practical. In: 12th Usenix Security Symposium (2003)
7. Boneh, D., Lynn, B., Shacham, H.: Short signatures from the weil pairing. In: Boyd, C. (ed.) ASIACRYPT 2001. LNCS, vol. 2248, pp. 514–532. Springer, Heidelberg (2001)
8. Cramer, R.: Modular Design of Secure, yet Practical Cryptographic Protocols. PhD thesis, University of Amsterdam (1996)
9. Di Crescenzo, G., Lipton, R.J., Walfish, S.: Perfectly secure password protocols in the bounded retrieval model. In: Halevi, S., Rabin, T. (eds.) TCC 2006. LNCS, vol. 3876, pp. 225–244. Springer, Heidelberg (2006)
10. Dodis, Y., Vadhan, S., Wichs, D.: Proofs of retrievability via hardness amplification. In: Reingold, O. (ed.) TCC 2009. LNCS, vol. 5444, pp. 109–127. Springer, Heidelberg (2009)
11. Duc, A., Dziembowski, S., Faust, S.: Unifying leakage models: from probing attacks to noisy leakage. In: Nguyen, P.Q., Oswald, E. (eds.) EUROCRYPT 2014. LNCS, vol. 8441, pp. 423–440. Springer, Heidelberg (2014)
12. Dziembowski, S.: Intrusion-resilience via the bounded-storage model. In: Halevi, S., Rabin, T. (eds.) TCC 2006. LNCS, vol. 3876, pp. 207–224. Springer, Heidelberg (2006)

13. Dziembowski, S., Pietrzak, K.: Leakage-resilient cryptography. In: FOCS, pp. 293–302 (2008)
14. Faonio, A., Nielsen, J.B., Venturi, D.: Mind your coins: fully leakage-resilient signatures with graceful degradation. In: Halldórsson, M.M., Iwama, K., Kobayashi, N., Speckmann, B. (eds.) ICALP 2015. LNCS, vol. 9134, pp. 456–468. Springer, Heidelberg (2015)
15. Faust, S., Rabin, T., Reyzin, L., Tromer, E., Vaikuntanathan, V.: Protecting circuits from leakage: the computationally-bounded and noisy cases. In: Gilbert, H. (ed.) EUROCRYPT 2010. LNCS, vol. 6110, pp. 135–156. Springer, Heidelberg (2010)
16. Feige, U., Fiat, A., Shamir, A.: Zero knowledge proofs of identity. J. Cryptology 1(2), 77–94 (1988)
17. Halderman, J.A., Schoen, S.D., Heninger, N., Clarkson, W., Paul, W., Calandrino, J.A., Feldman, A.J., Appelbaum, J., Felten, E.W.: Lest we remember: cold boot attacks on encryption keys. In: USENIX, pp. 45–60 (2008)
18. Juels, A., Kaliski, B.: PORs: proofs of retrievability for large files. In: CCS (2007)
19. Katz, J., Vaikuntanathan, V.: Signature schemes with bounded leakage resilience. In: Matsui, M. (ed.) ASIACRYPT 2009. LNCS, vol. 5912, pp. 703–720. Springer, Heidelberg (2009)
20. Kocher, P.C.: Timing attacks on implementations of Diffie-Hellman, RSA, DSS, and other systems. In: Koblitz, N. (ed.) CRYPTO 1996. LNCS, vol. 1109, pp. 104–113. Springer, Heidelberg (1996)
21. Kocher, P.C., Jaffe, J., Jun, B.: Differential power analysis. In: Wiener, M. (ed.) CRYPTO 1999. LNCS, vol. 1666, pp. 388–397. Springer, Heidelberg (1999)
22. Lindell, Y.: Parallel coin-tossing and constant-round secure two-party computation. In: Kilian, J. (ed.) CRYPTO 2001. LNCS, vol. 2139, pp. 171–189. Springer, Heidelberg (2001)
23. Micali, S., Reyzin, L.: Physically observable cryptography. In: Naor, M. (ed.) TCC 2004. LNCS, vol. 2951, pp. 278–296. Springer, Heidelberg (2004)
24. Miles, E., Viola, E.: Shielding circuits with groups. In: STOC, pp. 251–260 (2013)
25. Naor, M., Segev, G.: Public-key cryptosystems resilient to key leakage. In: Halevi, S. (ed.) CRYPTO 2009. LNCS, vol. 5677, pp. 18–35. Springer, Heidelberg (2009)
26. Okamoto, T.: Provably secure and practical identification schemes and corresponding signature schemes. In: Brickell, E.F. (ed.) CRYPTO 1992. LNCS, vol. 740, pp. 31–53. Springer, Heidelberg (1993)
27. Quisquater, J.-J., Samyde, D.: ElectroMagnetic analysis (EMA): measures and counter-measures for smart cards. In: Attali, S., Jensen, T. (eds.) E-smart 2001. LNCS, vol. 2140, pp. 200–210. Springer, Heidelberg (2001)
28. Shacham, H., Waters, B.: Compact proofs of retrievability. In: Pieprzyk, J. (ed.) ASIACRYPT 2008. LNCS, vol. 5350, pp. 90–107. Springer, Heidelberg (2008)
29. Wang, C., Wang, Q., Ren, K., Lou, W.: Privacy-preserving public auditing for data storage security in cloud computing. In: INFOCOM, pp. 525–533 (2010)

Author Index

Printed in the United States
By Bookmasters